Evaluating and Monitoring
the Health of Large-Scale Ecosystems

NATO ASI Series

Advanced Science Institutes Series

A series presenting the results of activities sponsored by the NATO Science Committee, which aims at the dissemination of advanced scientific and technological knowledge, with a view to strengthening links between scientific communities.

The Series is published by an international board of publishers in conjunction with the NATO Scientific Affairs Division

A	Life Sciences	Plenum Publishing Corporation
B	Physics	London and New York
C	Mathematical and Physical Sciences	Kluwer Academic Publishers Dordrecht, Boston and London
D	Behavioural and Social Sciences	
E	Applied Sciences	
F	Computer and Systems Sciences	Springer-Verlag Berlin Heidelberg New York
G	Ecological Sciences	London Paris Tokyo Hong Kong
H	Cell Biology	Barcelona Budapest
I	Global Environmental Change	

NATO-PCO DATABASE

The electronic index to the NATO ASI Series provides full bibliographical references (with keywords and/or abstracts) to more than 30000 contributions from international scientists published in all sections of the NATO ASI Series. Access to the NATO-PCO DATABASE compiled by the NATO Publication Coordination Office is possible in two ways:

- via online FILE 128 (NATO-PCO DATABASE) hosted by ESRIN, Via Galileo Galilei, I-00044 Frascati, Italy.

- via CD-ROM "NATO Science & Technology Disk" with user-friendly retrieval software in English, French and German (© WTV GmbH and DATAWARE Technologies Inc. 1992).

The CD-ROM can be ordered through any member of the Board of Publishers or through NATO-PCO, Overijse, Belgium.

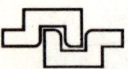

Series I: Global Environmental Change, Vol. 28

Evaluating and Monitoring the Health of Large-Scale Ecosystems

Edited by

David J. Rapport
Faculty of Environmental Sciences
Blackwood Hall, University of Guelph
Guelph, Ontario, Canada N1G 2W1

Connie L. Gaudet
Soil and Sediment Quality Section
Ecosystem Conservation Directorate
Environment Canada
Hull, Quebec, Canada K1A 0H3

Peter Calow
Institute of Environmental Sciences and Technology
Department of Animal and Plant Sciences
P. O. Box 601
University of Sheffield, Sheffield S10 2UQ, UK

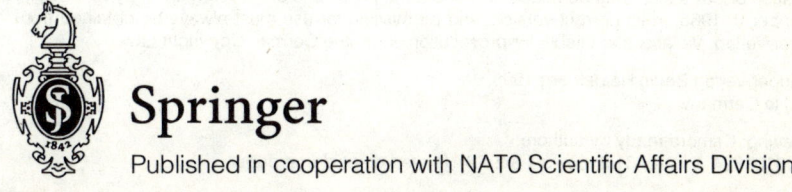

Published in cooperation with NATO Scientific Affairs Division

Proceedings of the NATO Advanced Research Workshop "Evaluating and Monitoring the Health of Large-Scale Ecosystems", held at Montebello, Quebec, Canada, October 10–15, 1993

ISBN 3-540-58805-1 Springer-Verlag Berlin Heidelberg New York

CIP data applied for

This work is subject to copyright. All rights are reserved, whether the whole or part of the material is concerned, specifically the rights of translation, reprinting, reuse of illustrations, recitation, broadcasting, reproduction on microfilm or in any other way, and storage in data banks. Duplication of this publication or parts thereof is permitted only under the provisions of the German Copyright Law of September 9, 1965, in its current version, and permission for use must always be obtained from Springer-Verlag. Violations are liable for prosecution under the German Copyright Law.

© Springer-Verlag Berlin Heidelberg 1995
Printed in Germany

Typesetting: Camera ready by authors
SPIN 10101824   31/3130 - 5 4 3 2 1 0 - Printed on acid-free paper

PREFACE

Ecosystem health offers a fresh perspective on the management of natural resources and the environment. While some of the root concepts can surely be traced back to Aldo Leopold and even earlier, it is only in the recent decade that a substantial body of work has emerged on this topic.

There is no question that a novel approach which is by its nature cross-disciplinary, bridging the health and biological sciences, will initially raise a number of questions particularly pertaining to the use of metaphors and the validity of the analogy. This volume however goes beyond merely the philosophical dimensions of the subject by covering a number of case studies which have given rise to the development of promising quantitative methods for diagnosis and rehabilitation of ecosystems under stress. The focus of most studies is on regional ecosystems i.e. ecosystems of large scale. As such, the methods and approaches should have wide appeal to government agencies charged with the responsibility of sustainable development of regional ecosystems and natural resources.

Health is one of those difficult concepts that everyone thinks they can define, until they come to try. We all have personal knowledge about health and illness and this makes the ecosystem analogy so potentially powerful. Yet it is also clear that the uncritical application of the concept could lead to overly simplistic approaches to analysis and management of ecosystem health.

The dilemma being faced is now to promote the positive aspects of the analogy while avoiding the pitfalls. This involves making transparent what is and can be intended by the ecosystem health metaphor. For example, all who have been involved in this Workshop reject the concepts of ecosystems as superorganisms and as homeostatic systems. On the other hand all of us recognize the similarity between what is done by clinicians in diagnosis and by ecologists assessing the state of ecosystems. Both involve an iteration between observations on, and sometimes hunches (call them hypotheses) about the state of the systems under consideration and more careful long-term observation and critical experimentation.

Health in the wider context associates societal values and goals (a matter for socio-political discussion and debate) with observations and predictions about ecosystem variation through space and time (matters for science). It clearly is not a matter of neutrality to observe declines in the provision of ecosystem services such as losses in productivity of fisheries and forests, water quality, biodiversity etc. Ecosystem health approaches ought to provide systematic methods to identify and assess risks to the integrity of ecosystem function, to detect early signs of a loss of resilience and to provide, where required, approaches to restructuring of damaged ecosystems. The challenge is in making this information available, in an intelligible and useful way, to policy makers, regulators and to the public in general so that rational decisions can be made about the protection of ecosystems.

The Editors hope that these contributions will provide fruitful avenues to address major questions on the preservation of the integrity of the earth's large-scale ecosystems - which will no doubt be a continuing challenge for decades to come.

This book emerged out of a NATO-funded Workshop that took place at the Chateau Montebello (Quebec, Canada) in October 1993. This involved a series of formal presentations that provide the basis of the chapters in this volume. All chapters were subsequently reviewed to reflect the workshop discussions and then further revised. They are arranged in five sections starting with definitions and measures and moving towards applications and management. Clearly there was much overlap and cross-referencing between sections.

The objectives of each session are summarized in the "Introduction" by the session chair. There was also a considerable amount of informal discussion that has been summarized in the "Rapporteur's Report" presented at the end of each section. Major conclusions and recommendations resulting from the workshop are included in the "Workshop Summary."

ACKNOWLEDGEMENTS

As well as receiving financial support from NATO (Special Programme on the Science of Global Environment Change), we are grateful for the additional funding made available by the following agencies:

Environment Canada
Canadian Forest Service
US-EPA Environmental Monitoring and Assessment Program
US-EPA Environmental Monitoring Systems Laboratory (Las Vegas)
Second European Conference on Ecotoxicology
National Board of Waters, Helsinki

Additional support was contributed by:

Biology Department, University of Ottawa, Ottawa, Canada
Desert Research Institute, Reno, Nevada
Maryland Institute of International Economics
CSIRO, Division of Water Resources, Canberra, Australia

We would like to especially thank Keith Desjardins who volunteered unstintingly of his time in handling the logistics for the workshop. It is a tribute to his organizational skills that the workshop was run in such an efficient and productive manner. His enthusiasm and ongoing support were instrumental to a highly successful workshop.

We also thank Geraldine Maini for designing the original logo for this Workshop.

CONTENTS

PREFACE

ACKNOWLEDGEMENTS

I. DEFINING ECOSYSTEM HEALTH

Introduction - *Robert Costanza*   3

1. Ecosystem Health: An Emerging Integrative Science
   David J. Rapport   5

2. Ecosystem Health - A Critical Analysis of Concepts
   Peter Calow   33

3. Qualitative and Quantitative Criteria Defining a "Healthy" Ecosystem
   François Ramade   43

4. The Relationship Between Ecosystem Health and Delivery of Ecosystem Services
   John Cairns Jr. & James R. Pratt   63

5. Environmental Health and Ecotoxicology: An Indispensable Link
   José Jerónimo Amaral-Mendes, Eric Pluygers & Juliania Fariña   77

Rapporteur's Report - *Edward Martinko*   95

II. QUANTITATIVE INDICES FOR ECOSYSTEM HEALTH ASSESSMENT

Introduction - *Peter Calow*   101

6. Ecological and Economic System Health and Social Decision Making
   Robert Costanza   103

7. New Approaches to the Assessment of Marine Ecosystem Health
   John S. Gray   127

8. Using Biological Criteria to Protect Ecological Health
   James R. Karr   137

9. Biological Changes in the German Bight of the North Sea
   as Indicators of Ecosystem Health
   Volkert Dethlefsen   153

10. Indices for Carcinogenicity in Aquatic Ecosystems:
    Significance and Development
    Paule Vasseur   179

11. Assessment of Ecosystem Health: Development of Tools and Approaches
    Peter-Diedrich Hansen   195

Rapporteur's Report - *Valery Forbes*   219

III. DIAGNOSTIC APPROACHES

Introduction - *Kenneth Sherman*   225

12. Degradation and Rehabilitation of Rivers:
    A Note on the Ecosystem Approach
    Wim Admiraal   227

13. An Influence Diagram Approach to the Diagnosis and Management
    of the Baltic Sea
    Mikael Hildén   241

14. Evaluation of New Techniques for Monitoring and Assessing
    the Health of Large Marine Ecosystems
    Robert Williams   257

15. Desertification: Implications and Limitations
    of the Ecosystem Health Metaphor
    Walter G. Whitford   273

Rapporteur's Report - *Tom Forbes*   295

IV. RECOVERY AND REHABILITATION OF LARGE-SCALE ECOSYSTEMS

Introduction - *David J. Rapport*   299

16. Paleolimnological Approaches to the Evaluation and Monitoring of
    Ecosystem Health:
    Providing a History for Environmental Damage and Recovery
    John P. Smol   301

17. Recovery and Rehabilitation of Mediterranean Type Ecosystem:
    A Case Study from Turkish Maquis
    Munir Ahmet Ozturk   319

18. Forest Recovery Following Pasture Abandonment in Amazonia:
    Canopy Seasonality, Fire Resistance and Ants
    Daniel C. Nepstad et al.   333

Rapporteur's Report - *Magda Havas*   351

V. METHODOLOGICAL ISSUES IN DESIGN AND ANALYSIS OF ECOSYSTEM HEALTH

Introduction - *Mikael Hildén*   357

19. Temporal and Spatial Variability as Neglected Ecosystem Properties:
    Lessons Learned from 12 North American Ecosystems
    Timothy K. Kratz et al.   359

20. Assessment and Monitoring of Large Marine Ecosystems
    Kenneth Sherman & Donna A. Busch   385

21. Remote Sensing and Ecosystem Health:
    An Evaluation of Time Series AVHRR NDVI Data
    David A. Mouat   431

Rapporteur's Report - *Connie L. Gaudet*   447

WORKSHOP SUMMARY

Peter Calow & David J. Rapport   453

Session I

DEFINING ECOSYSTEM HEALTH

INTRODUCTION

Chair: Robert Costanza

This session sets the stage for the workshop by addressing basic questions concerning what we mean by such concepts as "ecosystem health," "ecosystem integrity," "ecosystem medicine". Questions of the details of quantitative indices, diagnostic approaches, ecosystem restoration etc. are left for other sessions, but they must build on basic concepts discussed in this session.

The papers in this session present a range of perspectives on the issues and a major goal of this session is to come to some degree of consensus over basic definitions. Not that everyone has to agree, but rather we would like to determine what things we generally do agree upon and what things we do not. A major product of this session will be this list.

Critical Questions

- What constitutes health at multiple scales (ecosystem, landscape, global)?
- Integrating human health sciences and ecosystem sciences.
- Ecosystem-based frameworks for monitoring the state of the environment.
- Ecological economics and ecosystem health.
- What do we agree on - what do we not agree on - where do we go from here?

1. ECOSYSTEM HEALTH: AN EMERGING INTEGRATIVE SCIENCE

David J. Rapport
Tri-Council Eco-Research Chair
Blackwood Hall
University of Guelph
Guelph, Ontario, N1G 2W1
Canada

The world's ecosystems respond to the aggregate of both anthropogenic and natural stresses on their well-being, but our attempts to protect them are fragmented and unduly reductionist. While not intending to denigrate the reductionist approach"..."ecosystem quality control requires integrative science that employs a holistic view of multiple stresses, subsidies, and interactions in complex aquatic and terrestrial ecosystems. Integrative science requires that complex multivariate systems be considered in their entirety, not fragment by fragment.

John Cairns Jr. 1991

INTRODUCTION

Historical Context

The health metaphor has a long history within ecology and natural history. The use of the metaphor has evolved rapidly in the last several decades, as it has become far more obvious to both the public and politicians and decision makers that nature, increasingly, has become "disabled" as a consequence of human activity (World Resources Institute 1992).

Among early writings making reference to the health of nature, one finds the 1788 essay of James Hutton, a Scottish physician and geologist, who foreshadowed the Gaia Hypothesis, referring to the earth as a superorganism capable of self-maintenance (Hutton 1788 cited in Lovelock 1988). While ecosystems are not analogous to organisms or superorganisms - much of the Clementsian view to that effect has been discredited - they share common features with all complex systems, including mechanisms of self-regulation (although passively mediated) which are essential in maintaining system integrity and resilience.

In more recent times the ecosystem health concept was advanced by the naturalist, Aldo Leopold (Callicott 1992). In developing the concept of "land care," Leopold suggests that "...in medicine, symptoms of disease are manifest and that doctoring is an ancient art" and while "...the art of land doctoring is being practiced with vigour ... the science of land health is a job for the future" (Callicott 1992).

Leopold contributed to the development of the practice of "land health" by identifying key indicators of "land sickness". His list of common signs of land degradation included soil erosion, loss of fertility, hydrological abnormalities and occasional irruptions of certain species and mysterious local extinction of others, as well as qualitative deterioration in farm and forest products, the outbreak of pests and disease epidemics, and boom and bust wildlife population cycles (Leopold 1941). Thus, Leopold foresaw the development of a science of land health with the goal of "determining the ecological parameters within which land may be humanly occupied without making it dysfunctional" (Callicott 1992).

Problems of Definition

The concept of health has always presented somewhat of a quandary when it comes to definitions. It is one of those properties which is easier to identify the "absence of" than the "presence of". In the human health context the World Health Organization has provided a number of guidelines and definitions showing a clear evolution of the concept from a focus on narrowly conceived properties of physical well being, expanding to include aspects of mental well being, to the inclusion of psychological needs. Recently the World Health Organization defined health as "a state of complete physical, mental and social well-being and not merely the absence of disease" (Last 1988).

Lacking (as of yet) any international organization in the ecosystem health field with an equivalent role to the World Health Organization in the human health field, there are no official definitions of health at the ecosystem level. Rather, there is a plethora of attempts to define ecosystem health or desirable ecosystem states by individual researchers (e.g. Cairns Jr. 1992; Calow 1992, 1994; Costanza 1992, 1994; Karr 1991; Kay 1993; Kerr and Dickie 1984; Odum 1985; Rapport 1989a, 1992a; Schaeffer and Cox 1992; Schindler 1990; Smol 1992). These definitions range widely from very broad definitions which incorporate bio-physical, human and socio-economic components (e.g. Rapport 1992a) to definitions focusing primarily on the biophysical aspects (e.g. Costanza 1992) to definitions which focus on a single indicator within the biophysical domain (e.g. Kerr and Dickie 1984). Many definitions are based on "effects' or "impacts" of stress on ecosystems - focussing on cumulative impacts of stresses both temporally and spatially. Some definitions however are based upon the source of stress itself, focussing on risks associated with particular stresses (e.g. Minns 1992; Suter 1992). Other approaches make use of terms other than "health" to evaluate ecosystem transformation under stress, e.g. the use of the term "integrity" (Karr 1993; Kay 1993; Woodley et al. 1993). There are also very different approaches to reporting on the condition of ecosystems. Some rely on composite indices (e.g. Karr 1991); others make use of a suite of indicators (e.g. Odum 1985; Rapport et al. 1985); while others focus on a single indicator (e.g. Kerr and Dickie 1984).

As to examples of recent definitions, Costanza (1992) states that an ecological system is healthy and free from "distress syndrome" if it is stable and sustainable - that is, if it is active and maintains its organization and autonomy over time and is resilient to stress (Costanza 1992, p.248). Karr et al. (1986) state that a biological system can be considered healthy when its inherent potential is realized, its condition is stable, its capacity for self-repair when perturbed is preserved, and minimal external support for management is needed. Kerr and Dickie (1984) suggest evaluating ecosystem health by the size distribution of biota. A decline in the size of dominant species is one of the signs of ecosystem distress. Minns et al. (1990) provide a methodology for risk assessment with respect to acidification of inland lakes in eastern Canada. Minns et al. argue for integrated modelling to incorporate distress syndromes and counteractive capacity as part of a risk assessment approach. Odum (1985) suggests that stressed ecosystems are characterized by a reversal of trends found in ecosystem development. His analysis includes many of the signs of ecosystem distress. Schaeffer and Cox (1992 p. 159) state that health is achieved when functional ecosystem thresholds are not exceeded. Here thresholds are defined as "any condition (internal or external to the system) that, when exceeded, increases the adverse risk to maintenance of the ecological system". Schindler (1990) provides a detailed account of experimental results of acidification in freshwater systems, showing a sequences of changes or abnormal signs of ecosystem structure and function as acidification proceeds. Smol (1992 p.51) defines a healthy ecosystem as an "ecosystem that existed prior to cultural impact". Health may also be assessed in terms of resistance to disease. Steedman and Regier (1990) advance the notion of "ecosystem integrity" which is evaluated by a series of indicators of ecosystem breakdown, many of which are signs of ecosystem distress (e.g. increased dominance by 'r' selected species, less symbiotic interactions). and the loss of resilience or capacity to rebound from an external stress force.

Placing this wide array of definitions into common categories serves to show common features. Table 1 identifies three properties that are at the core of most definitions: 1) absence of distress syndrome; 2) resilience or counteractive capacity; and 3) risk factors.

Ecosystem Health as a Societal Goal

Fostering the practice of Ecosystem Health has never been more critical than at the present time. Global monitoring data suggest a marked deterioration (both quantitatively and qualitatively) in major ecosystems of the world (World Resources Institute 1992; World Watch Institute 1994; Tolba et al. 1992). And while effective solutions to mounting global and regional environmental crises depend primarily on stemming the tide of global population growth, they also require effective monitoring of the condition of the world's ecosystems and reliance on "early warning" indicators of loss of ecosystem resilience (Maini 1992). In short, this requires development of an ecosystem health approach at ecosystem, landscape, and biosphere levels.

Table 1. Categorization of approaches to assessing ecosystem health.

Author	Absence of Distress Syndrome	Counteractive Capacity / Resilience	Risk Factors
Costanza 1992	✓	✓	
Karr 1986	✓	✓	
Kerr and Dickie 1984	✓		
Minns et al. 1990; Minns 1992			✓
Odum 1985	✓		
Schaeffer and Cox 1992			✓
Schindler 1990	✓		
Smol 1992	✓	✓	
Steedman and Regier 1990	✓	✓	

The Health Paradigm

As evidence accumulates that many of the earth's ecosystems have been degraded as a result of chronic stress from human activity, it has become evident that a more holistic approach to ecosystem management is required. Rather than deal with one issue or one objective at a time, it is necessary to evaluate ecosystem condition with respect to multiple processes and societal values. It is the biophysical processes and societal values that collectively define the criteria of well-functioning ecosystems and ipso facto, ecosystem health.

An holistic approach to the question of ecosystem health must account for significant linkages between economic activity, biophysical ecosystem condition and human health. In integrating across the biological, social and health sciences, the approach transcends the separate disciplines and yields methods that guide the development of expert based knowledge systems fully reflective of the complexities of ecosystem behaviour under stress (Friend and Rapport 1991; Rapport et al. 1981, 1985; Rapport 1989a).

The process of assessing ecosystem health is complex. However, the essential aspects are: (i) selection and validation of a suite of indicators which, collectively, distinguish well-functioning systems from pathological systems; (ii) systematic protocols for diagnosis of probable causes of pathology; and (iii) methods for preventative and rehabilitative actions. The "health model" in fact goes beyond this in its preventative aspect. This may take the form of actively encouraging management practices which are directed to more efficient recycling of nutrients, promoting biodiversity etc., and are undertaken even in the absence of any signs of ecosystem pathology.

Cumulative stress from human activity often results, with various temporal lags and spatial displacements, in ecosystem breakdown, i.e. the system exhibits a loss of resilience, and an increased vulnerability to simplification as a result of both internal and external pressures (Odum 1985). This transformation is often accompanied by a loss in species highly valued by humans as well as losses in other "ecosystem services" and management options (Whitford this volume; Regier and Baskerville 1986; Cairns Jr. and Pratt this volume).

As the practice of ecosystem health and medicine transcends the health sciences and ecosystem sciences, it includes elements of models drawn from both fields. With respect to health sciences it includes both aspects of a modified "medical model" geared for early detection of the onset of ecosystem pathology and the "health model" where the emphasis is on both preventive actions and promotion of ecosystem health. With respect to ecosystem sciences, it draws upon a variety of approaches which have served to illuminate the complex nature of ecosystem response to stress, including hierarchy theory (O'Neill et al. 1986), systems analysis (Holling 1992; McKenzie et al. 1992; Steedman and Regier 1990), risk assessment (Suter 1993) and stress ecology (Hunsaker 1993; Odum 1985; Rapport et al. 1985; Smol 1994; Sprugel 1991).

The ecosystem health paradigm draws upon these diverse models and in effect constitutes a larger synthesis for the purpose of assessment. For example, systems approaches to complex ecosystem dynamics (Kay 1991), documenting biotic integrity (Karr 1991; Woodley et al. 1993), risk assessment approaches to ecosystem condition (Suter 1992), and applications from health and medicine to describing ecosystem change (Rapport 1989a,b, 1992a,b), incorporate common goals: early detection of system abnormalities; appropriate restoration or rehabilitation options; and maintaining system resilience.

Central to the health paradigm are value judgments implicit in the question of what is health. Clearly health assessments require norms, which while objectively measured, are not objectively established. By this I mean the characteristics of the system of interest - for example in humans: longevity, vitality, energy levels, etc., and for ecosystems: biodiversity, productivity, etc. - are those features valued by society, or deemed "desirable". The father of human ecology, Hawley (1950 p.7), succinctly stated the perennial dilemma in the evaluation of any complex system, namely: "How the effectiveness of a system is to be measured - as durability, productivity or efficiency, or all together - is open to debate".

In the ecosystem context, all of these dimensions are important (and interrelated). In fact one might view, within the ecosystem context, a hierarchy of considerations which at the pinnacle, might be the sustainability of ecosystem services and the preservation of flexibility and reversibility in management options.

ASSESSING ECOSYSTEM HEALTH

As shown in Table 1, most definitions of ecosystem health can be related to one or more of three basic methods of assessment:

(i) Ecosystem Distress Syndrome (diagnostics, ecosystem response);
(ii) Counteractive Capacity (system resilience stability/persistence); and
(iii) Risk Analysis (stress effects).

These assessment approaches combine elements of a modified "bio-medical model" (relating to diagnostics and interventions) and a "health model" (relating to preventative actions). Each approach has associated with it one or more specialized sets of indicators (generally groups of indicators).

Ecosystem Distress Syndrome

This approach assesses health by the presence or absence of a highly selected group of key indicators characteristic of ecosystems under stress (Odum 1985; Rapport et al 1985). An ecosystem is presumed healthy if it displays none of the signs associated with the Ecosystem Distress Syndrome (EDS). The concept of a "syndrome" is critical, and it is the collective signs of distress that signal ecosystem breakdown. One sign by itself may be insignificant. The syndrome comprises a suite of indicators (Odum 1985; Rapport et al. 1985). Many of these indicators were, as it turns out, identified decades earlier by Leopold as characteristic of "land sickness." EDS however, has been shown to have application not only to terrestrial ecosystems, but as well, with suitable modifications, to aquatic systems. Indicators included in Leopold's list and which are part of EDS include: reduced biodiversity (at levels of habitat, species and gene pools), loss of nutrient capital, reduction in primary productivity, shifts in biotic composition, resulting in increased dominance by exotics and "r" selected opportunistic species, increased amplitude of oscillations of component species, reduced size distribution, changes in energy flow, circulation of contaminants in biota and media, and others (Karr 1991; Kerr and Dickie 1984; Odum 1985; Rapport and Regier 1994; Rapport et al. 1985; Schaeffer et al. 1988).

EDS has proved instructive in gauging the extent of ecosystem degradation in both aquatic (e.g. Harris et al. 1988; Hilden and Rapport 1993; Rapport 1983b, 1989b) and terrestrial (e.g. Bird and Rapport 1986; Rapport et al. 1985) ecosystems. Changes in the Gulf of Bothnia (Baltic Sea) resulting from over four centuries of intensifying human use show clear evidence of EDS at spatial scales ranging from estuaries and bays to coastlines to the entire basin (Hilden and Rapport 1994; Rapport 1989b). In both the Baltic Sea and the Lower Laurentian Great Lakes, there have been numerous signs of EDS, including losses in species richness, shifts in biotic composition to favour opportunistic species, and changes in size distributions to favour smaller biota, increased nutrient concentrations, and bioaccumulation of toxic substances in biota. These changes resulted from a combination of stresses including nutrient loading, toxic loadings, habitat restructuring, and to some degree, harvesting (Harris et al. 1980; Rapport 1983b). An additional sign of EDS was documented for the Lower Great Lakes, namely increased oscillations of key fish populations just prior to becoming extinct (Regier and Hartman 1973).

In these examples, EDS is clearly evident only after the ecosystems have become severely degraded. At the onset of stress, in the very early stages of degradation, not only is the syndrome less evident, but indeed, some of the indicators may move in the opposite direction. For example, in the process of desertification in the Chihuahuan Desert (south-western USA) many common signs of distress such as reduced species richness do not appear - rather the opposite, species richness may initially increase rather than decline owing to an increased shrub diversity (e.g. the invasion of creosote bush) which creates favourable habitat for some avian species (Whitford 1994). However, as desertification becomes more pronounced eventually species richness declines. This suggests that the ecosystem distress syndrome, while serving well for "general screening" purposes, is insufficient for early detection of ecosystem breakdown. So-called 'early warning indicators' are found in the behaviours and population changes in key species and for toxic stresses, in biochemical and behavioral changes in sensitive organisms.

Counteractive Capacity (Resilience)

This approach defines ecosystem health in terms of capabilities for coping with stress rather than disabilities which may result from failure to cope. In the eco-health model, a key measure of the system performance is its capability to rebound from perturbations or disturbances which are a natural part of the system (e.g., drought in desert systems, annual stream scouring in aquatic systems, fire in the boreal forest ecosystem (Rapport 1989a, 1992a; Vogl 1980). The healthier the ecosystem, the greater its capabilities for recovery from disturbance (Rapport 1989a, 1992b). In essence, the approach involves comparing the resilience or "return time" of an ecosystem under stress with that of a similar type system (or the same system in an earlier period) under lower levels of cumulative stress.

Most ecosystems are perturbation dependent, that is they require disturbance for periodic revitalization of the system (Vogl 1980). Stressed ecosystems, whether from human or natural sources are more susceptible to additional stress. Several historical studies might be cited in support of this thesis. In Canada in 1980, over 22 million hectares of once forested lands were classified as "not sufficiently restocked" (Bird and Rapport 1986). These are lands which once bore merchantable timber, but failed to recover sufficiently after harvesting or some other source of disturbance (e.g. fire). That these areas tend to be concentrated in areas near pulp mills suggests that repeated cuttings weakened the forest's ability to regenerate (perhaps by leaching of soil nutrients).

A study of an upland area in Northern Ireland over historical periods offers evidence in support of a hypothesis of this nature. After repeated episodes of land clearing 5000 BC to 1000 AD, regeneration of the forest became progressively weaker leading to more open conditions. Similarly, in the Jutland heath in northern Germany on base-poor soils, the successive waves of regeneration got progressively weaker until the forest ultimately failed to return (Dimbleby 1978, p.137).

Risk or "Threats"

Both the ecosystem distress syndrome and counteractive capacity (described in sections above) are methods to assess ecosystem health based on ecosystem condition or behavioral responses to perturbations. The risk approach looks at the other side of the coin; that is it focuses on the stress side rather than the response side. Here emphasis is placed on estimating potential impacts of known sources of stress on receiving systems. Risk can then be calculated as potential damage to the receiving ecosystem (e.g. loss of productivity, species diversity, or other ecosystem services) from single or multiple stresses (Rapport 1992a, b).

Since ecosystem condition is in part determined by stress events, there is a need for well documented long term case histories which record not only the response history, i.e. the biophysical condition of the ecosystem, but equally important, the stress history, i.e. the various stresses that have likely impacted the ecosystem over time (Hilden and Rapport 1993). In documenting stress history, it is not only the type, but also the intensity, duration and timing of stress loadings that are critical (Hilden and Rapport 1993; Rapport et al 1985). This type of study provides the foundation for developing a risk approach to ecosystem health (Rapport 1992a,c).

By placing emphasis on the probable consequences to ecosystems from exposure to known sources of stress, the risk approach is well adapted for preventative ecosystem health care. The best examples thus far come from aquatic systems exposed to single dominant sources of stress, e.g. nutrient loadings, or acid deposition. Simulation models, developed on the basis of historical associations between stress loadings and ecosystem response, may be used to predict risks of impairment for other similar type ecosystem impacts by the same class of stress. For example Minns et al. (1990) modeled the probable impact of acid precipitation species richness in fish in eastern Canadian lakes. Drawing from a population of a large number of lakes of known acid buffering potential, it was shown that there was a high correlation between species richness and lake acidity - i.e., lower pH correlated with lower species richness in fish. In this case there was a sufficiently large database such that it was possible to establish a quantitative relationship between pH and species richness. From such retrospective studies and a knowledge of the impact of lake geochemistry, it is possible to predict risks to species loss from further acidification of the eastern Canadian lakes.

ECOSYSTEM BEHAVIOUR AND SOCIETAL VALUES

There is no denying that ecosystems exhibit very complex behaviour, of which there is only partial understanding. However, imperfect knowledge has not precluded development of methods for diagnosis and rehabilitation in the health sciences and it ought not preclude progress in ecosystem sciences.

Ecosystem Behaviour

The development of a practice of ecosystem health and medicine rests upon the following tenets:

a) Ecosystems are in a dynamic state of flux and likely have multiple states of equilibrium. These alternative states may be self sustaining and often serve to maintain system functional integrity (Rapport and Regier 1994; Regier 1993 personal communication; Sprugel 1991). Change takes place, simultaneously, on a variety of temporal (i.e. diurnal, meteorological, seasonal, successional, climatic, evolutionary, geological), and spatial (i.e. local, regional, landscape, biosphere) scales.

b) When natural disturbances are frequent and their impacts are small compared with the size of landscape units, a dynamic equilibrium may be maintained comprising all phases of the disturbance (Sprugel 1991). However, when the spatial scale of disturbance is large relative to the landscape unit, such that a single disturbance affects a large proportion of the landscape, the "Shifting Mosaic Steady State" (Bormann and Likens 1979) is unlikely, and instead there will be wide swings from one decade or century to the next in the proportion of landscape that exhibits one phase or another of disturbance. Examples include large-scale fires in the Pacific Northwest in pre-historic times and eastern forests under major hurricane tracks. Further, as a result of unique events, one might have disequilibrium conditions lasting over many centuries. A prominent well documented example of long-term dynamics concerns the appearance of a novel pathogen of eastern hemlock, *Tsuga canadensis*, which caused a sharp decline in the hemlock some 4800 years ago throughout its range, and it was almost two millennia before the hemlock returned to its pre-decline levels (Sprugel 1991).

c) Although loosely organized, ecosystems are nonetheless well adapted to recurrent naturally induced perturbations (e.g. floods, fire, drought). Ecology may in fact be defined as "the science of the principles of the self-organization of nature" (Faber et al. 1992). Even if recovery results in achieving a new and altered state of dynamic equilibrium, a healthy ecosystem will retain the functional integrity characteristic of mature systems (Pickett and White 1985; Rapport and Regier 1994; Regier 1993; Vogl 1980).

d) Alterations of the normal disturbance regime (e.g. reducing or increasing the frequency and/or intensity of the disturbance), or the introduction of various combinations of single or multiple chronic or acute stresses, if prolonged or intense, often results in eventual ecosystem breakdown (Rapport and Regier 1994; Rapport 1983b; Schindler 1990).

e) Ecosystems exhibit a natural tendency towards increasing integration, differentiation, mechanization and control (Regier 1993; Steedman and Regier 1990). Progressive integration entails the development of integrative linkages between different species of biota and between biota, habitat and climate; progressive differentiation entails the development of parasitic and symbiotic relationships; progressive mechanization entails gradual loss of individual flexibility; and progressive centralization entails development of structures which facilitate more centralized control within the ecosystem (e.g. "resource islands" in arid ecosystems, shoals or marshes in aquatic systems).

Societal Values

It is a much misunderstood aspect of the health sciences that one often has the impression that human health determinations are wholly objective - a simple matter of clinical test results. Quite the opposite impression may be held for ecosystem assessments - often thought to be wholly subjective, being largely dependent on social values.

In fact, societal values indeed underpin all health assessments. The seeming "objectivity" (from the layman's perspective) in human and veterinary arenas reflects general agreement on the values that govern these types of assessments. The notion advanced by Rene Dubos (1968) that health is *a modus vivendi*, enabling imperfect man to achieve a rewarding and not too painful existence" reflects a generally shared social value. There is widespread agreement that "health" be equated with vitality, long life, freedom from pain, etc. However, when Porn (1984) extends the definition to incorporate judgments by individuals with respect to their life goals, evaluations of health become much more of a personal judgment. In the final analysis, health can be viewed as a resource which enables individuals to satisfy their life aspirations.

Similarly, evaluation of ecosystem or landscape condition is only meaningful relative to societal values (Rapport 1992b). The challenge for the practice of ecosystem health is to specify desired or acceptable states clearly and unequivocally. A fundamental goal might be "sustainability", particularly with respect to natural resources, but more and more also with respect to cultural resources (Hartig and Zarull 1992).

Having socially determined objectives however, is not an "open door" for any criteria for ecosystem health, so long as the public is in favour of it. For example, the public may favour stocking the Great Lakes with a sports fishery. However as the Pacific Salmon and other sports fisheries are not self-reproducing this may be taken as a symptom of ill health rather than health - i.e. the system has to be continuously subsidized to persist.

Ecosystem health then is partly a matter of social values and partly a matter of the requirements for persistence or resilience of ecosystems. The fact that social values play a role has lead some to claim that any condition of the ecosystem might be found, at least by some, to be healthy. This argument ought to be rejected as Callicott (1992) suggests by raising the issue in the following terms:

What is wrong, objectively wrong, with urban sprawl, oil slicks, global warming,

> *or, for that matter, abrupt, massive, anthropogenic species extinction -other than that these things offend the quaint tastes of a few natural antiquarians? Most people prefer shopping malls and dog tracks to wetlands and old-growth forests. Why should their tastes, however vulgar, not prevail in a free market and democratic policy?* (Callicott 1992, p.46)

Here one can argue (as Callicott does) that the concept of ecosystem health serves as an important rescue mission. It provides a strong rebuttal to the argument that since nature is dynamic, since communities may be nothing more than a fortuitous collection of species, since extinction is commonplace throughout evolutionary history, there is nothing wrong with change of any sort! From an ecosystem health perspective, preferences of "consumer societies" are generally not compatible with maintaining viable ecosystems as the record of declines in quantity and quality of ecosystem services clearly shows (World Resources Institute 1992). As losses of vital whole ecosystems become an increasingly significant proportion of the landscape, the entire human life support system becomes at risk.

This is not to suggest that the opposite extreme, namely that only a pristine ecosystem, as it existed without humans, is a healthy system. Obviously humans and their activities are part of the system, and the challenge is to establish those criteria which the entire system including humans must meet in order to be sustainable. In striking a balance Aldo Leopold (1941) emphasized erring on the cautious side. In explicating the art of 'land doctoring' he suggested that "the land should retain as much of its original membership as is compatible with human land-use (and) should be modified as gently and as little as possible" (Callicott 1992, p.50). In this he greatly preferred the "gentle and restrained" land uses, in contrast to "violent and unrestrained modification" using a medical metaphor to drive home the difference, terming one the equivalent of "organic" and the other the equivalent of "mustard-plaster therapeutics" in the field of land-health.

DISCUSSION

Challenges to the Ecosystem Health Concept

Despite the promise of an ecosystem health approach as a necessary and essential tool for environmental management and sustainable development, a number of objections have been raised. These range widely, from objections to the use of the terms "health" or "ecosystem health", to the use of the metaphor itself - which like all metaphors can be carried too far (although this hardly negates its proper use), to specific critiques suggesting the difficulties of assessing ecosystem condition (of which there are many, but this applies to all approaches to assessing ecosystem condition).

Interestingly enough, many of the critics of ecosystem health have themselves contributed their own perspective on assessing ecosystem condition, which while using different terminology, incorporates the basic elements. Suter (1993) for example, while lambasting the eco-health approach, devises a risk assessment which forms an essential element of that approach. I think it is undeniable that almost every approach to environmental assessment has some core elements of the ecosystem health model embedded within it. That is, the various approaches collectively are (and have to be) concerned with assessing and evaluating changes in ecosystem condition, diagnosing probable causes, and suggesting corrective interventions. While the preferred terms vary (e.g. biotic integrity or ecological integrity, rather than ecosystem health), the underlying concepts are not incompatible with the health paradigm, rather in many cases they serve to advance it.

It would thus appear that many of the objections, to be examined below, rest on preferred epistemology, or a skeptical attitude towards the health sciences. At the root of the problem may be a reluctance to expropriate terms from a 'foreign' field, even if it can be demonstrated that these terms are highly appropriate and have the advantage of suggesting methodologies that are transferable between fields. Among the more specific critiques are the following:

(a) Inappropriate Metaphor - Objections to the effect that since ecosystems are far less homeostatic and less centrally controlled than organisms, the health metaphor is inappropriate when applied to the ecosystem level (Calow 1992; Kelly and Harwell 1989). A variant of this (Minns 1992) is that the use of the metaphor may be appropriate for some aspects (e.g. diagnosis) but can be "taken too far".

While one must agree that the use of any analogy or metaphor can be taken too far, this does not imply that one should, in any particular case, "throw out the baby with the bath water!" The criticism here appears to rest on a view that the health concept is only appropriate at the organismic level, and that since ecosystems are not analogous to organisms, it is totally invalid to use the metaphor at the ecosystem level.

I find several difficulties with this line of attack. Firstly, health concepts are not only applied at the organism level, but in both epidemiology and public health, where they are widely applied at population and community levels. Secondly, it is not necessary to adopt an organismic view of ecosystems to appreciate that they have a degree of integration and degree of capacity for buffering external pressures. Stress acting at both the level of ecosystems and organisms can serve to degrade the systems and make it dysfunctional (Rapport et al. 1985). Therefore overarching concepts of "health", "diagnosis", "assessment" are just as valid at the level of organisms as for the level of ecosystem.

(b) Lack of objective basis for ecosystem assessment - The underlying concern here is that "health" always implies an element of subjectivity (i.e. a social value judgment as to what is important) and therefore should not serve as the guiding concept for ecosystem assessment.

In reply, one might argue that while the criteria for "health" contain a subjective element (e.g. if biodiversity, productivity, water quality are valued "ecosystem services" and declines in these attributes collectively may be taken as a sign of unhealthy systems), each of these values can be related to an explicit and objective measure (assessment endpoints), and the system can be assessed according to changes in selected indicators. Further, since many ecosystems display a variety of potentially stable states (Holling 1993; Rapport and Regier 1994), assessment of system condition cannot be based solely on a single desired state (e.g. some historical "baseline"), but rather on an envelope of alternative self-sustaining states.

Reluctance to use the health metaphor may also hinge on a misconception on the part of ecologists that medicine is more of an exact science than is ecosystem analysis (Kelly and Harwell 1989). Here it may be of some comfort to ecologists to learn that this is not the case! The complex process of medical diagnosis rests on a holistic assessment not reducible to an exact science. For example, Siegel (1979, p.411) in referring to writings of Thomas Addis, concludes that it is not rigorous science but rather holistic judgment that ultimately shapes the opinion of the diagnostician on any particular case. This aspect is well described by Addis (quoted in Siegel 1979):

...the physician will ... for a time dismember his patient - isolate, for instance, his kidneys or his heart and observe their action under very specialized conditions - but in the end, he has to put these parts together again in his 'diagnosis'. ...This diagnosis is his total conception of the relationships between the patient as a person, the disease as a part of the patient, and the patient as a part of the world in which he lives.

(c) Not possible to establish norms which differentiate between healthy and pathological conditions - Objections to the effect that since is not possible to establish "norms" for ecosystems, health assessments which imply differentiating between the "normal" and "abnormal" are likewise not possible (Calow 1992; Kelly and Harwell 1989; Minns 1992). This recurring criticism suggests (quite correctly) that health assessments depend on establishing standards or "norms", but then goes on to suggest (quite wrongly) that norms cannot be established for ecosystems, and thus health assessments are impossible (Kelly and Harwell 1989, p.13). For example, while acknowledging that the "goal of environmental protection is to ensure the 'health' of ecosystems", Kelly and Harwell (1989, p.13) find the health metaphor unsuitable because:

...unlike indicators of adverse human health effects, there are no comparable integrative, simple measures or indices that show the effects of disturbances on ecosystems.

They go on to suggest (p. 13) a range of other complicating factors:

Attempts at an analogy of ecological health to human health have not been satisfying, in part because the exposure of ecosystems to stress is very complex...in part because ecosystems are more diverse and more complex than the human metabolic systems...and in part because ecosystems are much less internally coordinated and less able to respond to stress by controlled compensatory mechanisms that engender homeostasis of the systems...If ecosystems truly could be seen as superorganisms, where a few key components or processes reflected the state of being of the ecosystem-superorganism, then ecological response and recovery predictions in principle could be as reliable as human health check-ups and prognoses. But the reality is that predictive ecology lags far behind because ecological systems are less robustly defined, their dynamics thus being inherently less tractable, and their state not so easily fully characterized.

Their criticism suggests that the challenges of ecosystem assessment and diagnosis are as great or greater than human health challenges, but this by itself is hardly a negation of the health model! Further, while one might agree that ecology by its nature may never be fully predictive (but neither are human systems, and human medicine is hardly an exact science), it is gaining ground, and case studies such as those of Schindler (1988, 1990) and Hilden and Rapport (1993) lay the foundation for a better appreciation of the complex mechanisms governing the response of ecosystems to single and multiple stresses. These studies suggest the importance of temporal lags, spatial displacements, synergistic and antagonistic effects of stressors, ambiguities in the use of single indicators of system response, and a variety of biological, physical and chemical mechanisms in mediating the impacts of stress on the recipient system.

Minns (1992, p.110) in voicing a related concern, suggests that there is little consensus on expected values of various indicators that have been suggested as characterizing ecosystem health, and that the complex interactions among indicators are rarely addressed. In reply it might be stated that it has become possible to characterize ecosystem response to stress pressures with a suite of indicators. Recent empirical studies (e.g. Elmgren 1989; Hilden and Rapport 1993; Schindler 1990) not only pinpoint such a suite but also evaluate ecosystem condition against "norms" based on pre-stressed or historical states. Obviously if "norms" or "standards" could not be established, ecosystem assessment would not be possible - and the objection would not only prove fatal to the health paradigm but to all approaches to ecosystem assessment!

Schindler (1987) suggested that it may be years before sufficient data are accumulated to "confidently distinguish between natural variation and low-level effects of perturbations on ecosystems". However, by 1990 Schindler concluded from his studies that sufficient long-term data can differentiate pathological states from normal. While establishing "norms" for particular ecosystems will provide an ongoing challenge, new statistical approaches (Carpenter 1990), coupled with increased availability of long term data sets for major ecosystems, will provide the basic material to develop a clearer picture of the normal bounds for specific systems. Further, advances in paleoecology and ecological applications of remote sensing offer the promise of inexpensive long-term data sets.

While many ecosystems operate in a non-equilibrium range and exhibit a large degree of variability, long-term data sets permit quantification of the "normal" range of variability, which in many systems varies with ecosystem development and evolution

(Odum 1969, 1985). Interestingly enough, the crucial question of whether "norms" can be established for ecosystems was raised nearly half a century ago by Aldo Leopold. He suggested that principles of "land health" needed to be established with reference to norms, and that in this endeavour, wilderness "may serve as a base-datum for land health" (Callicott 1992, p.43). Not all systems of course can be returned to a pristine state. Thus for many systems, rehabilitation may involve restructuring so that an alternative, but equally acceptable state, is reached (Rapport and Regier 1994).

(d) Shortcomings of the medical model - Objections to the effect that the bio-medical model itself is flawed and/or limited. It is flawed because the stereotyped model is indeed very simplistic - asserting linear causality between an (single) agent of disease and illness. It is limited or inappropriate as the emphasis in western medicine has been by-and-large on "react and cure" whereas the emphasis sought by environmentalists is on "health care" and prevention.

With respect to disadvantages of applications of the simple bio-medical model or its practice, one can state that:

(i) The occurrence of disease has been found to often be attributable to a far more complex set of factors than implied by the simple biomedical model;
(ii) "Doctoring", particularly in the form that empowers practitioners (the anointed medical doctors or ecosystem scientists) and dis-empowers the patient, is not the desired social model for safeguarding nature;
(iii) Traditional medical practice has emphasized curative rather than preventative medicine - while clearly effective ecosystem management must rely on preventative actions (Maini 1992).

These inadequacies, fortunately are well recognized within the health sciences and critiques of the practice of medicine serve as a valuable guide for the practice of ecosystem health. Desirable modifications benefitting both the practice of human medicine and ecosystem medicine are:

(i) Replacing the simple bio-medical model with a complex multi-causal model. Many (if not most) diseases have been shown to be multi-causal. This is true even of biotic vector disease - where the effectiveness of the vector is modified by many other factors.
(ii) sharing responsibility between the "doctor" and "patient". In human medicine it is essential to consider the patient's own desires and needs determined by a consultation process. Similarly with respect to rehabilitation of ecosystems, involvement of 'stakeholders' is a necessary and essential part of the process of developing and implementing remedial action programs. This has in fact proved moderately successful in the efforts to rehabilitate areas of concern in the Great Lakes (Hartig and Zarull 1992).
(iii) Emphasizing preventative approaches. While there is no question about the desirability of a preventative approach in principle, there are questions within human medicine about how much "prevention" can realistically be undertaken. Some programs,

particularly those aimed at stopping smoking, or preventing the spread of epidemics, appear to be highly cost effective. Other programs aimed at extending life (through exercises, diet) may be less effective. In contrast, the area of ecosystem care and maintenance, there is no question but that preventative approaches are perhaps the only truly cost-effective interventions (Maini 1992).

Strengths of the Concept

There are at least three major advantages to the ecosystem health approach: a) providing a conceptual framework and methodology for ecosystem preventative and rehabilitative ecosystem health practice; b) integrating the ecological, socio-economic and human health aspects of ecosystem analysis; and c) communication to the public.

a) Providing a comprehensive conceptual framework for ecosystem preventative and rehabilitative ecosystem health practice - The ecosystem health framework provides a comprehensive methodology for assessment, diagnosis, and rehabilitation of large-scale ecosystems. Drawing upon methods developed and tested in the health sciences - at individual, population and community levels - it has already been demonstrated that there are straightforward applications to the ecosystem level - in both aquatic and terrestrial systems - e.g. the Baltic Sea (Elmgren 1989; Hilden and Rapport 1993; Rapport 1989b), the Laurentian Great Lakes (Harris et al. 1988; Rapport 1983b), desert ecosystems (e.g. Whitford 1994).

Clearly the evaluation of changes in ecosystems shares much in common with the practice of medicine. The identification of "vital signs" of ecosystem condition (Epstein 1993 personal communication; Rapport 1983b; 1989a) provides a necessary probe to track the health of these systems. The development of "early warning" indicators (Rapport 1992b), risk approaches (Suter 1992) and tests of "ecosystem fitness" (Rapport 1992c), parallels closely developments in health sciences. When it comes to diagnostics, here too systematic rule-in rule-out procedures can be readily shown to be applicable to ecosystem analysis (e.g. Hilden and Rapport 1993; USEPA 1993). For example, based on an extensive review of surface water conditions in eastern USA aquatic ecosystems where sensitive fish species had been lost, there was sufficient evidence for concluding that while no single indicator was sufficient, the weight of evidence supported the conclusion that acid deposition had contributed to these deteriorating conditions and that this condition would continue unless emissions decreased (NAPAP 1991).

b) Integration of socio-economic, biophysical and human health dimensions - Ecosystem health ultimately requires meeting human socio-economic needs within the requirements for biophysical integrity of ecosystems (Leopold 1941). The concept of ecosystem health thus requires integration across bio-physical, socio-economic and human health dimensions. An integrated view helps resolve certain paradoxes arising from more limited perspectives. For example, from a solely economic perspective, Canadian agriculture has been judged to be in excellent health - at least in terms of crop yields and (subsidized) farm incomes. However, from an ecological perspective the same system can be judged to be in poor health - with loss of soil fertility, soil erosion, etc. (Rapport 1993). An ecosystem health perspective would resolve this apparent conflict: clearly the whole system is not sustainable in the long run if the biophysical integrity is being compromised. While in the case of agriculture, it may be a matter for debate whether or not these intensively utilized systems can be sustained in the long term, clearly some management practices are less destructive of regenerative capacity than others.

c) Facilitating communication to the public - The health metaphor is ideally suited to foster communication between scientists, the public and politicians and administrators. The power of the metaphor rests on its capacity to convey the notion of ecosystem integrity and threats to ecosystem sustainability owing to easily made associations with the human condition. Here the concepts of pathology, early warning signs, costs of delayed action, advantages of preventative approaches, diagnostics, etc. are readily grasped in their applications to the care of environment at the regional ecosystem and landscape levels.

A Research Agenda

Ehrenfeld (1993 p.15) suggests that "health is a bridging concept connecting two worlds - it is not operational within science if you try to pin it down, yet it can help foster the necessary process of enabling scientists and non-scientists to communicate with each other". I would go somewhat further suggesting that health has become an operational concept in ecosystem assessment and has become a concept around which a new integrative science is evolving. Research challenges encompass (a) conceptual; (b) methodological; and (c) statistical areas.

(a) Conceptual

(i) Ecosystem health and values - The essential "hooks" for environmental assessment are societal values. For example, in assessing landscape health, one needs explicit criteria - which in part reflect societal values. These values ought to be established from consultation with all sectors of society, not only ecologists. Once specific criteria are agreed upon - commonly such criteria include the provision of basic ecosystem services such as biodiversity, sustainability, wildlife, aesthetics, productivity, water quality etc. - health determinations are amenable to scientific assessment methods.

(ii) Linkage of socio-economic, biophysical and human health - A major challenge here is to develop conceptual frameworks for the integration of indicators across seemingly very different domains: biophysical, socio-economic and human health. Attempts at integration have thus far been limited at best to two of these dimensions: e.g. socio-economic and biophysical (Friend and Rapport 1991); biophysical and human health (McMichael 1992). What is needed is a comprehensive framework - perhaps by resource sector, e.g. agriculture, forestry -which links all three areas, i.e. socio-economic, human health and bio-physical integrity. The challenge is not only to identify indicators within each of the relevant domains, but to explore their interdependence, both qualitatively and quantitatively.

(iii) Questions of scale - There are a number of critical areas here which need to be resolved. Fundamental to the assessment of ecosystem health is the appropriate temporal and spatial scale of analysis. Boundaries often appear arbitrary, but their choice may make a critical difference in the validity of the analysis. If chosen at too fine a scale, normal disturbance patterns could appear as catastrophes. If chosen at too coarse a scale, insidious processes might be overlooked. Similar considerations may apply at various temporal scales. What may appear "abnormal" on a short-term time scale, may be seen as recurrent events on a century-long time scale. Partly the scale problem is a question of identifying major driving forces governing the dynamics of the system and bounding the system so as to internalize the critical functions.

(b) Methodological

(i) Sensitivity and validation - Questions of sensitivity offer considerable scope for methodological and empirical work. By sensitivity, one refers to the response of the particular indicators to changes in the level of chronic or acute stress. Indicators are likely to vary considerably in their sensitivity to particular stresses, and to the same stress in different ecosystems (Rapport and Regier 1994; Rapport 1992ac). Sensitivity of a particular indicator is obviously dependent on the nature of the stress, levels of stress (both chronic and acute), and the nature of the recipient ecosystem. What needs better specification is which groups of indicators are most sensitive to which groups of stresses for which types of ecosystems. A related and equally important question is: for what purpose? - by which I refer to (again borrowing from the metaphor) the very different functions indicators might serve, i.e. general screening of ecosystem condition, diagnostic, early warning, or risk.

A key stumbling block to investigations of sensitivity of indicators in field situations is the lack of appropriate statistical methodologies. In order to infer causality - the necessary precondition for testing sensitivity - one normally requires statistical methods comparing 'experimental' with 'control'. Since in nature there are seldom sufficient 'replicates' for applying classical statistical techniques, new methodologies are required to effectively deal with 'one of a kind' events. Given the natural variability that is inherent in ecosystems, and the complex dynamics of the system, this proves to be a difficult but not impossible challenge. Some promising avenues in the development of statistical methods include "intervention analysis", time-series analysis, and various statistical models (Carpenter 1990).

A critical aspect of the use of indicators in the health sciences is validation, that is, careful testing of the supposed relationship between the indicator and the likely future state of the system. Scarce attention has yet been given to validation of indicators in ecosystem applications. What would be required are retrospective studies which confirm the subsequent development of the pathology after the appearance of selected indicator(s). For example, if several key indicators point to ecosystem breakdown, does the subsequent history of the system confirm that diagnosis? This research is complex, particularly if most ecosystem "ills" are suggested not by a single indicator, but by a group of indicators (e.g. the ecosystem distress syndrome, Rapport et al 1985). In special cases, where there is a single dominant stress, and the character of the recipient ecosystem is well defined, diagnostic "early-warning" indicators have been found - e.g. elimination of sensitive species to eutrophication, acidification etc. Here validation has in some cases essentially been done by careful studies of the etiology of the stress impact, or by careful retrospective studies of the type now commonly undertaken in paleoecology (Schindler 1988, 1990; Smol this volume).

(ii) Diagnostic protocols - An indispensable element of taking corrective action or repair for any complex system would be to first diagnose the source(s) of the problem. Here one can borrow much from systematic "rule-in", "rule-out" procedures, both successes and failures (e.g. false diagnosis) gained in the medical profession. For ecosystem applications, there are a number of challenges to establishing practical diagnostic protocols. These include: (i) The complex dynamic behaviour of the system, at best only partially understood, rendering early detection of 'abnormalities' difficult particularly in response to low level chronic stress; (ii) The fact that many stresses yield common signs of distress (Odum 1985; Rapport et al 1985) implies that stress-specific signs occur at earlier stages of ecosystem breakdown, but at these earlier stages detection of abnormalities is generally more difficult; (iii) Often a large number of different stress factors impinge upon the same ecosystem. Thus diagnostic protocols for ecosystems would in many cases "rule in" a number of potential stress factors, resulting in considerable uncertainty as to the identification of the major stress forces (e.g. Francis et al 1979; Harris et al 1988).

These difficulties notwithstanding, it ought to be possible to establish generic protocols for the diagnosis of ecosystems under stress. What is crucial in this process is to record the "patient history" particularly at the very early stages of ecosystem breakdown. Well documented case histories of the type conducted by Hilden and Rapport (1993) and Schindler (1990) are invaluable for establishing the sequence of events that can be detected at ecosystem levels from the early onset of low-level stress to the final stages of chronic intense stress. These studies suggest that careful documentation of temporal and spatial consequences of stress yields information sufficient to establish probable causal linkages. Further, these analyses lead to hypotheses of mechanisms which link spatial/temporal patterns of ecosystem response with known spatial/temporal patterns of stress. What remains to be established in most cases is the formal application of diagnostic methods to ecosystem assessments. This has in fact been suggested in a recent framework paper for further developments of the framework for the US EPA's Environmental Monitoring and Assessment Program (USEPA 1993).

(iii) Risk assessment - This area already shows promising developments and the methods are advancing rapidly (Suter 1992). When the stress history of a particular system is well documented, such that both the intensity of stress and consequent ecosystem responses are well documented, it is a relatively straightforward matter to model probable consequences of stress pressures. Minns et al (1990) provide an excellent case study of how this was accomplished for inland lakes in Canada. This of course is the simplest situation wherein a single stress (acid precipitation) impacts inland waters (eastern Canadian lakes) for which the geochemistry and buffering capacity has been well documented. Far more difficult will be the development of a risk methodology to apply to multiple stressed systems. Whether it is possible to establish a "Health Smart" system for ecosystems of the type now being deployed for human health remains to be seen.

(iv) Taxonomy of ecosystem ills - Taxonomy is one of the critical areas in the development of the biological and health sciences. This is particularly true for the development of an ecosystem health practice. Case studies of systems around the world are the essential database for evaluations of new cases. Naturally, judgments on likely causes and potential interventions will be made on the basis of experience with similar type situations. This is precisely where taxonomy is needed: to classify both the ecosystem and the types of stress impinging on it - so as to identify "similar type situations". This requires development of taxonomies, not only of ecosystems and stresses acting upon them, but also of ecosystem ills or pathologies. While each case history has its own unique aspects, there ought to be patterns of distress associated with particular classes of stresses, and such patterns, if they are found, would form the basis for a newly constructed taxonomy of ecosystem ills.

The taxonomy of ecosystem ills is, at present, at a very early stage of development. Broad distinctions have been made, e.g. between phenomena associated with eutrophication, acidification, desertification, salinization, overharvesting, etc. However there has been little differentiation of symptoms within these broad categories, e.g. different forms of desertification, depending on climate and stress(es). Such distinctions may prove critical in deciding the appropriate amount and types of interventions.

(v) Risks of misdiagnosis - One of the undisputed parallels between the health sciences and ecosystem sciences is the phenomena of misdiagnosis. There are sufficient examples in both areas to warrant close scrutiny. For example, the die-back of *Fucus* (a macrophyte) along the Finnish Coast of the Baltic in the late 1970s and early 1980s was initially taken as an early warning symptom of ecosystem collapse owing to cultural eutrophication (Rapport 1989b). Several years later, however, *Fucus* partially recovered while there was no reduction in nutrient loadings. Thus the die-back could not be attributed solely to cultural eutrophication. It was then discovered that the enhanced algae growth was likely a result of a upwelling of nutrients from salt-water intrusions which occur very irregularly in the Baltic. Similar examples can be cited for other ecosystems- e.g. the die-back and subsequent recovery of sugar maple in Ontario and Quebec -

originally thought to be part of irreversible damage of acid precipitation.

(c) Statistical

One of the recurring difficulties in the analysis of changing conditions in large scale ecosystems is finding appropriate statistical methods for the analysis of events which are, by their very nature, non-replicable. Minns (1992, p.110) suggests the lack of replication and controls is a key deficiency in the analysis of the behaviour of large-scale ecosystems. He contends that ideal measures of stress impacts require "...application and removal of the stress of perturbation while measuring the performance of the ecosystem. This is ideal where experiments are performed but large-scale ecosystem alterations are usually made without regard to adequate controls and replication".

There are certain long-term studies carried out on natural ecosystems which are in essence large scale experiments with proper and adequate controls. One of the best examples is the work of Schindler (1988, 1990) on experimentally acidified boreal lakes. However for most large-scale ecosystems, "controls" are not possible, and new statistical methods are required. There are a number of promising avenues here, some of which involve applications of Baysean Models, impact analysis, and time-series analysis (e.g. Carpenter 1990). These methods are designed for essentially non-replicable, but repeatable events, where 'repetition' here refers to data from other systems under similar stress pressures, or data from the same system for an earlier time frame.

While analysis of long term data seldom yields results with the statistical confidence found in laboratory experiments, generally speaking, such situations provide a basis for a reasonable judgments on ecosystem health. John Cairns, Jr. (1991, p.55) cautions that in such cases it is not possible to have the kind of 95% or 99% certainty as achieved in carefully controlled lab experiments with only one major variable.

CONCLUDING REMARKS

Sustainable Development - Ecosystem Health as the Bottom Line

The concept of 'sustainable development' is at best an ambiguous one, and at worst a contradiction in terms. Given the second law of thermodynamics, there is a continuous transformation from higher ordered states to lower ordered states, with the consequence that entropy (degradation) is increasing overall. In open systems this law does not apply straight forwardly. However, given the pressures already acting upon the earth's biosphere and major ecosystems, sustainable development if measured in conventional economic terms i.e. increases in global consumption/production and energy use does not appear as a realistic prospect.

If ecosystem health is, after all, the bottom line for any future choices, development must be redefined so that it is not measured in conventional economic terms, but in evolutionary terms - where evolution towards long term viability (sustainability) would entail reductions in the stress loadings humans place on the biosphere, through some combination of green consumerism, green technology, reduced population growth, reduced consumption per capita, and increased equity. Given the economic, political and ecological interdependence and the impossibility of effectively blocking impacts from the "rest of the world" on a particular nation, sustainable development goals must be achieved by all countries or those that have not achieved this will continue to pose threats to those that have.

Opportunities for Understanding System Behaviour Under Stress

While it may be poetic license to speak of ecosystems as "dead" or in a "state of collapse", it is apt to speak of degrading systems, on a world-wide basis (Brown et al. 1989; Postel 1989; World Resources Institute 1992). Signs of degradation are now validated for many ecosystems, and recurrent patterns of degradation have been documented. These case studies, including a number of contributions to this volume, provide the basic data for systematic evaluation of ecosystem health and threats to system integrity.

To further this work calls for integration and synthesis of the biological, social, and health sciences. Within the biological sciences, advances in ecosystem science will draw upon many fields, both at the micro and macro level. Such fields contribute to understanding the complex dynamics that drive ecosystems to altered states in response to stress and disturbance (Rapport and Regier 1994). Ecosystem management will be furthered by more systematic means of identification of ecosystems at risk, and systematic means of diagnosing likely causes. Ultimately safeguarding the earth's ecosystems will depend on evolving a preventative ecosystem health practice - taking measures to reduce threats before ecosystem resilience is compromised and management options and ecosystem services are irretrievably lost.

ACKNOWLEDGEMENTS

This work was supported in part by a Tri-Council Eco-research Chair Award to the author and the University of Guelph, an Eco-research Grant in Agro-Ecosystem Health, and an EPA Innovative Research Award. The Tri-Council comprises the Medical Research Council, the Social Science and Humanities Research Council, and the Natural Sciences and Engineering Research Council of Canada.

REFERENCES

Bird P.M., Rapport D.J. (1986) State of the Environment Report for Canada. Canadian Government Publishing Centre 264 pp.
Bormann F.H., Likens G.E. (1979) Pattern and Process in a Forested System.

Springer-Verlag, New York
Brown L.R., Flavin C. and Postel (1989) A world at risk. In Brown et al. (eds.) State of the World. A Worldwatch Institute Report on Progress Toward a Sustainable Society. W.W. Norton and Co., New York
Cairns Jr. J. (1991) Will integrative science develop with sufficient rapidity to mitigate global environmental degradation? Speculations in Science and Technology 15(1):54-59
Callicott J.B. (1992) Aldo Leopold's metaphor. In Costanza R, Norton B., Haskell B. (eds.) Ecosystem Health - New Goals for Environmental Management. Island Press, Washington DC
Calow P. (1992) Can ecosystems be healthy? Critical consideration of concepts. J. of Aquatic Ecosystem Health 1:1-5
Calow P. (1994) Ecosystem health - a critical analysis of concepts. In Rapport D.J., Calow P., Gaudet C. (eds.) Evaluating and monitoring the health of large-scale ecosystems. Springer-Verlag, New York (in preparation)
Carpenter S.R. (1990) Large-scale perturbations: opportunities for innovation. Ecology 71:2038-2043
Clements F.E. (1916) Plant succession: analysis of the development of vegetation. Publ. Carnegie Institute 242:1-512
Costanza R. (1992) Toward an operational definition of health. In Costanza R., Norton B., Haskell B. (eds.) Ecosystem Health - New Goals for Environmental Management. Island Press, Washington DC
Costanza R. (1994) Ecological and economic system health and social decision making. In Rapport D.J., Calow P., Gaudet C. (eds.) Evaluating and Monitoring the Health of Large-scale Ecosystems. Springer-Verlag, New York (in preparation)
Dimbleby G.W. (1978) Prehistoric man's impact on environments in north-west Europe. In Holdgate M.W., Woodman M.J. (eds.) The Breakdown and Restoration of Ecosystems. Plenum Press, New York
Dubos R. (1968) Man, Medicine and Environment. Praeger, New-York
Ehrenfeld D. (1992) Ecosystem health and ecological theories. In Costanza R., Norton B., Haskell B. (eds.) Ecosystem health - New Goals for Environmental Management. Island Press, Washington DC
Ehrenfeld D. (1993) Ecosystem health. Orion (Winter):12-15
Elmgren R. (1989) Man's impact on the ecosystem of the Baltic Sea: energy flows today and at the turn of the century. Ambio 18:326-331
Faber M., Manstetten M., Proops J (1992) Toward an open future: ignorance, novelty and evolution. In Costanza R., Norton B., Haskell B. (eds.) Ecosystem Health - New Goals for Environmental Management. Island Press, Washington DC
Francis G.R., Magnuson J.J., Regier H.A., Talhelm D.R. (1979) Rehabilitating Great Lakes ecosystems. Great Lakes Fishery Commission Tech. Rep. 37, Ann Arbor, Michigan
Friend A.M., Rapport D.J. (1991) Evolution of macro-information systems for sustainable development. Ecological Economics 3:59-76

Harris H.J., Harris V.A., Regier H.A., Rapport D.J. (1988) Importance of the near shore area for sustainable redevelopment in the Great Lakes wiith observations on the Baltic Sea. Ambio 5:261-163

Hartig J.H., Zarull M.A. (1992) Towards defining aquatic ecosystem health for the Great Lakes. J. Aquatic Ecosystem Health 1:97-108

Hawley A.H. (1950) Human Ecology: A Theory of Community structure. Roland Press, New York

Hilden M., Rapport D.J. (1994) Four centuries of cumulative cultural impact on a Finnish river and its estuary: an ecosytem health approach. J. Aquatic Ecosystem Health (in press)

Holling C.S. (1985) Resilience of ecosystems: local surprise and global change. In Malone T.F., Roederer J.G. (eds.) Global Change. Cambridge Univ. Press, Cambridge

Holling, C.S. (1992) Cross-scale morphology, geometry and dynamics of ecosystems. Ecological Monographs 62(4):447-502

Holling C.S. (1993) New science and new investments for a sustainable biosphere. (draft manuscript)

Hunsaker C.T. (1993) Ecosystem assessment methods for cumulative effects at regional and global levels. In Hildebrand S.G., Cannon J.B. (eds.) Environmental Analysis: the NEPA Experience. CRC Press, Inc, Boca Raton, Florida

Hutton J. (1788) Theory of the earth or An investigation of laws observable in the composition, dissolution and restoration of land upon the globe. Trans. Roy. Soc. Edinburgh 1:209-304

Karr J.R. (1991) Biological integrity: a long-neglected aspect of water resource management. Ecolog. Appl. 1:66-84

Karr J.R. (1993) Measuring biological integrity: lessons from streams. In Woodley S., Kay J., Francis G. (eds.) Ecological Integrity and the Management of Ecosystems. St. Lucie Press, Delray Beach, Florida

Karr J.R., Fausch K.D., Angermeier P.L., Yant P.R., Schlosser L.J. (1986) Assessing biological intergrity in running waters: a method and its rationale. Illinois Natural History Survey, Champaigne, Illinois, Special Publication 5

Kay J.J. (1991) A non-equilibrium framework for discussing ecosystem integrity. Environmental Management 15(4):483-495

Kay J.J. (1993) On the nature of ecological integrity: some closing comments. In Woodley S., Kay J., Francis G. (eds.) Ecological Integrity and the Management of Ecosystems. St Lucie Press, Delray Beach, Florida

Kelly J.R., Harwell M.A. (1989) Indicators of ecosystem response and recovery. In Levin S.A., Kelly J.R., Harwell M.A., Kimball K.D. (eds.) Ecotoxicology: Problems and Approaches. Springer-Verlag, New York

Kerr S.R., Dickie L.M. (1984) Measuring the health of aquatic ecosystems. In Carins V.W., Hodson P.V., Nriagu J.O. (eds.) Contaminant Effects on Fisheries. J. Wiley and Sons, New York

Last J.M. (ed.) (1988) A dictionary of epidemiology, 2nd ed. Oxford University Press,

Oxford (edited for the International Epidemiological Association)
Leopold A. (1941) Wilderness as a land laboratory. Living Wilderness 6 (July):3
Lovelock J. (1988) The Age of Gaia. Oxford University Press, Oxford
Maini J.S. (1992) Sustainable development of forests. Unasylva 43(2):3-7
McKenzie D.H., Hyatt D.E., McDonald V.J. (eds.) (1992) Ecological Indicators, v.2. Elsevier, Essex
McMichael A.J. (1992) Ecological disruption and human health: the next challenge to public health. Aust. J. Pub. Hlth. 16:3-5
Minns C.K (1992) Use of models for integrated assessment of ecosystem health. J. of Aquatic Ecosystem Health 1:109-118
Minns C.K., Moore J.E., Schindler D.W., Jones N.L. (1990) Assessing the potential extent of damage to inland lakes in eastern Canada due to acidic deposition IV: predicted impacts on species richness in seven groups of aquatic biota. Canadian J. Fish Aquatic Sci. 47:821-830
NAPAP (National Acid Precipitation Assessment Program) (1991) 1990 Integrated Assessment Report. Washington DC: National Acid Precipitation Assessment Program, Office of the Director
Odum E.P. (1985) Trends expected in stressed ecosystems. BioScience 35:419-422
Odum E.P. (1969) The strategy of ecosystem development. Science 164:262-270
O'Neill R.V., DeAngelis D.L., Waide J.B., Allen T.F.H. (1986) A Hierarchical Concept of Ecosystems. Princeton University Press, Princeton
Pickett S.T.A., White P.S. (1985) The Ecology of Natural Disturbance and Patch dynamics. Academic Press, San Diego
Postel S. (1989) Halting land degradation. In Brown et al, (eds.) State of the World. A Worldwatch Institute Report on Progress Toward a Sustainable Society. W.W. Norton and Co., New York
Porn I. (1984) An equilibrium model of health. In Nordenfelt L, Lindahl B. (eds.) Health, disease and causal explanations in medicine. Reidel, Durdecht
Rapport D.J. (1983a) Ecosystem medicine. In Calhoun J.B. (ed.) Environment and population: problems of adaptation. New York, Praeger pp:96-98
Rapport D.J. (1983b) The stress-response environmental statistical system and its applicability to the Laurentian Lower Great Lakes. Statistical Journal of the United Nations ECE 1:377-405
Rapport D.J. (1989a) Symptoms of pathology in the Gulf of Bothnia (Baltic Sea): ecosystem response to stress from human activity. Biological Journal of the Linnean Society 37:33-49
Rapport D.J. (1989b) What constitutes ecosystem health? Perspectives in Biology and Medicine 33:120-132
Rapport D.J. (1991) Myths in the foundations of ecology and economics. Biological Journal of the Linnean Society 44:185-202
Rapport D.J. (1992a) Evaluating ecosystem health. J. Aquatic Ecosystem Health 1: 15-24
Rapport D.J. (1992b) Defining the practice of clinical ecology. In Costanza R., Norton

G., Haskell B. (eds.) Ecosystem health - New Goals for Environmental Management. Island Press, Washington DC

Rapport D.J. (1992c) Evolution of indicators of ecosystem health. In Mckenzie D.H., Hyatt D.E, Mcdonald V.J (eds.) Ecological Indicators. Elsevier Applied Science 1:121-134

Rapport D.J (1993) Approaches to reporting on ecoytem health. Commissioned paper for the National Round Table on the Environment and the Economy, Ottawa, Canada

Rapport D.J. Friend A.M. (1979) Towards a comprehensive framework for environmental statistics: a stress-response approach. Statistics Canada (11-510), Ottawa, Canada

Rapport D.J., Regier H.A. (1994) Disturbance and stress effects on ecological systems. In Patten B.C., Jorgensen S.E. (eds.) Complex Ecology (Memorial volume in honor of G. VanDyne). Prentice Hall (in press)

Rapport D.J., Regier H.A., Hutchinson T.C. (1985) Ecosystem behaviour under stress. The American Naturalist 125:617-640

Rapport D.J., H.A. Regier, C. Thorpe (1981) Diagnosis, prognosis, and treatment of ecosystems under stress. In Barrett G.W., Rosenberg R (eds.) Stress Effects on Natural Ecosystems. John Wiley and Sons, New York

Regier H.A. (1993) The notion of natural and cultural integrity. In Woodley S., Kay J., Francis G. (eds.) Ecological Integrity and the Management of Ecosystems. St Lucie Press, Delray Beach, Florida

Regier H.A., Baskerville G.L. (1986) Sustainable redevelopment of regional ecosystems degraded by exploitive development. In Clark W.C., Munn R.E. (eds.) Sustainable Development of the Biosphere. Cambridge University Press, Cambridge

Regier H.A., Hartman W.L. (1973) Lake Erie's fish community: 150 years of cultural stresses. Science 180:1248-1255

Schaeffer D.J., Henricks E.E., Kerster H.W. (1988) Ecosystem health: I. Measuring ecosystem health. Environmental Management 12:445-455

Schaeffer D.J., Cox D.K. (1992) Establishing Ecosystem Threshold Criteria In Costanza R., Norton B., Haskell B. (eds.) Ecosystem Health - New Goals for Environmental Management. Island Press, Washington DC

Schindler D.W. (1987) Detecting ecosystem response to anthropogenic stress. Can. J. Fish. Aqua. Sci. 44(supplement):6-25

Schindler D.W. (1988) Effects of acid rain on freshwater ecosystems. Science 239:149-159

Schindler D.W. (1990) Experimental perturbations of whole lakes as tests of hypotheses concerning ecosytem structure and function. Oikos 57:25-41

Siegel I.M. (1979) The nature of the diagnostic process: a comparison with judicial decision making. Perspectives in Biology and Medicine pp:410-414

Smol J.P. (1992) Paleolimnology: an important tool for effective ecosystem management. Journal of Aquatic Ecosystem Health 1(1):49-59

Smol J.P. (1994) The importance of paleolimnological approaches to the evaluation and monitoring of ecosystem health: providing a history for environmental damage and recovery. In Rapport D.J., Calow P., Gaudet C. (eds.) Evaluating and Monitoring the Health of Large-scale Ecosystems. Springer-Verlag, New York (in prep.)

Sprugel D.G. (1991) Disturbance, equilibrium and environmental variability: what is 'natural' vegetation in a changing environment? Biological Conservation 58:1-18

Steedman R.S., Regier H.A. (1990) Ecological bases for an understanding of ecosystem integrity in the Great Lakes Basin. Proceedings of a Workshop on Integrity and Surprise, June 14-16 1988. International Joint Commission, Windsor, Ontario and Great Lakes Fishery Commission, Ann Arbor, Michigan

Suter II G.W. (1992) Ecological Risk Assessment. Lewis Publishes, Chelsea, Michigan

Suter II G.W. (1993) A critique of ecosystem health concepts and indexes. Environmental Toxicology and Chemistry 112:1533-1539

Tolba M.K. El-kholy O.A., El-Hinnawi E., Holdgate M.W., McMichael D.F., Munn R.E. (1992) The World Environment 1972-1992. Chapman and Hall, London

USEPA (1993) EMAP: the Environmental Monitoring and Assessment Program. Assessment Framework (draft)

Vogl R.J. (1980) The ecological factors that produce perturbation-dependent ecosystems. In Cairns Jr. J. (ed.) The Recovery Process in Damaged Ecosystems. Ann Arbor Science, Ann Arbor

Whitford W.G. (1994) Desertification: implications and limitations of the ecohealth metaphor. In Rapport D.J., Calow P., Gaudet C. (eds.) Evaluating and Monitoring the Health of Large-scale Ecosystems. Springer-Verlag, NewYork (in prep.)

Woodley S., Kay J., Francis G. (1993) Ecological Integrity and the Management of Ecosystems. St. Lucie Press, Delray Beach, Florida

World Resources Institute (1992) World resources 1992-93: a guide to the global environment. Oxford University Press, Oxford

World Watch Institute (1994) State of the World 1994. Earthscan Publications Ltd., London

2. ECOSYSTEM HEALTH - A CRITICAL ANALYSIS OF CONCEPTS

Professor Peter Calow
Department of Animal and Plant Sciences
University of Sheffield,
PO Box 601
Sheffield, S10 2UQ
United Kingdom

INTRODUCTION

If we are going to protect ecosystems against adverse effects emanating from human activities, we need to be able to distinguish between what is "normal" and "abnormal" for them. By analogy with humans, this is increasingly referred to as ecosystem health. The analogy can be intended in a variety of ways: for example, that there is a direct similarity between the health states of humans and ecosystems (strong sense of the analogy); that there are sufficient similarities to justify drawing a parallel (weaker sense of the analogy); that there is a similarity in the way that medical practitioners and ecologists approach diagnosis and treatment. In terms of the general promulgation of environmental protection, the use of imagery drawing an analogy with the human condition is likely to have more impact than more abstract concepts.

Yet the definition of "health" both for humans and for ecosystems has proved somewhat elusive (e.g. Costanza et al. 1992). Minimally, though, as indicated above, it has to imply that there is some norm, deviations from which represent a deterioration in health. If we could not identify a norm for which the descriptor "healthy" applied, it would not make much sense to have the descriptor in the first place.

Here, I explore the kinds of norms that we can expect to find in ecological systems and consider how they relate to the health analogy. As I have argued elsewhere (Calow 1992) there are certain kinds of norms, that follow from the strong sense of the health analogy, that are not appropriately applied to ecosystems. Even the norms associated with the less strong forms of the analogy face a number of problems, such that, in the final analysis, justification for using the analogy may be either mainly promotional or in terms of anthropocentric management practices.

ASSUMPTIONS THAT MIGHT BE ATTACHED TO THE HEALTH CONCEPT

A definition of health is elusive, even for people (Costanza et al. 1992). There do, nevertheless, appear to be a couple of unavoidable, core assumptions that are packaged into the health concept. One is a presumption that a state of normality can be defined, however loosely, for the system under consideration. And the other is the presumption that this state of normality can be specified in terms of the properties of the systems as a whole, to which the parts contribute. This is known as a holistic approach.

Different forms of the ecosystem health concept vary in the definition of the norm and in the way they envisage that the parts are involved in maintaining the whole; i.e. in the strength to which the holistic philosophy is applied. One strong sense of holism is that the norm is an optimum state, which is actively defended by appropriate and active control of the parts. This implies programming and goal-directedness achieved by active feedback control.

An alternative interpretation is that the norm represents an equilibrium of parts that arises simply because they are associated, through various kinds of interaction (food/feeder; competitor; etc.), in a system. Any complex system of parts that interact either strongly or weakly can develop equilibrium states that are more or less stable - i.e. capable of resisting change, and returning to normal after it - depending upon the properties of the system. Clearly, in principle ecosystem norms and health states might be defined in terms of these kinds of equilibria and/or stability. And indeed the concept of stability has emerged as a central feature of a definition of ecosystem health that has arisen out of recent discussions (Haskell et al. 1992).

Another definition of ecosystem health might be in terms of the parts, rather than the wholes; i.e. that a healthy system is one in which the component populations and individuals can be "healthy". For example, Rapport (1989) has suggested that disease prevalence is one sign of ecosystem distress. A special, anthropocentric, form of this is when it is defined in terms of the health and general well-being of humans. For example, increasingly ecosystems are being considered as providing services in terms of their contributions to such things as the maintenance of water cycles, the maintenance of the ozone layer, the provision of food and recreational opportunities, and as objects of aesthetic value (Westman 1977). The extent to which they are able to provide these services could be used as a definition of ecosystem health (Rapport 1992).

Finally, the health analogy might be intended not so much in terms of the state of the system but in terms of the way ecosystems are studied and managed. Just as medicine is an imperfect science, so also is ecology. Diagnosis in medicine is often achieved by a matching and synthesis of information about body processes and structures (the parts) against both objective and subjective expectations. In other words diagnosis is achieved by a balance of scientific knowledge and experience that can be enhanced by knowledge of the history of the patient.

A summary of this range of definitions of ecosystem health and their treatments of norms and relationship of parts to whole is given in Table 1.

ECOSYSTEMS ARE NOT ORGANISMIC [DEFINITION 1]

One definition of health is: an optimum state that can be actively defended (Calow 1992; Suter 1993). For organisms the optimum states can be related to Darwinian fitness; and the active control systems that have evolved to defend them, such as the immune systems, have done so under the influence of natural selection. This active control requires a cybernetic organisation with programmed "set points" and feedback control systems.

Such an organisation is an integral part of what is meant by organismic. To evolve in this way requires competition for limited resources between the units of selection, reproduction and a genetic memory. Ecosystems are not organismic in this sense, since they do not reproduce themselves as unitary wholes, do not have a common genetic memory and are therefore not subject to natural selection as unitary systems. Their component parts may have cybernetic organisation, but the rules of natural selection are such that the evolution of these is unlikely to lead to a "balanced economy" at the ecosystem level. This is because natural selection on individuals and populations will favour genotypes that maximise the command of resources even if this is at the expense of the rest of the system. The alternative is a group-selection interpretation of evolution (Dawkins 1982) that does not seem likely to be applicable for most ecosystems (Maynard Smith 1984).

If the health analogy is intended to suggest that ecosystems are subject to optimal control then it is not only misleading but seriously flawed (Calow 1993). Williams (1992) has made a similar case against the representation of the biosphere as an organism in the GAIA models.

DO ECOSYSTEMS EXIST IN STABLE STATES? [DEFINITION 2]

Ecosystems may not have optimal states but, like all complex systems, they may have stable states towards which, other things being equal, they should tend. Hence, the health norms could be and indeed have been defined in these terms. For example, a general definition of ecosystem health that emerged out of recent workshops (Haskell et al. 1992) embodies this concept:

An ecological system is healthy and free from "distress syndrome" if it is stable and sustainable - i.e. if it is active and maintains its organisation and autonomy over time and is resilient to stress.

This appears to be based on an earlier definition formulated by Karr (1991).

What properties do we expect to be associated with ecosystems in stable equilibria? The answer to this question is not straightforward (May 1981), but a combination of theory and observation suggest, in general terms, that ecological stability, the capacity to resist disturbance and recover after it, tends to be associated with a low diversity system, in which species relative abundances follow a log-normal distribution, connectance (Pimm 1982) is low, food chains are short (Lawton 1989) and there is a balanced energy economy in which the output of energy is equivalent to the input.

One of the reasons that it may not be straightforward, though, to associate particular ecosystem properties with stability is that the extent to which particular organisations are stable will depend upon their environmental context (May 1981). Thus stable environments are likely to enable the development of more diverse and connected communities. These are likely to be intrinsically fragile, but the fragility will not be manifest under normal circumstances. On the other hand disturbing environments are likely to lead to the development of less fragile, intrinsically more robust, systems, with lower diversity and connectance and possibly shorter food chains.

Paradoxically, then, anthropogenic disturbance, by causing lower diversity and potentially impacting connectance and food chain length, could lead to systems that have improved resistance and resilience. And this complicates a definition of ecosystem health that uses stability, resistance and resilience as key properties. Changes that occur under disturbance need not necessarily lead to a system that has reduced stability, in the sense of being more susceptible to subsequent disturbances in terms of resisting their effects and recovering after them.

There is also a developing view in ecology that not all or indeed not most ecosystems may be in stable equilibrium states (e.g. Yodzis 1986). This presents another problem for normative definitions of ecosystem health. These non-equilibrium models emphasise both spatial and temporal variability and focus on the importance of history and chance. Natural disturbances generate ecological space that might be filled according to a predictable sequence of species, as different species have different strategies for exploiting resources. This is the basis of succession (cf. Horn 1981). Alternatively, patches may be filled according to local availability of species, their colonising ability and chance (Yodzis 1986). Hence, here the process and its outcome is less predictable.

The extent to which communities and ecosystems are driven by these deterministic or stochastic processes is likely to vary from system to system and which dominates or occurs most frequently is likely to be something that can only be decided empirically. On the other hand, certain systems are subject to frequent disturbance and may well be more obvious candidates for stochastic controls. Amongst these, flowing-water systems are obvious examples (Townsend 1989). Interestingly, it is these that have often been used to transport wastes and that have therefore been vulnerable to polluting stresses. Because they are subject to natural disturbances (see above) and may potentially be characterised by stochastic dynamics, it could prove difficult to pick up abnormal dynamics within them.

Yet at a very basic functional level of energy economy stable equilibrium states exist, because they are based upon the laws of thermodynamics. This requires that for any stable system energy input to an ecosystem must equal output. Otherwise changes might occur as a result of the accumulation or depletion of material and energy within the system. Such imbalances might be caused by anthropogenic disturbances (e.g. input exceeding output through excessive organic loading), or by output being greater than input as might occur if, for example, the system as a whole had to do more metabolic work to combat the stress (Calow 1991). These changes would lead to predictable shifts in ecological efficiency ratios, something that has been subject to theoretical analysis (e.g., Odum 1985) but has not frequently been used as a measure of ecosystem health.

Apart from problems of measurement (Gray, this volume), another problem with these functional measures is that they do not seem to be coupled very tightly to particular species compositions. There appears to be much functional redundancy in most species assemblages (Schindler 1987) as would be anticipated on theoretical grounds from the way communities are structured (above; and see also O'Neil et al. 1986). So functional stability need not signal structural stability; and structural instability need not lead to functional changes. Hence, it is probably better to focus on structure rather than function as a measure of stress, since it is likely to be more sensitive.

In conclusion, the entire edifice of the ecosystem health analogy is based on the presumption that ecosystem norms exist. However, a major conclusion of this section is that normative ecology, in so far as it depends upon equilibrium and/or stable states, does not always apply (Ehrenfeld 1992).

A PRAGMATIC APPROACH TO ECOSYSTEM HEALTH [DEFINITION 4]

Whether or not norms can be precisely defined in the way intended by definition 2 of Table 1 is open to some doubt. Moreover, predictability of community structure if not functioning seems likely to be context dependent; i.e. more easily defined locally. At the same time, even inexperienced observers can recognise ecosystem degradation when it is sufficiently bad, and especially in the light of local experience.

This therefore emphasises the importance of monitoring programmes directed at interrogating particular systems. However, even with this approach, norms are still presumed and usually defined comparatively by reference to similar systems thought or known not to be impacted (e.g. for structure see Wright et al. 1989; for a combination of function and structure see the ecosystem integrity approach of Karr 1991) or the same system at some earlier time. There is not necessarily any scientific understanding of why the normal states are as they are; minimally, they are taken as given. Rather the scientific approach is used in designing the sampling programme and critically assessing the results.

The major problem with the design of these monitoring programmes is that it can never be assumed that the "control" is necessarily in the same state that the observed would have been in in the absence of impact. There is probably no absolute solution to this problem. However, the expeditious application of careful sampling design and good sense can improve confidence in identifying deterioration in quality and ascribing causes. Moreover, it is at least possible in principle to supplement the observational approach with hypothesis testing in a more rigorously controlled experimental setting. Thus hypotheses could be developed about the cause of a divergence from expectations and explored, for example by ecotoxicological assays that seek to consider if known contaminants could, at the concentrations observed, lead to changes observed in structure and function. This, for example, is the basis of the toxicity identification procedures (TIE) that are being developed in ecotoxicology (Doi 1994) and again is analogous to the way medical diagnosticians do their work, checking diagnoses based on examinations with laboratory analyses.

NORMS IN TERMS OF SERVICES [DEFINITION 3]

Ecosystems can be represented as if they provide services for their components. Thus in this sense a "healthy ecosystem" might be represented as one that contains "healthy" populations either in general or with respect to a particular reference species. "Health" here presumably means the capacity of populations to persist in the circumstances presented. But a complication is that most populations can adapt to disturbance. Thus Menzel (1977) demonstrated, in artificial ecosystems, that there was only a transient inhibition of phytoplankton primary production when exposed to toxicants and recovery occurred within a couple of weeks. And Blanck, Wangberg and Molander (1988) have used pollution-induced community tolerance (PICT) of this kind as an indicator of prior pollution exposure. Such "community responses" might occur as a result of the replacement of sensitive with insensitive species and/or sensitive with insensitive genotypes within species. There is certainly evidence for the evolution of tolerance to anthropogenic disturbance, by selection of stress resistant genotypes, in plants (Bradshaw and Hardwick 1989) and animals (Calow 1989). The tolerant genotypes might have different physiological and/or life-history traits relative to less tolerant genotypes (Calow 1989). It is, of course, possible that these adaptations are disturbance-specific (Blanck et al. 1988) and that an enhanced capacity for tolerance may trade-off with other components of fitness (Calow 1991).

Increasingly it is recognised that ecosystems also provide services for humans; for example, in the form of atmospheric support, weather modulation, production of biomass, nutrient cycling, recreation, aesthetics etc. (Westman 1977). We might therefore define ecosystems' health in terms of their contribution to those services (Rapport 1992); in other words in terms of our own health and well being. This moves away from the definition of ecosystem health in terms of naturally defined norms, to anthropocentric ones - and there is then a direct relationship between human and ecosystem health. Moreover, we can view the services as goal states, and aim to achieve them through active monitoring and management. The system can now be treated properly as a cybernetic one, where the definition raised earlier and rejected - in which an optimum state is actively defended - becomes appropriate. This is because the active control is imposed by ourselves from outside. Here the remaining concern must be whether the ecosystem management that is implemented is sustainable in the long term.

CONCLUSIONS

The holistic philosophy implicit to some definitions of ecosystem health, (Nos. 1 and possibly 2 in Table 1) is hard to defend. Ecosystems are not superorganisms in which parts are actively controlled for the sake of the well being of the whole (definition 1). Nor is it obvious in natural ecosystems that parts will always interact in a way that leads to equilibrium and/or stable states (definition 2). On an ecological scale these may be disrupted by environmental disturbances and variability; and on an evolutionary scale there is an expectation that species will be selected for maximising there own command of resources rather than balancing their needs with those of the whole ecosystem and this is potentially destabilizing.

Table 1. Some definitions of ecosystem health and their key elements.

	Norm that is Presumed	Relationship of Parts to Whole	Form of Definition
1	optimum state	parts actively directed to well-being of whole	holistic/organismic
2	equilibrium/stable state	parts equilibriate, resist change and return to norm after it	holistic/systems
3	health/well-being parts	whole provides appropriate conditions for parts	
	health/well-being people	whole services people	anthropocentric
4	defined pragmatically by combination of scientific and subjective judgement	performance of parts judged against subjective/objective expectations	pragmatic

Yet there are patterns to nature, and even inexperienced observers can pick up signs of ecosystem degradation when they are sufficiently serious. This suggests an approach to ecosystem management and protection that is similar in some respects to medical diagnosis (definition 4 in Table 1). It involves interrogation of particular systems by monitoring one or several of their characteristics - integrating this information and making judgements about quality on the basis of both an objective and subjective assessment of the observations. Identification of problems and, indeed, general understanding of ecosystem structure and function can be improved by an iteration between the field monitoring programmes and experimental analysis. Ideas should be expressed as hypotheses about the nature and cause of deterioration that can be rigorously and critically tested. There is a similarity here between medical and ecological diagnosis.

Finally, there is the anthropocentric view that ecosystems are healthy when they service human health in its broadest sense. This is probably the basis for all ecological management practice. It also provides a basis for valuing ecological services that can be used in the development of an economics approach to management of the environment and management of our affairs within the environment (Pearce 1993).

REFERENCES

Blanck H., Wangberg S-A., Molander S. (1988) Pollution-induced community tolerance - a new ecotoxicological tool. In Cairns J., Pratt J.R. (eds.) Functional Testing of Aquatic Biota for Estimating Hazards of Chemicals. ASTM-STP 988. American Society for Testing and Materials, Philadelphia pp:219-230

Bradshaw A.D., Hardwick K. (1989) Evolution and stress - genotypic and phenotypic components. Biological Journal of the Linnean Society 37:137-155

Calow P. (1989) Proximate and ultimate responses to stress in biological systems. Biological Journal of the Linnean Society 37:173-181

Calow P. (1991) Physiological costs of combating chemical toxicants; ecological implications. Comparativ Biochemistry and Physiology 100C:3-6

Calow P. (1992) Can Ecosystems be healthy? Critical consideration of concepts. Journal of Aquatic Ecosystem Health 1:1-5

Calow P. (1993) Ecosystems not optimised. Journal of Aquatic Ecosystem Health 2:55

Costanza R., Norton B.G., Haskell B.D. (eds.) (1992) Ecosystem Health - New Goals for Environmental Management. Island Press, Washington D.C., California

Dawkins R. (1982) The Extended Phenotype. Freeman, London

Doi J. (1994) Complex mixtures. In Calow P. (ed.) Handbook of Ecotoxicology, Volume II. Blackwell Scientific Pubs., Oxford pp:289-310

Ehrenfeld D. (1992) Ecosystem health and ecological theories. In Costanza R., Norton B.G., Haskell B.D. (eds.) Ecosystem Health - New Goals for Environmental Management. Island Press, Washington D.C., California pp:135-143

Haskell B.D., Norton B.G., Costanza R. (1992) Introduction: What is ecosystem health and why should we worry about it? In Costanza R., Norton B.G., Haskell B.D. (eds.) Ecosystem Health - New Goals for Environmental Management. Island Press, Washington D.C., California pp:1-19

Horn H.S. (1981) Succession. In May R.M. (ed.) Theoretical Ecology (2nd Ed). Blackwell Scientific Pubs., Oxford pp:253-271

Karr J.R. (1991) Biological integrity: a long-neglected aspect of water resource management. Ecological Applications 1:66-84

Lawton J.H. (1989) Food webs. In Cherrett J.M. (ed.) Ecological Concepts. B.E.S.

Symp. Blackwell Scientific Pubs., Oxford pp:43-78

May R.M. (1981) Patterns in multi-species communities. In May R.M. (ed.) Theoretical Ecology (2nd Ed). Blackwell Scientific Pubs., Oxford pp:197-227

Maynard Smith J. (1984) The population as a unit of selection. In Shorrocks B. (ed.) Evolutionary Ecology. Blackwell Scientific Pubs., Oxford pp:193-202

Menzel D.M. (1977) Summary of experimental results: controlled ecosystem pollution experiment. Bulletin of Marine Science 27:1-7

Odum E.P. (1985) Trends expressed in stressed ecosystems. Bioscience 35:419-422

O'Neill R.V., De Angelis D.L., Waide J.B., Allen, T.F.H. (1986). A Hierarchical Concept of Ecosystems. Monographs in Population Biology 23. Princeton University Press, Princeton, New Jersey

Pearce D. (1993) Economic Values and the Natural World. Earthscan Pubs. Ltd., London

Pimm S.L. (1982) Food Webs. Chapman and Hall, London

Rapport D.J. (1989) Symptoms of pathology in the Gulf of Bothnia (Baltic Sea): Ecosystem response to stress from human activity. Biological Journal of the Linnean Society 37:33-49

Rapport D.J. (1992) What is clinical ecology? In Costanza R., Norton B.G., Haskell B.D. (eds.) Ecosystem Health - New Goals for Environmental Management. Island Press, Washington D.C., California pp:144-156

Schindler D.W. (1987) Detecting ecosystem responses to anthropogenic stress. Canadian Journal of Aquatic Sciences 44 (Suppl. 1):6-25

Suter G.W. (1993) A critique of ecosystems health concepts and indexes. Environmental Toxicology and Chemistry 12:1533-1539

Townsend C.R. (1989) The patch dynamics concept of stream community ecology. Journal of the North American Benthological Society 8:36-50

Westman W.E. (1977) How much are nature's services worth? Science 197:960-964

Williams G.C. (1992) GAIA, nature worship and biocentric fallacies. The Quarterly Review of Biology 67:479-494

Wright J.F., Armitage P.D., Furse M.T., Moss D. (1989) Prediction of invertebrate communities using stream measurements. Regulated Rivers: Research and Management 4:479-494

Yodzis P. (1986) Competition, mortality and community structure. In Diamond J.M., Case J.J. (eds.) Community Ecology. Harper and Row, New York pp:480-491

3. QUALITATIVE AND QUANTITATIVE CRITERIA DEFINING A "HEALTHY" ECOSYSTEM

François Ramade
Laboratory of Ecology and Zoology, Bat 442, University of Paris-Sud
F - 91405 Orsay
France

WHAT IS ECOSYSTEM HEALTH ?

As Costanza et al. have previously pointed out (1992), the terms "health" and "integrity" are used widely in policy documents and scientific publications regarding environmental protection. However, although the term "healthy" ecosystem has been in use for a long time, its representativeness regarding ecological reality and therefore relevance may still be contended. Defining a "healthy" ecosystem is a more complex task than might be expected.

The analogy between human health and ecosystem integrity seems at first glance very coarse since the complex ecological systems possess very specific properties because of their higher level of organization - taken apart their homeostasis properties are shared by all living entities from the simplest one, the cell, to the whole biosphere. Accordingly, in order to define the welfare of large ecosystems, we would likely rely on more specific concepts related to the structure and function of complex systems which stand as the focal point of the present work.

Before addressing this topic here it will be useful to review the history of the term "healthy" as applied to ecosystems. Two major factors contribute to the widespread acceptance of the "ecosystem health" term. The first stems from concerns which relate to the environmental issues involved in the protection of human populations and therefore public health. Accordingly, specialists from the biomedical disciplines have taken the lead in developing conservation policies by extending medical models to the protection of the natural environment (Norton 1991).

Another explanation for the introduction of the term "ecosystem health" stems from Lovelock's Gaïa hypothesis (1972). One of the major assertions of Lovelock's basic concepts of biosphere lay in the paradigm that the earth biosphere as a whole can be equated to an individual living organism. Therefore, it may be assumed that ecosystems and more complex systems such as landscapes are more or less "healthy" - as healthy as an individual organism - depending on the extent of the anthropogenic degradations to which they are exposed.

The concept of "ecosystem health" has been strongly questioned by a number of ecologists, most of them having abandoned the organismic theory of ecological systems. This position however, was not antagonistic to some concepts of the Gaïa hypothesis such as the regulation of geochemical processes at global scale by the living communities (Margulis and Lovelock 1989).

Notwithstanding the above critique the idea of "ecosystem health", which covers a wide array of affairs from the optimal one to the decline and even the death of communities, is still widely applied by some ecologists especially the ones involved in ecological restoration. For example, J. Berger uses the term of ecological health in the introduction to "Environmental Restoration" (1990). As Ehrenfeld (1992) advocates, we shall rely thereafter on this term, keeping in mind that in order to avoid "damaging a useful, indeed necessary idea, it cannot be associated closely with a particular model of ecosystem structure and function".

ECOSYSTEM THEORY AND ITS RELEVANCE FOR ASSESSING ECOSYSTEM HEALTH

The present concept of ecosystem is that of a non-equilibrium system or rather a system in a metastable equilibrium so that environmental changes - even discrete ones - will disrupt the steady state ecological balance and lead to another equilibrium (Frontier 1992). Natural disturbances such as fire or hurricanes are more and more perceived by ecologists not so much as catastrophic events than as fundamental processes, the occurrence of which are necessary to maintain the ecosystem at a given stage of development. This is particularly obvious when an intermediate successional stage is concerned; for instance in the case of open mediterranean woodland or that of tropical savannah where the cyclical burning of vegetation stands as a prerequisite for maintaining its characteristic grass and shrubs cover. Even in climax ecosystems such as tropical rainforests, there is evidence showing that forest regeneration depends primarily on tree falls and (or) of flooding due to episodic hurricanes (Colinveaux 1989).

The early theory claiming a causal link between diversity and stability in ecosystems is presently considered obsolete. In fact it has been demonstrated that simple communities of low diversity consisting of euryoecic species (tundra for example) show a far stronger ability to damp major disturbances than evolved climax ones like those of rainforests. However, such an observation does not invalidate the homeostasis concept which still applies as long as a given threshold of disturbance is not exceeded. It may be inferred from these considerations that a healthy ecosystem is a system one which maintains its capacity for self-organization.

Another difficulty arising from ecosystem theory in defining ecosystem "health" is the integration of time and space scales. To what extent does the occurrence of a change in a so called indicator of ecosystem integrity provide early warning of the decline of the system as a whole? This obviously raises the problem of scale both from temporal and spatial viewpoints.

Another practical difficulty arises when one is considering large ecosystems; namely the boundaries problem. This is of importance from a managerial as well as from a scientific standpoint. After all, little attention has been paid to biogeographical and even jurisdictional boundaries as far as environmental goals are concerned. The case of acid rain abatement in Europe is a good example. Progress in defining more comprehensive management actions requires a better understanding of how scientifically adequate ecosystem boundaries can be used in integrating the work of a number of agencies which manage parts of large ecosystems (Norton 1992).

Keeping in mind these general considerations regarding ecosystem concepts, we shall now attempt to sort out some basic principles which will allow us to define more precisely an "healthy" ecosystem. Norton (1991) has put forward five axioms that could be considered major terms of reference in order to complete this task:

1) The axiom of <u>Dynamism</u>. Flux of matter and energy is a basic property of ecosystems so that at any given time an ecosystem is in a temporary steady state of a dynamic process. Moreover, as a consequence of successional processes, ecosystems develop and eventually age over time although when the climax stage is reached, they will keep their metastable equilibrium as long as average abiotic conditions do not change. This is especially true for example in tropical rainforests in South East Asia which where almost unaffected by quaternary glaciations.

2) The axiom of <u>Relatedness</u>. In an ecosystem all ecological processes are interrelated so that a significant disturbance affecting only one process will impinge on the whole (Figure 1).

3) The axiom of <u>Hierarchy</u>. From the most complex to the simplest, processes unfold in units within units which may be ranked along a decreasing scale of space and time. Additionally, this hierarchy is often fractal with each subunit being similar to the preceding one at a lower scale (Frontier 1992).

4) The axiom of <u>Creativity</u>. Ecological processes are creative" since they are the very origin of biological productivity. The energy flow through ecological systems is the vehicle for this productivity which in turn provides self organization and supports an increasing biomass and organizational complexity (i.e. biodiversity) a physical measure of which is the level of negentropy reached by the system.

5) The axiom of <u>Differential Fragility</u>. As with natural disturbances, ecological systems may in their autonomous homeostasis processes buffer human caused disruptions, until a given threshold level of disturbance is reached after which a collapse of the whole system occurs usually quite abruptly.

From a functional viewpoint, taking into account the preceding axioms, a "healthy" ecosystem could be simply defined by reference to the criteria of sustainability, namely as an ecosystem which retains its level of species richness, diversity and productivity provided that the natural environmental parameters stay themselves unchanged. This definition applies to a majority of cases related to anthropogenic disturbances with the exception of successional communities.

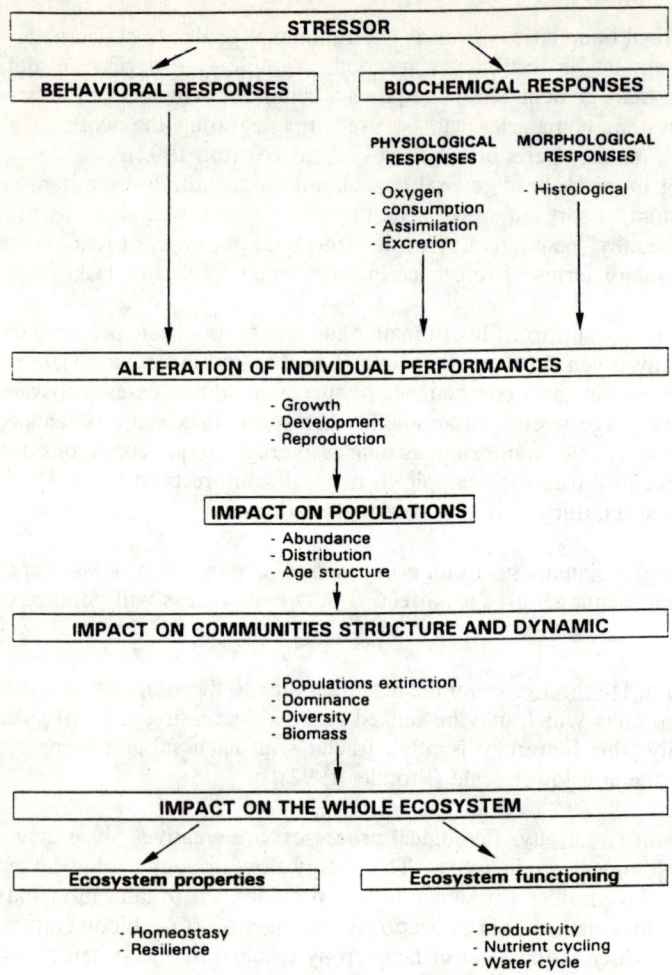

Figure 1. Scheme of the overall impact of a stressor acting initially on a given species but impinging on the whole ecosystem.

Criteria Regarding the Physicochemical Component of Ecosystems

A number of "elementary" qualitative and quantitative criteria may be used as a reference point for assessing the "health" of large ecosystems including their abiotic and biotic compartments. With respect to the abiotic component of an ecosystem, terrestrial and aquatic habitats have to be addressed separately due to their specific physico-chemical characteristics. Terrestrial ecosystems may suffer, for example from various environmental degradations, the most obvious being alterations of soils and even worse, soil erosion. With respect to aquatic ecosystems both freshwater and marine, their state of "health" depends primarily on the absence of pollution. They may be additionally exposed to other degradations such as the spilling of suspended matter or other disturbances impacting both the water column and the sediment. Accordingly a "healthy" aquatic ecosystem has to be devoid of such alterations.

Criteria Regarding the Ecosystem's Communities

The living part of an ecosystem - namely its biocoenosis - stands as an extremely complex community of organisms in a dynamic steady state. This is controlled by the mutual interaction of its living components as well as by its abiotic component, especially climatic, and broadly speaking its physico-chemical ones. The overall result of this equilibrium translates into structural and functional characteristics.

Structural criteria defining a "healthy" ecosystem - An undisturbed community may be defined as a community which possesses the maximum species richness and diversity allowed by its level of evolution into the serial gradient depending on its geographical location and (or) on its edaphic components. Another major criterion structurally defining an "healthy" ecosystem relies on the key species occurrence and relative abundance as key species rank among the first of the various living components of the community to be affected by an anthropogenic disturbance.

Functional criteria defining a "healthy" ecosystem - The disturbance of an ecosystem can impinge on fundamental ecological processes such as decomposer activity, nutrient cycling, biogeochemical cycles and productivity. Detrimental effects may occur even if structural changes such as a decline of the most sensitive populations are still unobserved or discrete. Therefore, the best criteria defining an "healthy" ecosystem are unquestionably functional. These may be found among those which define the optimum for the fundamental ecological processes - considering the overall environmental characteristics of the ecosystem (climatic, edaphic factors for example); the major quantitative criteria defining an "healthy" ecosystem may include the relevant local values of primary and secondary productivity, soils or water microbial efficiency in the nutrient cycling, overall condition of the nitrogen or phosphorous biogeochemical cycle as well as the hydrological cycle (in terrestrial ecosystems).

STRESS EFFECTS ON COMMUNITY STRUCTURE AND DYNAMICS

Natural communities consist of complex assemblages of hundreds to hundreds of thousands of species which are in a dynamic equilibrium and which interact with the complex physico-chemical components of their ecosystems. The biota specific to a given ecosystem play a major role in the fundamental ecological processes and also modify their physical and chemical environment. Conversely, the latter influences the composition and diversity of the community. Therefore, it is of the utmost importance for the assessment of the state of "health" of an ecosystem to appraise the effects of anthropogenic stresses on the community structure and dynamics. Such studies will require specific methodologies in order to predict the potential effects of a pollutant on living resources.

STRESS RELATED CHANGES IN DIVERSITY AND SPECIES RICHNESS: THE INDICATOR SPECIES

The reduction in density and species richness of chronically polluted habitats is the primary factor altering community structure. Indicator species are also useful for the assessment of disturbances on communities. Despite a number of criticisms regarding their use, they can provide valuable information on the early effects of various stress chemicals, either physical or of biotic origin.

The properties required for a good bioindicator are the ability to display a response to discrete changes in the environment induced by chemicals and therefore be hypersensitive and reliable in its response. The other desirable characteristics are the ability to respond more specifically to a given chemical family of pollutants, to give a fast response and to be of easy use in monitoring. If these prerequisites are met, the change in value of carefully selected indicators would allow a measurement both of the rate and of the value of change induced by a chemical in a community as a whole.

SPECIES DIVERSITY: THE USE OF BIOECOLOGICAL INDICES

The concept of ecological diversity integrates both the number of species (richness) and their relative frequency in a given biota. Decreased diversity has been used to assess gross environmental degradation in ecosystems for several decades. Among these which are the most frequently applied is Margaleff's (1956, 1968):

$$H = (1/N) \; Log_2 \; N! - \sum_{I=1}^{S} Log_2 \; n_i!$$

where N = total number of individuals; S = total number of species; n_i = number of individuals of the *i*th species.

The major limitation of this index stems from the fact that it requires a census of all the individuals from the stressed community. Therefore, for practical reasons ecologists prefer other indices that may be applied to samples, because communities cannot usually be entirely numbered. The index of Llyod, Zad and Karr (1968) has been routinely applied to the study of pollution in the Seine River (Figure 2):

$$H' = C/N \left(\log_{10} N - \sum_{i=1}^{S} n_i \log_{10} n_i \right)$$

where C is the number of classes of frequency expressed in bits (for 10 classes, C=3,3219).

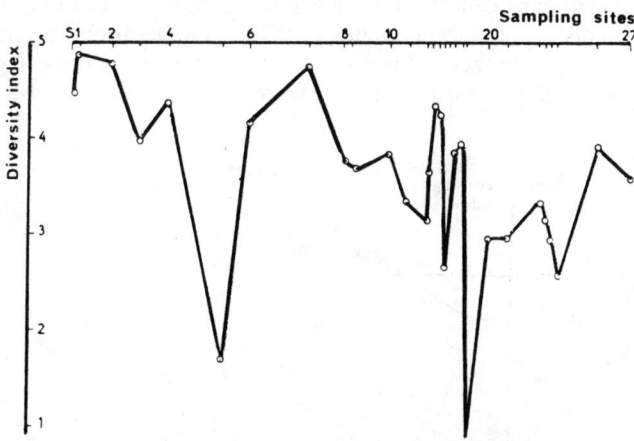

Figure 2. Variation of the diversity index of the community of benthic diatoms from the Seine river as a result of sewage discharges (arrow) (after Costes in Ramade 1992).

However, Shannon's index (Shannon 1963) has been by far the most widely used in the study of disturbed ecosystems especially those disturbed by pollution:

$$H' = - \sum_{i=1}^{S} (n_i/N) \; Log_2 \; (n_i/N)$$

Nevertheless, various limitations hamper the effectiveness of the diversity index for assessing the effects of chemicals on community structure.
Shannon's index, for example, gives the same weight to systematic units of the same abundance, whatever their taxonomic level and *a fortiori* affinities.

In order to avoid these limitations Osborne, Davis and Linton (1980) have proposed the use of a hierarchical diversity index devised by a formula expanded from Pielou's to include three taxonomic levels (familial, generic and specific):

$$HDI = H'_{(F)} + H'_{(G)} + H'_{(S)}$$

where $H'_{(F)}$ is the familial component of the total diversity, $H'_{(G)}$ the generic component and $H'_{(S)}$ the specific component of the total diversity.

The most universal criticism of the application of the Shannon diversity measure to assess the impact of anthropogenic disturbances on ecosystems structure is the misleading interpretation of data from depauperate communities resulting from the large influence of the evenness component (Godfrey 1978). Another problem related to the use of a diversity index is the absence of linearity of a community response to a given gradient of increasing stress (pollution for example) (Figure 3).

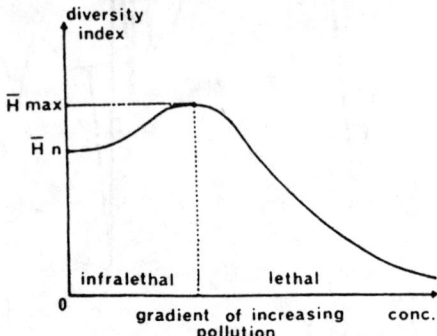

Figure 3. Effect of a gradient of increasing stress (pollution for example) on the diversity index of a community (after Ramade 1993).

During an initial stage, at sublethal exposure, some dominant species among the most pollutant sensitive will decrease their abundance, increasing the evenness and therefore the diversity index value. It is only at the onset of lethal exposure which triggers the disappearance of species less pollutant sensitive that the index will drop.

Boyle et al. (1990) have carried out a theoretical study in order to assess the validity of the use of 16 indexes for studies on freshwater ecosystems. They started with three communities that differed in their species richness but with the same abundance-value distribution curves. Their overall conclusion was as follows: although the Shannon index properly represents changes in species richness, the index does not generally afford a good representation of changes occurring in the stressed communities and even can lead to misleading conclusions.

Accordingly, diversity indices reflect changes in ecosystems structure only during periods of severe stress. In moderately altered ecosystems, changes in dominance strongly affect equatability thereby hampering the effectiveness of diversity indices in distinguishing degraded communities from unstressed ones. An improvement of this index use can be obtained in applying it on a taxonomic group previously identified and carefully studied for its value as an indicator from a given disturbance - especially pollution. Lichens afford a good illustration of such a use of diversity index for the assessment of the impact of air pollution by SO_2 on forest ecosystems and for the prediction of the ecotoxicological effects associated with the observed levels. For example, an Index of Atmospheric Purity (IPA), was defined, derived from biocoenotic surveys revealing that the species diversity of lichens communities were related to the level of SO_2 pollution (Ramade 1987):

$$IPA = (S/100) \left[\sum_{i=1}^{S} (Q \times f)\right]$$

where S = number of lichen species in the area sampled; f = frequency of each species; Q = index of toxiphobia of each species.

EFFECTS OF DISTURBANCE ON THE FREQUENCY DISTRIBUTION OF SPECIES

The means by which species populations are controlled is one of the most important issues related to the ecological niche. In degraded habitats, the stress resulting from the occurrence of a given disturbance- whether chemical or physical- will necessarily impact the extent of the resources space occupied by each species, according to the level of tolerance or sensitivity of the species. As a consequence, the balance existing between the various components of the community will be disturbed since the occurrence of pollutants or any other human disturbance will force modifications of the interspecific competition thereby leading to the vanishing of the most sensitive populations. The frequency distribution of species will be therefore more or less distorted by the chronic pollution of an ecosystem.

The use of importance value distribution curves will provide an interesting opportunity, from a methodological standpoint, to assess the response of a given community to a given stressor. For example, in our own research (Ramade et al. 1983), we compared the importance value distribution curves computed from experimental data related to two lentic communities of benthic invertebrates: one from ponds contaminated by insecticides and the other from meadow ponds used as a control (Figure 4). The latter fits well with Preston's distribution whereas the community from field pond fits an intermediary position between Preston's model and a log-linear one. Generally speaking, communities strongly disturbed by any pollutant are prone to fit a log linear distribution.

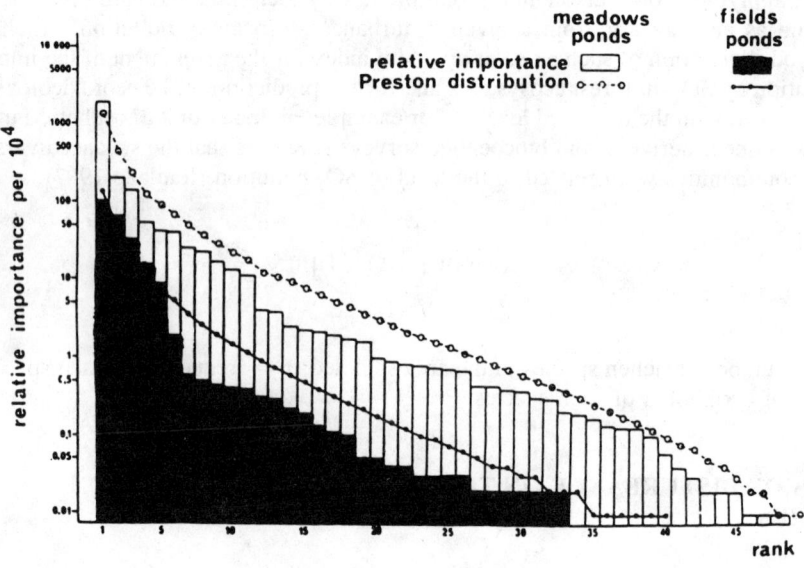

Figure 4. Comparison of the relative abundance distribution curve of a stressed macroinvertebrates benthic community from freshwater ponds chronically polluted by pesticides (i.e. field ponds) with undisturbed ones taken as control (meadow ponds) (after Ramade et al. 1985).

PRINCIPLES PERTAINING TO THE ASSESSMENT OF DISTURBANCE EFFECTS ON COMMUNITIES AT A FUNCTIONAL LEVEL

An ecosystem is a complex system resulting from the association of a living community and of its physico-chemical environment. It may be considered both from a structural and a functional standpoint. As a consequence, it may be assumed that an anthropogenic disturbance which will induce changes in the community structure will impinge on its functions, notwithstanding possible correlative changes in the physico-chemical structure of the degraded habitats due to a modification of the activity of its whole biota or especially of its decomposer biomass.

In spite of extensive studies on acid rain impinging upon large lakes (e.g. Anderson and Olson 1986) and on temperate forests, or on the ecological degradation of the Great Lakes ecosystems (e.g. Harris et al. 1990), little information exists yet about most aspects of the effects of anthropogenic stresses on large ecosystems. There is still a critical need for investigations on the appraisal of effects on species diversity, organismal genetics, community stability, physical resources, energy flow, nutrients cycling, from compounds of widespread use like pesticides (Pimentel and Edwards 1982).

It has been frequently argued that indicators of the structure of ecosystems stand as a rather inaccurate method to appraise the effective state of health of a whole ecosystem. In order to provide an answer to these criticisms, Ulanowicz (1992) has devised an index which takes into account the functional characteristics of ecosystems by relating the overall state of activity of an ecosystem to the organization of its trophic network. As Ulanowicz (1992) points out, the various criteria that define an "evolved" ecosystem according to Odum (1969) may be grouped into four categories: greater species richness (and diversity), higher niche specialisation, more developed cycling and feedback, and greater overall activity. Accordingly, Ulanowicz has devised a single index, the so-called "network ascendency" which integrates into a single algorithm all four of these criteria. Ascendency is the product of two factors, one that gauges the level of the system activity and another that appraises the level of trophic organization.

If T_{ij} is the transfer from compartment i to j ($i,j = 1,2,3, ...n,$) of some system products, assuming that exogenous inputs derive from the zero compartment and exogenous outputs flow to the compartment $n+1$, then:

$$T = \sum_{i=0}^{n} \sum_{j=1}^{n+1} T_{ij} \qquad (1)$$

Since T defines the size of the system, any increase in T will be defined as growth. The organization of the network is more complex to quantify. Using the Shannon-Weaver index of uncertainty, itself related to the quantity of information carried by the system, it is possible to measure how the flows T_{ij} relate to each other:

$$H = \sum_{i=0}^{n} \sum_{j=1}^{n+1} (T_{ij}/T) \text{Log}(T_{ij}/T) \qquad (2)$$

Knowing how the flows are interrelated in the network (trophic webs for example), its uncertainty is reduced by an amount so-called average mutual information:

$$I = \sum_{i=0}^{n} \sum_{j=1}^{n+1}(T_{ij}/T) \text{Log}[(T_{ij}/T) / (\sum_{k=1}^{n+1} T_{ik})(\sum_{m=0}^{n} T_{mj})] \qquad (1)$$

It has been demonstrated that $H > I > 0$ for any given network of exchange T_{ij}. Any growth in the organization I of an ecological system can be equated to development. Ulanowicz (1992, 1986a) has called the product of T and I the "system ascendency" (A) for it is quite obvious from an ecological standpoint that in the absence of major anthropogenic (or natural) stress, succession proceeds towards mature stages accordingly in the direction of increasing ascendency. In order to avoid bias in the application of ascendency to the measure of ecosystem health given the problems of scale, a descaled version such as $I = H/T$ has been proposed which is correlated to the stage of development and therefore to ecosystem health. On the other hand, in order to solve another problem related to the necessity of taking into account the distinction between inputs and outputs, Ulanowicz (1992) has further proposed a more complex index S, the so called "Scope for ascendency" which distinguishes between inputs and outputs in the system.

Whatever the usefulness of indexes measuring the ecosystem function (e.g. Ulanowicz), a number of practical criteria must be taken into account as a prerequisite in order to assess the effective state of health of an ecosystem. Clinical ecology (Rapport 1992) requires not only theoretical considerations but also that a number of investigations be carried out in the field both in "pristine" natural ecosystems and in those disturbed by human impact. The present state of affairs shows that there is still a considerable need for research to be initiated in this respect.

SUCCESSION AND RECOVERY

There are few ecological data regarding changes in successional patterns in ecosystems exposed to chemicals or to any other kind of anthropogenic disturbances. Broadly speaking, permanent exposure to a stressors especially chemicals maintains the biota in an early successional stage where only some opportunistic, euryoecic, pollutant-tolerant and short-lived species of high reproductive potential (r strategist) are able to

Another kind of disturbance is related to the exploitation of a juvenile ecosystem by a mature one at their ecotone. A good example is afforded by the coupling of an urban technological system which may be considered as mature with another ecosystem which is an earlier successional stage such as agroecosystems. The result of this kind of stressing action is that the whole productivity of the exploited ecosystem is diverted to maintain a high level of evolution (negentropy) of the mature urban system while the exploited one -the agroecosystem- is maintained at a poorly differentiated stage, the only feed-back being the injection of auxiliary energy into the latter in order to sustain its productivity.

CRITERIA FOR ASSESSING EFFECTS OF DISTURBANCE ON THE FUNDAMENTAL ECOLOGICAL PROCESSES OF THE ECOSYSTEM

Extreme anthropogenic alterations exert a drastic impact on energy flow and biogeochemical cycles in ecosystems. The stress resulting from pollution provides demonstrative examples of such effects. Though internal feedbacks and control can minimize overall disturbances as various forms of redundancy exist inside each biota at lower levels of ecosystem organization, tolerant species being able to replace sensitive ones, the decline of some dominant - or key species - may strongly impinge on the ecosystem functioning as a whole.

Effects on Biomass and Productivity

The action of chronic stress (not only from pollution) on photosynthesis ranks among the major potential effects on fundamental ecological processes as it impinges on community productivity. Discrete actions can affect the primary productivity of terrestrial or aquatic ecosystems. For example, SO_2 may affect the primary productivity of the Scot pine (*Pinus sylvestris*) forest ecosystems at only 10 ppb, even though no morphological or anatomical damage can be detected at this concentrations (Grodzinsky et al. 1984).

Photosynthetic activity in phytoplanktonic algae may be inhibited at doses lower than 1 ppb of some pesticides. Research carried out in our own laboratory brought evidence of a strong decrease of productivity in phytoplankton and filamentous algae in freshwater habitats contaminated by various herbicides especially Chlortoluron, Triazine and Neburon coming from adjacent fields (Goacoulou and Echaubard 1987).

Effects on secondary productivity though less thoroughly explored are also potentially important in the assessment of effects at the ecosystem level. Among the most investigated is the study of the action of acid rains and pesticides on invertebrates and on freshwater fisheries exposed to forest spraying or acidification (Ide 1967). For example, we have demonstrated a strong decrease in benthic macroinvertebrate productivity as a result of contamination of freshwater habitats by organochlorine insecticides run off from surrounding fields although the level reached in water and sediments was well under the thrive.

acute toxic concentrations (Figure 4). A sharp decrease in the abundance and biomass of zooplankton has been observed in acidified lakes (Almer et al. 1974; Stenson and Oscarson 1985). A sharp decline has also been observed in salmon fisheries as a consequence of organochlorine insecticide pollution. For example, the New Brunswick salmon streams were badly affected by DDT spraying intended to control the spruce budworm. It was demonstrated that the decrease in the secondary productivity was not so much a consequence of the insecticide toxicity for fishes but of the collapse of the aquatic insect populations upon which young salmon normally feed (Ide 1967). Effects of river acidification on the productivity of salmon fisheries has also been well documented and careful studies have shown a strong decline in direct relationship with the level of acidification experienced (Leivestad and Muniz 1976; Watt et al. 1983)

Some attempts have been made to model the effects of a chemical on some major components of ecosystems structure and (or) function in order to develop a new methodology of ecosystem risk analysis. For example, O'Neill et al. (1982) have devised a method of extrapolating laboratory toxicity data to aquatic ecosystem effects such as decreased productivity or reduction in fish biomass. This implies translating laboratory data into changes in the parameters of an ecosystem model known as the Standard Water Column Model (SWACOM). The translation is effected through knowledge of toxicological mode of action. The approach is achieved by a simulation of the effects of a toxic substance across different trophic levels - therefore on the relationship between nutrients, phytoplankton, zooplankton, and fishes. Various scenarios are run and each scenario affects populations interactions and alters both the level and the nature of the risk to ecosystem processes. Moreover, the method analyses the uncertainties associated with laboratory measurements and extrapolation, and risk estimates are given in terms of probabilities.

Relying on this SWACOM method, O'Neill et al. found a good representation of the effects of phenol and quinoline in the stimulated aquatic ecosystem. However Cairns, Smith and Orvos (1988) have stressed that validating experimental simulation results is essential to predict effects of chemicals on natural systems because many so called ecosystem simulation models are low in environmental realism and because vast differences may occur between systems. Consequently, field studies on natural ecosystems are invaluable for achieving this ecological validation.

Research on the effects of stressors carried out in full scale ecosystems may prove long, complex, sometimes inaccurate due to the vast number of variables, and expensive. Accordingly, the mesocosm approach has been developed for predicting the effects of human changes on ecosystems, particularly those of pollutants. (e.g. Mauck et al. 1976; Vosheel 1989). The problem which still impedes the mesocosm methodology for the assessment of effects of stressors stems from their cost, which prevents their wider use in applied ecology. Research is in progress in our own laboratory (Caquet et al. 1992) as well as in various European countries in order to develop mesocosms which would prove cheaper but would still give a realistic representation of natural ecosystems.

Effects on Decomposers and Nutrient Cycling

Another major parameter of ecosystem functioning is related to nutrient cycling. Though less explored, the impairment of decomposer activity by a chemical or physical stress may lead to major alterations and even to the collapse of the whole ecosystem. For example it was observed that the effects of air pollutants on saprophagous insects that feed on dead litter in forest ecosystems disrupts the nutrient cycle, as the consumption of dead organic matter by the soil arthropods is much slowed (Sheehan 1984). Others studies have shown effects of the same order regarding leaf decomposition in the waters of acidified lakes.

The study of potential effects of a given pollutant on microorganisms involved in the decomposition of dead organic matter in the various ecosystems ranks among the major parameters required for a sound assessment of the ecological risk of a chemical at the ecosystem level. The appraisal of the potential effects on the biogeochemical cycles of major nutrients stands also as a prevalent issue regarding risk assessment at the ecosystem level.

Some attempts have been made to rely on the impairment of the nitrogen cycle for the assessment of the potential ecotoxicological effects of a chemical. For instance, Mathes and Shultz-Berendt (1988) have assessed the effect of Aldicarb on nitrification in agroecosystems. Starting from these previous studies, Mathes and Weideman (1990) proposed an integrated approach to the effects assessment of chemicals on terrestrial ecosystems including an appraisal of the effects on primary production by comparison between a part of the ecosystem exposed to a given chemical and another part of the same ecosystem used a control. Other major ecosystem parameters important to its functioning that were taken into account for this assessment were litter decomposition and the level of impairment of the Nitrogen cycle. This kind of approach involves several prerequisites but especially a suitable indicator system. The criteria requested for selection of these indicators are summarized in Figure 5.

The spatio-temporal scale of exposition has to be considered as well as the temporal variability. The selection of reference organisms must incorporate both short-lived and long-lived species.

CONCLUSION

There is no simple solution to a quantitative and quick assessment of ecosystem "health". The various criteria discussed and evaluated here provide an unidimensional evaluation which may in no way be taken as an absolute appraisal of this state of "health". It has been argued (Costanza 1992) that the overall functional properties of an ecosystem could be reduced to three components: vigour, V (an expression of productivity and metabolism), organization O (structure, biodiversity) and resilience, R. A formulation of an overall and comprehensive system health index (*HI*) could be proposed from these major components quoted above which could provide also a measure of sustainability: $HI = V*O*R$.

Whatever the soundness of the considerations discussed previously in this paper, the assessment of ecosystem health both from structural and functional standpoints needs further basic studies.

Among the various criteria required to assess sustainability are the identification of key-species and the improvement of the knowledge of functional bioindicators as a tool of assessment of potential effects of anthropogenic disturbances on communities. The need for accurate assessment of ecosystem "health" also requires further research on the impingement of various anthropogenic perturbations, especially chemical pollution, on fundamental ecological processes such as primary and secondary productivity as well as at the decomposers level and on nutrient cycles - not to mention research regarding their effects on biogeochemical cycles.

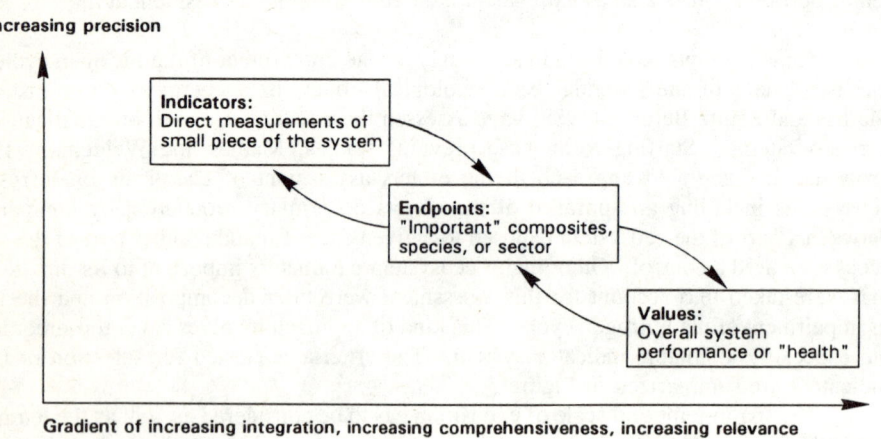

Figure 5. Relationship between criteria of ecosystem health and the increasing complexity and relevance of assessment allowed (after Costanza 1992).

On the other hand since the effects of anthropogenic stress, especially the ones of chemicals, are not usually restricted to an ecosystem, but impinge on large areas that involve both terrestrial and aquatic habitats, an additional part of the assessment of effects requires information from the field of landscape ecology. Indeed, the assessment of effects on ecosystem health needs ultimately to be addressed at a regional scale since it affects the landscape physiognomy and the ecological fundamental processes at complex systems level.

REFERENCES

Almer B., Dickson W., Eckstrom C. (1974) Effects of acidification on Swedish lakes. Ambio 3(1):30-36

Berger J.J. (ed) (1990) Environmental Restoration. Island Press, Washington DC

Boyle T.C., Smillie G.M., Anderson J.C., Beeson J.R. (1990) A sensitivity analysis of nine diversity and seven similarity indices. Journ. Water Poll. Control. Feder. 62:749-762

Cairns J. Jr., Orvos D. (1989) Ecological consequence assessment: predicting effects of hazardous substances upon aquatic ecosystems using ecotoxicological engineering. In Ecological Engineering. An Application to Ecotechnology. John Wiley and Sons, New York 411pp.

Cairns J. Jr, Smith E.P., Orvos D. (1988) The problem of validating simulation of hazardous exposure in natural systems. Barnett C.C., Holmes W.M. (eds.) Proc. Summer Conf. Soc. Comp. Simul. Int. San Diego, California pp:448-454

Caquet T., Thybaud E., Le Bras S., Jonot O., Ramade F. (1992) Fate and biological effect of Lindane and Deltamethrin in freshwater mesocosms. Aquatic Toxicology 23:261-278

Colinveaux P.A. (1989) The past and future of the amazon. Scient. Am. 260(5): 68-74

Costanza R., Norton B.G., and Haskell B.D. (1992) Ecosystem Health - New Goals for Environmental Management.. Island Press, Washington and Covelo, California, 270pp.

Ehrenfeld D. (1992) Ecosystem health and Ecological theory. In Costanza R., Norton B.G., Haskell B.D. (1992) Ecosystem Health - New Goals for Environmental Management. Island Press, Washington DC pp:135-143

Frontier S. (1992) Théorie des écosystèmes. In Frontier S., Pichot-Viale D. (eds.) Ecosytèmes: Structure, Fonctionement, Évolution. Masson, Paris pp: 279-365

Goacoulou J., Echaubard M. (1987) Influence des traitements phytosanitaire sur les biocoenoses limniques: le phytoplancton. Hydrobiologia 148:269-280

Godfrey P.J. (1978) Diversity as a measure of benthic macroinvertebrates community response to water pollution. Hydrobiologia 57:111-122

Grodzinski W., Weiner J., Maycock P.F. (1984) Forest Ecosystems in Industrial Regions Springer Verlag, Berlin 223pp.

Harris J., Sager P., Regier H.A., Francis G.C. (1990) Ecotoxicology and ecosystem integrity: the Great Lakes examined. Envir. Sci. Technol. 24(5):598-603

Ide F.P. (1967) Effects of forest spraying with DDT on aquatic insects of salmon streams in New Brunswick. J. Fish. Res. Board. Canada 24:769-805

Jackson D.R., Watson A.P. (1977) Disruption of nutrient pools and transport of heavy metals in a forested watershed near a lead smelter. J. Environ. Qual. 6:331-338

Keenleyside M.H.A. (1967) Effects of forest spraying with DDT in New Brunswick on food of young atlantic salmon. Journ. Fish. Res. Board of Canada. 24(4):807-822

Lovelock J.E. (1972) Gaia as seen through the atmosphere. Atmospheric environment 6:575-578

Llyod M., Zad J.H., Karr J.R. (1968) On the calculation of information, theoretical measures of idversity. Am. Mid. Nat. 79 (2):257-272

Margaleff R. (1956) La teoria de la informacion en ecologia. Mem. Real Acad. Ciencias, Barcelona 32:373-449

Margaleff R. (1968) Information theory in ecology Gen. Syst. 3:36-71

Margulis L., Lovelock J.E. (1989) Gaia and geognosy. In Rambler, Margulis, Fester, Global Ecology: Towards a Science of Biosphere. Academic Press pp:1-30

Mathes K., Schultz-Berendt V.M. (1988) Ecotoxicological Risk assessment of chemicals by measurements of nitrification combined with a computer simulation model of N-Cycle. Toxicity Assessment: An Intern. Journ. 3:1271-286

Mathes K., Weidemann G. (1990) A baseline-ecosystem approach to the analysis of ecotoxicological effects. Ecotox. Environ. Safety 20:197-202

Niederlehner B.R., Pratt J.R., Buikema A.L., Cairns J. (1986) Comparison of estimates of hazards derived at three levels of complexity. In Community Toxicity Testing Amer. Soc. Test. Mat., TP 920:30- 48

Norton B.G.(1992) A new paradigm for environmental managment. In Costanza R., Norton B.G., Haskell B.D. (eds.) Ecosystem Health - New Goals in Environmental Management. Island Press, Washington and Covelo, California. pp:24-41

O'Neill R.V., Gardner R.H., Barnthouse G.W., Suter S., Hildebrand S.G., Gehrs C.W. (1982) Ecosystem risks analysis:a new methodology. Envir. Toxicol. Chem. 1:167-177

Osborne L.L., Davies R.W., Linton K.J. (1980) Use of hierarchical diversity indices in lotic community analysis. Journ. Appl. Ecol. 17:567-580

Pascual J.A., Peres J. (1992) Effects of forest spraying with two application rates of Cypermethrin on food supply and on breeding success of the Blue Tit (*Parus Coeruleus*). Envir. Tox. Chem. 11:1271-1280

Pimentel D., Edwards C.A. (1982) Pesticides and ecosystems. Bioscience. 32(7):595-600

Ramade F. (1977) Ecotoxicologie. Masson, Paris 214 pp.

Ramade F. (1987) Proposal of ecotoxicological criteria for the assessment of the impact of pollution on environmental quality. Toxicol. Environ. Chem. 13:189-203

Ramade F. (1987) Ecotoxicology. John Wiley and Sons, Chichester and New York 262pp.

Ramade F. (1992) Précis d'Ecotoxicologie. Masson, Paris 97pp.

Ramade F. (1993) Methodologies for evaluating the ecotoxicological impacts of chemicals on ecosystems. Sgomsec-Scope nø 10 (in press)

Ramade F., Echaubard M., Le Bras S. Moreteau J.C.(1983) Influence des traîtements phytosanitaires sur les Biocoenoses limniques. Acta Oecologica 4:3-22

Rapport D.J. (1992) What is clinical ecology? In Costanza R., Norton B.G., and Haskell B.D., Ecosystem Health - New Goals in Environmental Management. Island Press, Washington and Covelo, California. pp:144-156

Shannon C.E. (1963) The mathematical theory of communication In Shannon C.E., Weaver W. (eds.) The Mathematical Theory of Communication. University of Illinois Press, Urbana, Illinois 117pp.

Ulanowicz R.E. (1992) Ecosystem health and trophic flow networks. In Costanza R., Norton B.G., Haskell B.D. (eds.) Ecosystem Health - New Goals for Environmental Management. Island Press, Washington DC pp:190-206

4. THE RELATIONSHIP BETWEEN ECOSYSTEM HEALTH AND DELIVERY OF ECOSYSTEM SERVICES

[1]John Cairns Jr. and [2]James R. Pratt

[1]University Center for Environmental and Hazardous Materials Studies and Department of Biology, Virginia Polytechnic Institute and State University Blacksburg, Virginia 24061-0415
USA
[2]School of Forest Resources, Pennsylvania State University, University Park, Pennsylvania 16802-4302
USA

INTRODUCTION

Ecosystem services is a term applied to any functional attribute of natural systems that is easily perceived by policy makers as beneficial to human society. If a society were highly environmentally literate, it would probably accept the assertion that every ecosystem function is, in the long term, beneficial to human society. However, because environmental literacy is less robust than it should be for the present decisions society must make, it would be best to use only those services that are likely to be persuasive at the present level of literacy. A biologically impoverished natural system delivers services of poorer quality and dramatically reduced quantity and may even perform some functions (such as concentrating persistent toxic substances) that are harmful to human health. As one reviewer noted, a non-impoverished system may also concentrate toxic substances. However, it seems probable that biotic impoverishment may lead to reduced functional capabilities, which, in this case, would mean reduced ability to transform obnoxious wastes into something less so. This attribute, which ecologists regard as an ecosystem function, would be regarded by laypersons as an ecosystem service which otherwise might have to be carried out in waste treatment plants or other treatment systems. However, the correspondence of ecosystem health and the delivery of environmental services is poorly documented. The rapid rate of ecological destruction and the concomitant rapid increase in human population make a drastic reduction in ecosystem services per capita inevitable by the year 2000 and in the period immediately thereafter when the human population is expected to exceed 10 billion. This discussion explores some of the ways in which a more robust understanding of the relationship between ecosystem health and ecosystem services can be developed.

The significant problems we face cannot be solved at the same level of thinking we were at when we created them.

Albert Einstein

COMMUNICATING THE CONSEQUENCES OF POOR ECOSYSTEM HEALTH

Skinner (1938) argued persuasively that human behavior is selected or determined by its consequences and that substantial numbers of people cannot be expected to change their behavior as a result of information or advice alone, especially when the information is about a distant future. He further stated that people might follow advice when the information from the advice giver has led to beneficial consequences in the past; however, this situation requires that people experience the reinforcing consequences of prior compliance with similar advice givers or similar rules. Such operant learning or response selection by reinforcing consequences is difficult or impossible when reinforcement lies in the future or punishing consequences are unclear, uncertain, or remote. Lack of clarity, uncertainty, and remoteness are all common characteristics in making judgments about complex, multivariate systems such as ecosystems. Geller (1994) espouses a well-known, but rarely used, educational principle: *tell them and they'll forget --demonstrate and they'll remember -- involve them and they'll understand.* In short, education/awareness sessions and informational packages that did not involve the participants or provide incentives or punishing consequences are not successful in motivating newspaper recycling (e.g. Geller et al. 1975) or water conservation (Geller et al. 1983). Geller (1992) has classified intervention programs into multiple tiers or levels. Level 1 interventions are least intrusive and target the maximum number of people for the least cost per person (e.g. behavioral prompting through signs and public service announcements). Higher level and more influential intervention processes require increased costs in terms of materials and personnel. For example, a feedback or incentive/reward process changes the behavior of more individuals but is much more costly. The following quote from Geller (1994) is instructive.

> *Unless business can make money from environmental products or politicians can get elected on environmental issues, or individuals can get personal satisfaction from experiencing environmental concern, then individuals and organizations will simply do whatever competes with environmentalism if they see the payoff as greater.*
>
> C. Seligman (personal communication, March 8, 1990)

Figure 1 identifies three basic types of actively caring (AC). The importance of this figure is that behavior-focus AC is intervention attempting to influence another individual's behavior in desired directions (e.g. giving, rewarding, or correcting feedback, demonstrating or teaching desirable behavior, developing or implementing a behavior change intervention program). The literature on "social traps" (Brockner and Rubin 1985; Costanza 1987; Cross and Guyer 1980; Platt 1973; Teger 1980) is very consistent with the Geller materials cited here.

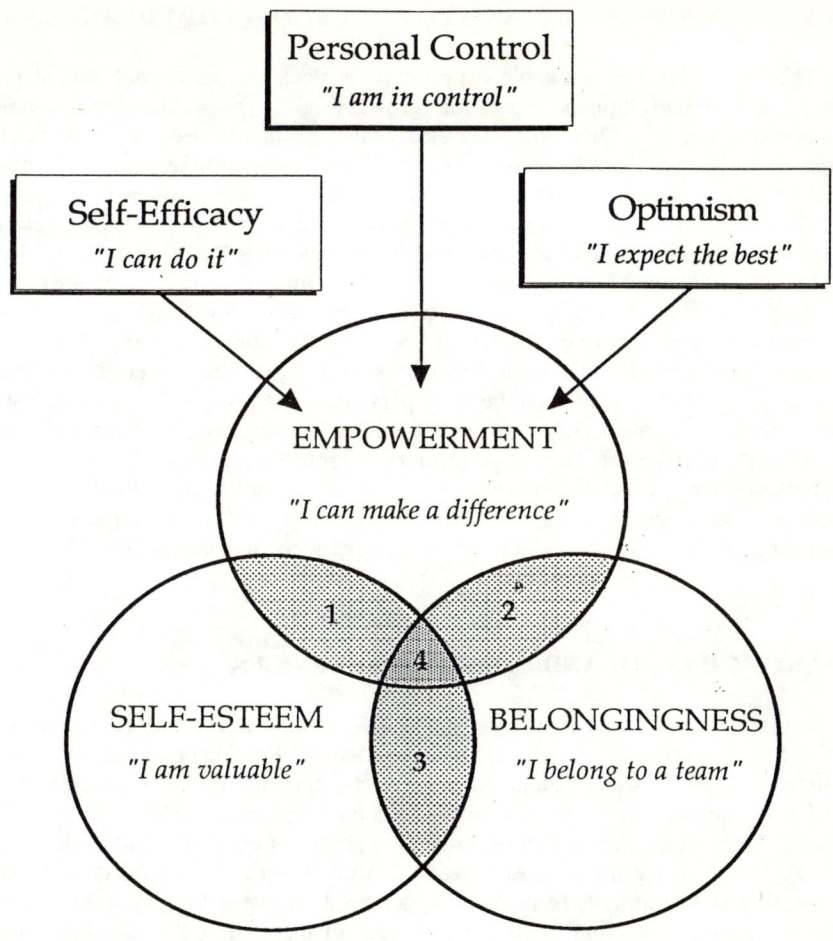

1. I can make *valuable* differences.
2. We can make a *difference*.
3. I am a *valuable team* member.
4. We can make *valuable differences*.

Figure 1. The actively caring model (adapted from Geller 1991).

MAJOR WARNINGS TO HUMANITY ABOUT ECOSYSTEM HEALTH

There have been two notable expressions of deep concern about the interactive effects of accelerated human population growth and human-induced environmental destruction and change. They were released nearly simultaneously, and, although the contents differ, they both make nearly identical assumptions: (a) the present relationship of human society to the environment is not suitable for the sustained beneficial use of the environment and (b) the consequences of not altering the relationship will be deleterious, perhaps catastrophic, to human society. The World Scientists Warning to Humanity was signed by over 100 Nobel Laureates and over 1600 scientists that included many members of the U.S. National Academy and its counterparts in other countries. The other document is a joint statement by the Royal Society of the London and the U.S. National Academy of Sciences (1992). Both documents were signed by leaders in mainstream science, and, therefore, they should be quite persuasive to those who respect mainstream science. However, large numbers of people will either be unaware of this information or essentially indifferent to it. The primary reason for this indifference is almost certainly the rather low level of environmental literacy, even among individuals with one or more advanced degrees. As a consequence, the Society of Sigma Xi and a number of collaborating organizations and institutions sponsored a meeting on this subject (Blackburn 1992).

ECOSYSTEM HEALTH AND ECOSYSTEM SERVICES

In this discussion, the term ecosystem health is used to indicate robust ecological condition (e.g. high resistance to disturbance, resiliency following disturbance). The term ecosystem services is value laden, since it refers only to functions perceived to be beneficial to humans. On a local level, ecosystem services include such functions as nutrient cycling and the production of food. The ability of ecosystems to cycle nutrients is often exploited for the final stages of waste treatment from sewage plants or the small-scale assimilation of nutrients from septic systems. Local stream ecosystems also retain eroded sediments, protecting downstream water quality. Food production includes harvesting from both agroecosystems and native ecosystems (e.g. harvesting of fishes). At the regional or landscape level, ecosystems are exploited for fiber and fuel wood. Usually, local depletion is quite rapid, so sustained use usually occurs over a large area (e.g. timber management for wood products). At the biosphere level, the combined functioning of ecosystems maintains the gas balance of the atmosphere within broad boundaries. A close correlation should exist between ecosystem health and ecosystem services. While this may indeed be true, this question is too important to be judged by anecdotal evidence and intuition. Therefore, a number of hypotheses deserve rigorous testing:

(1) There is a close correspondence between ecosystem health and the production of ecosystem services.

(2) Deterioration of ecosystem health does not affect ecosystem services uniformly.
(3) Decline of ecosystem services important to human society can be predicted accurately from measures of ecosystem health.
(4) Ecosystem services of importance to human society may vary markedly even for ecosystems in robust health or condition.
(5) For ecosystems in robust health with highly variable delivery of ecosystem services, these changes can be predicted from climatic, life cycle, or other similar information.
(6) Ecosystem services decline linearly with declines in ecosystem health.
(7) Some ecosystem services are important globally and, therefore, must be maintained at a global level.
(8) Some ecosystem services are local and regional, and management at those levels will ensure appropriate delivery of services.

The list of hypotheses given in this discussion refer to relationships between ecosystem health and ecosystem services, where services are defined as those functions which are perceived by humans to be useful. In this context, the relationship between function and services is then one of cultural environmental literacy and perception. It is quite possible to use the term ecosystem function in the hypotheses and elsewhere in the text under these circumstances, but the choice was made to use services because the term ecosystem services is more likely to be understood by and attract interest of non-ecologists, whereas the term ecosystem functions is cast in the more disciplinary mode. This seems inappropriate for the context of this discussion, although it is conceptually correct.

These are not strongly focused or easily tested hypotheses such as one might expect to find in a grant proposal. There is an understandable fear on the part of some scientists that these general statements might undercut strongly discipline-oriented sciences. However, the National Research Council (1992) report used as two of the criteria suggested for determining whether an ecological restoration had been successful as: (1) the ability for self-maintenance or self-perpetuation and (2) integration into the larger ecological landscape in which the restored patch occurs.

Each of these is an important consideration and each is difficult to measure. Does this mean that they should not be stated? If one is interested in the philosophy of science, they should be stated even though they are not cast within rigid disciplinary boundaries. With regard to ecological restoration, Bradshaw (1987) indicated that this rapidly developing field is the acid test for theoretical ecology. Science progresses by considering awkward questions, and scientists should not hesitate to do so even though there may be temporary adverse consequences. Webster's Third International Dictionary defines hypothesis as a proposition tentatively assumed in order to draw out its logical or empirical consequences and so test its accord with facts that are known or may be determined. Readers can undoubtedly devise hypotheses in addition to these, but, if the validity of these hypotheses could be determined, it would enormously benefit the development of a management plan to ensure sustainable delivery of ecosystem services.

BIODIVERSITY AND ECOSYSTEM SERVICES

Much attention has been given in recent years to the topic of biodiversity, particularly in tropical rain forests. Inadequate attention, however, has been given to the relationship between biodiversity and the ability of ecosystems to function properly or, in the context of this discussion, the ability to deliver ecosystem services. Two of the most notable books on this subject are by Wilson (1989) and the National Research Council (1992). These two volumes summarized the problems and put them in perspective. Abundant evidence in both books (and in many of the citations given there) indicates that the diversity of species varies dramatically from one part of the world to another. The number of species of insects in a single tropical rain forest tree may be astonishingly high (e.g. Ervin (1988) found 3000 species of beetles in a single plot in the Amazonian forest, more than had previously been know for all of Brazil), while the number of species in some greatly simplified ecosystems such as deserts is quite low. Areas with both high and low numbers of species are each capable of delivering ecosystem services. It is also fairly well established that functional rates (e.g. detritus processing) are notably different in low and high species areas. However, this difference may be due to factors other than the number of species, such as temperature, rainfall, underlying geological formations, etc., and there may be cause/effect interactions in complex ways among these. The concept of ecological redundancy is very important in this regard.

ECOLOGICAL REDUNDANCY

For many years, a confusing idea has persisted in applied ecological literature called, variously, functional redundancy or ecological redundancy (Walker 1992). This idea is rooted in equilibrium theory and suggests that ecosystems have homeostatic mechanisms (because they are cybernetic systems) which allow them to maintain function in the face of stress-induced structural changes. There have been few experimental tests of this idea, but it is especially germane to a discussion of ecosystem health and services.

The services an ecosystem renders might be considered similar to its ecological functions. For example, ecosystems process detritus; consequently, adding detritus to the system in the form of sewage takes advantage of this ecosystem service, and, as long as the additional sewage does not alter the provision of the basic and added ecological services, the ecosystem might remain healthy. Interestingly, the addition of sewage to aquatic ecosystems such as streams is well known to have adverse effects, typically resulting in the loss of pollution sensitive taxa and their replacement by pollution tolerant forms. Invariably, the species richness of the system diminishes (Schaeffer et al. 1988), even though the ecosystem service is largely unaffected. Since even low levels of sewage input have adverse effects on stream ecosystems, environmental legislation in most western countries regulates dischargers to minimize the problem described in this example. The fallacy in the concept of ecological redundancy lies in the separation of structure and function (Cairns and Pratt 1986). Ecosystems might be considered functionally redundant for several compelling reasons; three are examined briefly here.

First, the number of species in ecosystems might be increasing. This idea, the hypothesis of evolution as entropy, suggests that increasing diversity in systems is a result of a trend toward maximum disorder. There is no solid evidence to suggest that species richness has consistently increased through time, especially since key functional components of ecosystems, such as bacteria, cannot be effectively studied from the fossil record.

An alternative explanation for functional redundancy is the observation that ecosystems have apparent homeostasis (Odum 1985). That is, even a stressed system may maintain primary production, biogeochemical cycles, or nutrient recovery in the early stages of stress. This suggests that systems are capable of maintaining function when the structure is changing. Observations of this apparent functional conservatism may indicate that most processes are substrate limited. That is, the rate at which most processes occur is limited by whatever component of the process is in short supply, commonly known as Liebig's law of the minimum. In freshwater ecosystems, for example, primary production is often limited by phosphorus. If stress within the ecosystem reveals that primary production is unaffected, it might be concluded that some primary producers (generally algae from a diverse community) are redundant. However, this process is substrate limited, and stress within the system does not typically change the supply of the limiting substrate. Because most organisms are substrate limited, they are not operating at capacity, so a reduced number of processors (i.e., fewer numbers or kinds of algae present after stress) simply increase their individual processing rates to a higher, but still substrate limited, capacity. The result is a measurement of apparent homeostasis when, in fact, community members have been lost and processing rates within the tolerant forms have increased (Levine 1989).

A third explanation for the apparent observation of functional redundancy is more mundane: researchers may be unable to measure ecological function precisely. This explanation is perhaps the simplest, since it requires the fewest assumptions about how ecosystems operate. The ability to measure ecological functions precisely is less than the ability to measure ecological structures (Pratt 1990), and functional changes are considered to be poor predictors of change in aquatic ecosystems (Schindler 1987). In this case, the apparent observation of functional conservatism can be explained simply by the limitations of the measuring techniques.

On a few occasions, measurements of functional alterations in stressed communities have been attempted using intensive measuring methods (Van Voris et al. 1980). A comparable experiment in aquatic systems (Pratt and Rosenberger 1993) has also been done. In the former case, soil core microcosms were treated with heavy metals, and changes in both macronutrient retention and power spectrum of microcosms carbon dioxide flux showed coupling of functional changes. In this case, no obvious structural changes occurred (the plants lived, but no investigation of meso- or microbiota was done). The latter experiment was a similar test in aquatic microcosms and showed that structural changes and changes in the power spectrum of continuously collected pH data were closely linked. Gross structural measurements (biomass, species richness, and composition) showed no effect of toxic material at the lowest doses, but the power spectrum measurement showed marked changes in the pattern of daily pH change. The power spectrum identified the dominant frequency (e.g. 24 hr cycle) of pH change. The daily pH cycle is analogous to the daily production and respiration cycles. At higher doses, no remnant of the expected diurnal functional pattern was evident, and structural changes were pronounced. In other words, the function of the system was changing as the species structure changed. No redundancy was evident in either system that allowed maintenance of function, but measuring the functional change was problematic, requiring several thousand measurements to observe the signature of the functional response. Therefore, it seems reasonable to conclude that the inability to measure functional changes is a result of measurement techniques, not of underlying redundancy in ecosystems. This is a classic case of a type II statistical error (accepting a null hypothesis when it is, in fact, false).

There is a danger in claiming or considering functional redundancy as a means of rationalizing species loss. The loss of, or change in, species in human-influenced ecosystems may lead to changes in ecosystem services (functions) that are valuable but not easily predicted using current measurement methods. Species interactions result in ecological functions, so it is possible to infer system function from structural information (e.g. Karr et al. 1986), but it is impossible to infer ecological structure from functional information. In addition to the loss of species made possible by conceiving of systems as functionally redundant, individuals within species may also appear to be redundant, allowing the loss of additional genetic diversity under human-induced stress. Changes in structure result in changes in function and, therefore, ecosystem services.

While the number of examples of ecosystem alteration resulting in changes in services is numerous, only a few are presented here for clarification. The deciduous forest of northeastern North America once contained American chestnut and American elm trees as important, if not dominant, members of a tree community that included about two dozen members at a given site. Were these trees functionally redundant? A number of other tree species are also producing living tissue and providing habitat for other organisms. The loss of these trees from the forest has not resulted in a wholesale collapse of the system, but rather the ability of this ecosystem to provide services has diminished. At a utilitarian level, the wood of the chestnut is no longer available for building supporting beams of homes, which it was so prized for in the past century.

A second example might be the wholesale alteration of stream ecosystems by fisheries managers who have presided over the repeated introduction of exotic species (e.g. brown trout) and the concomitant extirpation of native species. Fisheries management in streams is now based primarily on annual subsidies since natural reproduction cannot, apparently, keep pace with the need to rear fish bamboos. Of course, fisheries managers are not wholly responsible for the alteration of stream ecosystems - for example, an increase in recreational fishing has altered ecosystems and they may be more or less vulnerable to increased stress from this use. In the case of recreational fishing, one ecosystem service (production of fish) has taken precedence over other functions to the extent that the ecosystem service must be perpetually subsidized. Organismal resources have been lost from these manipulated ecosystems because of structural change (species replacement); in addition, genetic structure (hatchery trout are less genetically variable than native trout) has been lost and ecosystem functions (services) have been altered.

THE STANDARDIZATION TRAP

In the late 1940s and early 1950s, receiving system (the ecosystem into which wastes were discharged) standards were reasonably common, though limited in extent. For example, dissolved oxygen concentration, pH, conductivity, suspended solids, and a variety of other chemical/physical measurements were made. Biological oxygen demand (BOD) was a very common measurement, and, although incubations were carried out in the laboratory, the intent was to predict what would happen in a stream. The BOD was customarily carried out with a complex inoculum of microorganisms, and the measurement was intended to simulate functional processes occurring in natural systems. Even then and for some years before, biological measurements commonly used "indicator species" or measurements of the condition of commercially, recreationally, or ecologically important species or, in relatively rare cases, entire communities (e.g. Patrick 1949). Chemical/physical measurements were easily standardized, as was the BOD; however, biological/ecological measurements were difficult to standardize. For example, much literature dating to the first half of the century referred to "indicator species," but such species were not always found in the ecosystem of concern. Additionally, even then, taxonomy and systematics (the identification of these organisms) was not in favor at most academic institutions in the United States and, to a considerable extent, in other developed countries as well. The attention had shifted at that time to molecular biology and other laboratory-based disciplines and subdisciplines. The marketplace for persons capable of identifying species had worsened. Nevertheless, substantial numbers of ecologists still existed, but, regrettably, as Harte et al. (1992) have demonstrated, most of these were interested in population biology rather than systems ecology. Additionally, a stigma was attached to applied ecology in which examination of pollution problems was considered, at best, second-class science. This has been partially reversed by the development of the new field of conservation biology, which is more acceptable to theoreticians, and continues to place considerable emphasis on anthropogenic effects on ecosystems. Now, in the 1990s, there is a return to receiving system standards. However, this time around, the standards are biological goals, not chemical ones. Individual goals for each possible contaminating chemical failed to take into account the cumulative effects of many stresses present at the same location and could not assess the success of restoration efforts because even when individual chemicals are gone, damage can persist. "Biocriteria" provides a means of evaluating whether protective actions are working by quantitatively evaluating the structure and function of the resident aquatic community (Bascietto et al. 1990). They are a quantifiable expression of the goals of the Clean Water Act and amendments of restoring and maintaining the biological integrity of the nation's waters. While biological goals were originally thought to be too variable to be of much regulatory use, regionalizing biological goals while standardizing field methods for collecting biological data has made biocriteria a tractable regulatory tool (Hughes and Larsen 1988). Karr (1991) provides a review of biocriteria currently employed in the United States.

Enormous amounts of money were spent by federal, regional, and local levels of government in the 1970s and 1980s to improve waste treatment systems. However, the benefits to ecosystems of these improvements have never been rigorously examined even though certain "end-of-the-pipe" improvements were made. This is curious because the value of such a vast expenditure of funds should have been validated or confirmed in the ecosystems that were purportedly to benefit.

One reason this was probably never done is standardization. Waste treatment systems can be standardized and standardized chemical/physical tests can be used to check their performance. However, the complex mixture of microorganisms that are responsible for the chemical transformation so important to the treatment process cannot be examined in detail. Ecosystem measurement, particularly in aquatic ecosystems such as rivers, lakes, and wetlands, has received comparatively little funding support and attention. This may be because such systems are not easily standardized and understanding the data requires considerably more professional judgment (since they are complex, multivariate systems) than does the examination of the performance of the less complex, technological treatment systems.

Another reason may be that, about the same time, enormous attention was being given to toxicity testing with a limited array of test species, primarily aquatic. This involved selecting a relatively small array of species that could easily be maintained under laboratory conditions and could be cultured so that their size and other characteristics were relatively uniform. In addition, some types of response or lack of response to the toxicant or other form of stress could be attained in a relatively short period of time by people with relatively modest training and in test systems that were fairly inexpensive, especially compared to chemical tests of exotic compounds. In fact, an enormous percentage of the toxicity testing was done with only three species - a fish (fathead minnow), a macroinvertebrate (often *Ceriodaphnia*), and a microscopic plant (*Selenastrum*). Not only could the organisms be raised in the laboratory under carefully controlled conditions (regulating size, appearance of different life-history stages etc.), but the test containers, the dilution water, and other test conditions could be relatively easily standardized as well. However, all of this standardization meant reduction of variability and, thus, less and less environmental realism. Somehow, the fact that these environmentally unrealistic test containers and limited array of test species were being used to predict what happened in complex, highly variable systems escaped the attention of the regulators and decision makers. Thus, because of the drive towards standardization, the connections to natural, highly variable complex systems not amenable to standardization were markedly diminished.

The reason for emphasizing these historic events is that the same tendency could easily develop in measurements for ecosystem health. This could occur because of the well-established tendency of scientists and engineers to depend on standardization and standard methods and to reject methods not amenable to standardization.

CONCLUDING REMARKS

It is astonishing, given the present human population size and rate of growth and the present amount of ecological destruction and the continuing rate of this destruction, that no substantive attention has been given to the relationship between ecosystem health and the delivery of ecosystem services per capita. Harte et al. (1993) have identified a major source of this problem by pointing out that most of the publications in ecological journals were at the population level. They might have added that an even smaller number (if any) were devoted to the relationship between ecosystem services and human population size, affluence, and use of technology.

This situation revolves around two other major factors. (1) The level of environmental literacy is astonishing low in academic institutions, among political leaders, and, most important, in the public at large. Persuasive evidence of this exists with every Earth Day that attempts to acquaint all of these groups with the most basic information on human population growth and the impact of the growth itself and the lifestyles of individuals upon the environment. (2) Most science is reductionist and only a small percentage is integrative; this is surely responsible for the fact that the most obvious relationships in a landscape or global perspective are frequently ignored.

Cautious optimism, however, is justified for a number of reasons. (1) the rapidly expanding interest in both the academic community and the general public in healing ecological damage through ecological restoration; (2) the increased recognition that applied and theoretical ecology do not occur along a linear gradient, but rather are more appropriately described in terms of the proportionality of each in a particular investigation (see Figure 2A,B --we are indebted to John Harte of the University of California at Berkeley for the ideas in this graphic); (3) some manufacturers are beginning to consider the environmental costs of extraction of raw materials; the energy and environmental costs in manufacturing a product; and the environmental, social, and economic costs of reincorporating both the product and the waste products into the environment to eliminate deleterious effects; (4) major attempts are ongoing to improve environmental literacy, such as the Talliores Agreement and a symposium sponsored by the Society of Sigma Xi (Blackburn 1992).

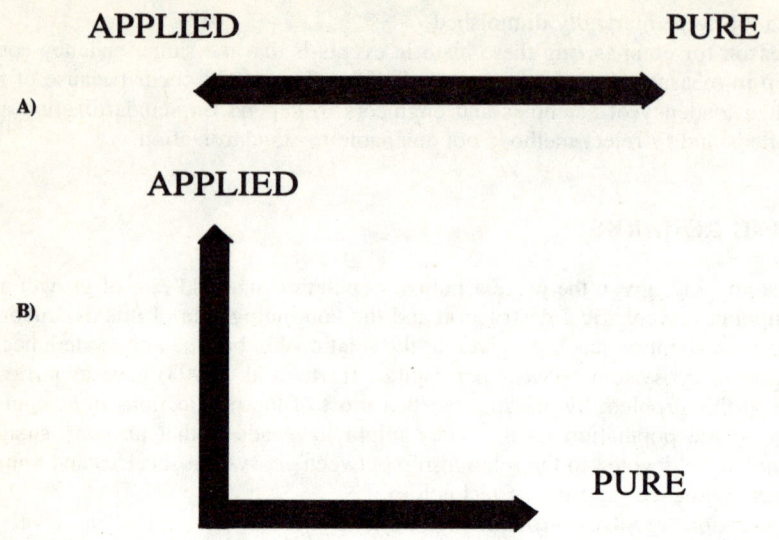

Figure 2. A. The common view of pure and applied research as having different directional components. B. The probable relationship between applied and pure research. Each project is probably a mixture of both.

REFERENCES

Bascietto J., Hinckley D., Plafkin J., Slimak M. (1990) Ecotoxicity and ecological risk assessment. Environ. Sci. Technol. 24:10-16

Blackburn C. (ed.) (1992) New Perspectives on Environmental Education and Research: A Report on the University Colloquium on Environmental Research and Education. Sigma Xi, The Scientific Research Society, Research Triangle Park, North Carolina 58 pp.

Bradshaw A.D. (1987) Restoration: an acid test for ecology. In Jordan W.R. III, M.W., Gilpin, J.D. Aber (eds.) Restoration Ecology. Cambridge University Press, New York pp:23-29

Brockner J., J.Z. Rubin (1985) Entrapment in escalating conflicts: a social psychological analysis. Springer-Verlag, New York

Cairns J. Jr., Pratt J.R. (1986) On the relationship between structure and function. Environ. Toxicol. Chem. 5:785-786
Costanza R. (1987) Social traps and environmental policy. BioScience 37:407-412
Cross J.G., Guyer M.J. (1980) Social Traps. University of Michigan Press, Ann Arbor, Michigan
Erwin T.L. (1988) The tropical forest canopy: the heart of biotic diversity. In E.O. Wilson (ed.) Biodiversity. National Academy Press, Washington DC pp:123-129
Geller E.S. (1991) If only more would actively care. J. Appl. Behav. Ana.l 24:607-612
Geller E.S. (1992) Applications of Behavior Analysis to Prevent Injury from Vehicle Crashes. Cambridge Center for Behavioral Sciences, Cambridge, Massachusetts
Geller E.S. (1994) The human element in integrated environmental management. In Cairns J. Jr, Crawford T.V., Salwasser H. (eds.) Implementing Integrated Environmental Management. UCEHMS, Virginia Tech., Blacksburg, Virginia pp:5-26
Geller E.S., Chaffee J.L., Ingram R.E. (1975) Promoting paper-recycling on a university campus. J. Environ. Syst. 5:39-57
Geller E.S., Erickson J.B., Buttram B.A. (1983) Attempts to promote residential water conservation with educational, behavioral, and engineering strategies. Pop. Environ. 6:96-112
Harte J., Torn M., Jensen D. (1992) The nature and consequences of indirect linkages between climate change and biological diversity. In Peters R.L., Lovejoy T.E. (eds.) Global Warming and Biological Diversity. Yale University Press, New Haven, Connecticut pp:325-343
Hughes R.M., Larsen D.P.(1988) Ecoregions: an approach to surface water protection. J. Water Pollut. Control Fed. 60:486-493
Karr J.R. (1991) Biological integrity: a long-neglected aspect of water resource management. Ecol. Appl. 1:66-84
Karr J.R., Fausch K.D., Angermeir P.L., Yant P.R., Schlosser I.J. (1986) Assessing biological integrity in running waters: a method and its rationale. Ill. Nat. History Survey Sp. Publ. 5 Champaign, Illinois
Levine S.N. (1989) Theoretical and methodological reasons for variability in the responses of aquatic ecosystem processes to chemical stresses. In Levin S.A., Harwell M.A., Kelly J.R., Kimball K.D. (eds.) Ecotoxicology: Problems and Approaches. Springer-Verlag, New York pp:145-180
National Research Council (1992) Conserving Biodiversity. Report of the Panel of the Board on Science and Technology for International Development. National Academy Press, Washington DC
Odum E.P. (1985) Trends expected in stressed ecosystems. BioScience 35:419-422
Patrick R. (1949) A proposed biological measure of stream conditions based on a survey of Conestoga Basin, Lancaster County, Pennsylvania. Proc. Acad. Nat. Sci. Phila. 101:277-341
Platt J. (1973) Social traps. Am. Psychol. 28:642-651
Pratt J.R. (1990) Aquatic community response to stress: prediction and detection of

adverse effects. In Landis W.G., van der Schalie W.H. (eds.) Aquatic Toxicology and Risk Assessment, 13th Symposium. American Society for Testing and Materials, Philadelphia, Pennsylvania pp:16-26

Pratt J.R., Rosenberger J.L. (1993) Community changes and ecosystem functional complexity: a microcosm study. In Gorsuch J.W., Dwyer F.J., Ingersoll C.G., LaPoint T.W. (eds.) Environmental Toxicology and Risk Assessment, 2nd Symposium. American Society for Testing and Materials, Philadelphia Pennsylvania pp:88-102

Royal Society of London and the U.S. National Academy of Sciences (1992) Population Growth, Resource Consumption, and a Sustainable World. National Academy of Sciences, Washington DC

Schaeffer D.J., Herricks E.E., Kerster H.W. (1988) Ecosystem health I: measuring ecosystem health. Environ. Manage. 12:445-455

Schindler D.W. (1987) Detecting ecosystem response to anthropogenic stress. Can. J. Fish. Aquat. Sci. 44(Suppl 1):6-25

Skinner B.F. (1938) The behavior of organisms. Appleton-Century Crafts New York

Teger A.I. (1980) Too Much Invested to Quit. Pergamon Press, New York

Union of Concerned Citizens (1992) World scientists' warning to humanity. Nucleus 14(4):1-3,12

Van Voris P., O'Neill R.V., Emanuel W.R., Shugart H.H. Jr. (1980) Functional complexity and ecosystem stability. Ecology 61:1352-1360

Walker B.H. (1992) Biodiversity and ecological redundancy. Conserv. Biol. 6:18-23

Wilson E.O. (1989) Biodiversity. National Academy Press, Washington DC

5. ENVIRONMENTAL HEALTH AND ECOTOXICOLOGY: AN INDISPENSABLE LINK

José Jerónimo Amaral-Mendes[1], Eric Pluygers[2] and Juliana Fariña[3]

[1]University of Évora, 7000 Évora
Institute of Pathological Anatomy, Unversity of Coimbra, 3000 Coimbra
National Institute of Health
1600 Lisboa
Portugal

[2]Honorary Head, Oncology Dept., Höpital de Joliment
1700 La Louvière
Belgium

[3]Department of Pathological Anatomy, Faculty of Medicine,
Complutense University, Pl. Cristo Rey, 28040 Madrid
Spain

INTRODUCTION

Well-being and increased life expectancy are to a great extent due to the progress of science and the modern technologies, in particular chemistry. At the same time, this progress is responsible for a number of new health problems, not only among workers in industry but also, and more seriously, in the population at large.

Cancer rates and the risk of allergies are increasing in association with the marked increase in the use of chemical substances. Large-scale releases of toxic pollutants become more frequent with the increasing probability of more long-lasting exposures. The borderline between working and living environments becomes less clear, with the result that the differential diagnosis between occupational and common diseases is made more difficult.

The basis for policy on health and the environment is the recognition that, in principle, almost every aspect of the environment potentially affects health for good or ill. This applies not only to specific agents (microorganisms, other biological entities, physical forces and agents, and chemicals), but also to elements of the urban and rural environment: homes, workplace, leisure facilities and the main components of the natural world (the atmosphere, soil, water, and many parts of the biosphere). A proper managed environment is therefore essential, not only to improve health but indeed to ensure human survival. It is increasingly necessary to ensure that the environment created by humans is planned and managed strongly and imaginatively enough to maximize its potential benefits to health and well-being. Hazard control is not enough if not coupled with the management of all aspects of the environment.

This paper is an attempt to address the interface between environmental health and ecotoxicology. The relationship between the two fields is complex and needs a transdisciplinary approach involving several aspects. Ecotoxicology, as opposed to classical toxicology, will be dealt with in more detail in the text. As far as environmental health is concerned, there is an increasing need to foster the development of a new field, ecosystem health and medicine based on application of the model of medical practice to ecosystem management. Both fields need biomarkers to help make such assessments. For both areas specific standard tests are required. However, in human medicine there is a vast database on the variations to be expected. Thus, a medical doctor can rapidly identify measurements outside the normal range.

There are two areas of interest: medical practice (or more broadly environmental health) will serve as a model to develop diagnostic protocols and tools to assess the adequacy and sensitivity of biomarkers characterizing ecosystem distress syndromes. On the other hand, environmental human health will benefit from the application of ecological concepts. The question is, as Calow (1993) puts it, "Even if the health analogy is philosophically flawed are there still pragmatic advantages to be gained from it?"

However the analogies cannot be carried too far. Environmental health, in the medical sense, is clearly related to public health activities, involving physicians and non-medical experts alike. The field is hygiene and prevention. Environmental health cannot be equated with environmental medicine, a domain of strict medical competence, essentially clinical, aimed at the detection and treatment of established disease.

Another concept that could be misleading is clinical ecology. Environmental illness was until very recently referred to as environmental hypersensitivity. This emerged in the early 1950s, contemporary with and largely as the result of the then popular concept of clinical ecology (Randolph 1982). Many adherents of clinical ecology redesignated it as environmental medicine. The concept was the result of a group of American physicians who were frustrated with the limits of "orthodox" medicine and, in particular "orthodox" allergy, in addressing problems present in some of their patients. Clinical Ecology/Environmental Ecology is still a controversial issue in the medical world. A large bibliography is available; for a review see Bell (1982) and for a critique see Gatien (1991).

A TENTATIVE HISTORICAL PERSPECTIVE IN EUROPE

In May 1977, the Thirtieth World Health Assembly resolved that the main social target of governments and the World Health Organization (WHO) in the coming decades should be "the attainment by all citizens of the world by the year 2000 of a level of health that will permit them to lead a socially and economically productive life" (WHO 1987).

In 1980, the WHO Regional Committee for Europe approved its first common health policy (WHO 1980). This "Health for All" (HFA) strategy called for a fundamental change in the health strategies of countries and outlined four main areas of concern: environmental risk factors; lifestyles; reorientation of the health care system;

and the political, managerial, technological, manpower, research, and other support necessary to bring about changes in these areas. A new era in health development was inaugurated.

In 1984, the Regional Committee agreed to endorse 38 targets in line with the European HFA strategy (WHO 1985). This stimulated debates at the country level on national health policies and helped the 32 Member States of the Region to set targets that reflect their specific needs and priorities. HFA rests on the basic principle that the coordinated action of all sectors is vital.

To promote these issues in Member States, the following main roles have been identified for EURO (the European Regional Office of the WHO) to play:

(a) Helping to make existing knowledge better known by "Identifying" innovative and improved approaches;
(b) promote priority health research;
(c) act as a catalyst in promoting rational health policy development towards HFA 2000 principles;
(d) improve cooperation and coordination between international organizations active in the health field.

Arising out of this, there were a number of subsequent developments:

1. In 1984, the Organization for Economic Cooperation and Development (OECD) Council updated sixteen guidelines of the volume "OECD Guidelines for Testing of Chemicals" (OECD 1981).

2. A Resolution of the Council of Ministers of the EEC and the representatives of the Governments of the Member States, meeting within the Council, of 29 May 1986, approved a programme of action for the European Communities (EC) on toxicology for health protection (EC 1986).

3. The Commission of the EC implemented a research and development Programme (1986-90) in the field of environmental protection (Council of EEC 1986a). A call for Research Proposals was published in 1986 (Council of EEC 1986b), covering all research areas of the Programme except for health effects of pollutants, which was covered by a special call for proposals published in April 1988 (Council of EEC 1986c). This contained two research themes: biomonitoring of human populations exposed to genotoxic environmental chemicals; and nephrotoxic early effect indicators from exposure to pollutants.

4. In 1987, the WHO Regional Office for Europe in its document "Environment and Health in Europe. A Regional Strategy" (Tarkowski 1987), described the main thrusts of a medium-term environmental health programme based on the European HFA Strategy. The programme was to be carried out in cooperation

with Member States as a continuation of the collaboration that had started more than 20 years before. Of the 38 targets endorsed by the Regional Committee in 1984, nine were concerned primarily with environmental health. They cover multi-sectoral policies, monitoring and control mechanisms, water, air, food safety, hazardous wastes, housing and settlements, the working environment, and accident prevention.

5. In 1987, The World Commission on Environment and Development (WCED) chaired by Gro Harlem Brundtland, the Prime Minister of Norway, produced a report "Our Common Future" (WCED 1987) which "reviewed the relationship between environment and development, and methods of assuming the progress of mankind while respecting the environment so as to make it possible to bequeath it in a healthy condition to future generations". Although the relationship with health was not considered in detail in the WCED report, concern for health underlay most of it. It therefore seemed opportune, three years later, for an independent body to make an assessment of the health consequences of environment changes, especially in anticipation of the United Nations Conference on Environment and Development held in Brazil in 1992.

6. The EC Council and country-member representatives approved the resolution of 31st May 1988, concerning the efforts to implement the community programme in the field of toxicology for health protection (Council of EEC 1988). Resolution 3 underlined:
    a. immunotoxicology studies
    b. nephrotoxic evaluation effects
    c. *in vitro* assay evaluation programmes
    d. general methodology for the evaluation of reproductive effects.
    Resolution 5 underlined the problem of extrapolation to man of animal data, as well as the quantitative carcinogenic evaluation risk. There was also a resolution concerning the improvement of prevention of acute intoxication in man (90/C 329/03 of December 3, 1990, O.J. C329, December 31); and conclusions concerning toxicology and sanitary protection (92/C 148/01 of May 15, 1992, O.J. C148, June 12).

7. In December 1989, the First European Conference on Environment and Health was held at Frankfurt-am-Main, Germany. The Conference brought together ministers and senior representatives from the environment and health administrations of 29 European countries and from the Commission of the European Communities. The "European Charter on Environment and Health" (WHO 1989) was adopted by the final session of the Conference, and was considered a further extension of the European Health policy targets adopted by 32 Member States of the European Region of WHO in 1984. The Charter also incorporates the basic philosophy of the World Commission on Environment and

Development (WCED) and represents a major step forward in the development of both public health and environment policies at a time when political change is greatly enhancing cooperation among Member State throughout Europe. In the words of Dr. J.E. Asvall, WHO Regional Director for Europe (WHO 1989): "it is already clear that governments are making use of the Charter as a basis for practical action in the interests of all our citizens now and in the future".

8. In 1989, the WHO/EURO published "Environment and Health. The European Charter and Commentary" (WHO 1990a). This is an exhaustive explanatory text originally provided as background for the delegations to the Conference. It expands and focuses on all aspects of the relationship between the environment and human health, as a concept referred to by WHO as "Environment Health". The Charter by which the governments of Europe have adopted a united position on the basic principles, mechanisms and priorities for further developing environmental health programmes, should, in itself, help to ensure that the subsequent action of those governments are undertaken confidently and decisively. However, international collaboration is hindered by several factors:
   (a) misunderstanding about the scope of the subject, since in some countries the term environmental health is not used or understood;
   (b) conflicting views on the nature and extent of the harm caused by environmental factors and on the benefits of more stringent controls;
   (c) severe limitations on the methods available for estimating harm;
   (d) the difficulty shared by organizations responsible for various aspects of environmental health in securing the collaboration of certain government sectors and society at large.

9. Immediately after the European Conference, a major development in the implementation of the Charter took place, in that an agreement on the establishment of a "European Centre for Environment and Health" was signed with the governments of Italy and The Netherlands. In order to increase WHO's capability to work with the countries of the European Region, the Governments of Italy and of The Netherlands offered resources and facilities for the establishment of the Centre with three units: Istituto Superiore di Sanitá in Rome, RIVM in Bilthoven, and the Centre Headquarters in Copenhagen. The Centre will work in cooperation with other international bodies such as the United Nations Environment Programme, the Economic Commission for Europe and the Commission of the European Communities. One of the priority tasks will be the organization of databases on all aspects of environment and health and to forecast trends. Another priority will be preparation of a comprehensive report "Concern for Europe's Tomorrow", that will collect and analyze existing information in order to provide a country-by-country survey of environment and health status, together with a forecast for the changes over the next 20 years.

The Centre will also develop a programme of technical cooperation with countries of the Region, having specific in-house expertise in air and water pollution, food safety, radiation protection, epidemiology and toxicology.

10. In early 1990, a WHO Commission on Health and Environment was appointed by the Director-General of WHO as a fully independent body. Composed of 22 members, the Commission was chaired by Simone Veil, France, member of the European Parliament. The Report of the Commission (WHO 1992) was considered by the Director-General the main WHO contribution to the Rio Conference in 1992, a contribution that underlines the importance of health in environmentally sound and sustainable development. It addressed the following:
    (a) Health and environment;
    (b) Integrating development, the environment and health;
    (c) Global challenges to health and the environment (population, poverty, resources use, macroeconomic frameworks);
    (d) Food and agriculture;
    (e) Water;
    (f) Energy;
    (g) Industry;
    (h) Human settlements, urbanization and basic services;
    (i) Transboundary and international issues (acid precipitation, the ozone layer, greenhouse gases, oceans, biodiversity);
    (j) Strategy and Recommendations.

11. In 1991, a resolution of the EC Council and Health Ministers meeting within the Council (Council of Minsters EEC 1991) required that the Commission, with the help of member states collect the available knowledge and experience of the relationship between health and environment.

12. In 1993, EC Council and country-member representative's of the February Meeting adopted a resolution concerning a community programme on policy and action related to environment and sustainable development (Council of Ministers EEC 1993).

THE CHALLENGES TO ENVIRONMENTAL HEALTH

The field of environmental health is more and more concerned with the effects of technology on health. The future in this field will be determined by the direction of scientific and technological development, in the industrialized societies and in the developing countries of the third world as well.

The objectives of health professionals must be to, understand, predict and prevent unwanted biological effects of the northern industrialization and technological development and make efforts to design biological, pharmacologic, behavioral and technical methods for control and ultimately treatment of such effects. The challenges lie in four major areas (Green 1987):

(a) Addressing the existing unsolved problems of environmental health

The impact of twentieth century events on population growth, human settlement, aging populations, lifestyles, and economic inequity, is having staggering socio-psychological effects. As a result, social discontent is expressed in reactionary anti-social behaviour such as terrorism, intolerance, racism, and widespread revolutionary movements. These problems may be greater than the ability to provide adequate nutrition to the rapidly growing number of people;

(b) Preventing the spread of diseases of industrialization around the world

Reduction in agricultural land, through urbanization and factory construction, natural resources depletion, transportation practices, the control of human, agricultural and industrial wastes, are problems demanding local and global biological control, as well as the prevention of chemical poisoning of air and water supplies. These changes will place enormous burdens on efforts to provide employment and a better income for the ever increasing numbers of their citizens. The average annual per capita earnings of 15 percent of population is $ 21,000 dollars, while 85 per cent of the rest of population earn only $ 1,000 dollars (Daly 1993). Attempts to reach an equitable balance may pose threats to the environment, and to human health. As Daly emphasizes "World free trade has long been presumed good, a presumption that has been the cornerstone of the existing General Agreement on Tariffs and Trade (GATT). That presumption should be reversed and should favour domestic production for domestic markets. When convenient, balanced international trade should be used, but it should not be allowed to govern a country's affairs at the risk of environmental and social disaster".

(c) Learning to anticipate the health impacts of new technologies in the post-industrialized societies

The increasingly rapid pace of technology development, and the demographic trends reached in most societies of the world, will continue to demand adequate populations to provide skilled labour. This will inevitably further develop between now and the beginning of the next century, with perhaps greater acceleration than that already seen and even more significant changes in the social structure of the labour force and in employment levels. For proof of these transformations it will suffice to quote just two phenomena, the consequences of which in social and health terms are already known. First, is the fact that in the majority of western countries and Japan, more than half the working population is no longer employed in agriculture or industry but in other sectors generally defined as "services" (Grieco 1987). Second, is the general fall in employment levels, as is reflected in the recent initiatives taken but WHO and the ILO aimed at stimulating research on "unemployment disease";

(d) Ethical considerations and education programs designed to provide the disciplinary strengths in the component sciences that underlie the evolving technologies

There will be a need to emphasize the health effects of chemical exposures as well as the health implications of computer technologies, robot labour, communication, information management and artificial intelligence. New programs will be needed to manage public health issues in near space environments, on land, as well as in the fast developing area of sea exploration.

Practitioners of environmental health will shift emphasis on skills required to contain contamination to acquisition of predictive and preventive management skills through health risk assessment of emerging technologies. From an ethical point of view, a serious concern should be placed upon the strong social rejection of the modern technological society by religious groups, of which islamic fundamentalism is the best example. Of paramount importance are the third world problems of poverty, illiteracy, high birth rates, compulsive industrial monocultures, natural resource depletion (e.g. forestry and fisheries), ethnic and minority group oppression, the overall migration phenomenon from poor countries to more industrialized societies, and the huge problem of refugees fleeing war or political unrest.

Environmental health in already industrialized countries is still faced with unresolved issues; housing, food safety, industrial disasters, the area of accidents and lastly chemicals.

Housing - The primary purpose of housing is to provide adequate protection from adverse forces and conditions of the natural environment, and to constitute the basis of the family social unit. While construction and technology have solved all the basic problems of housing safety, most of the world's population does not yet have access to these technologies and lives in primitive conditions. From a health point of view, even in advanced housing, there is a potential and significant health problem i.e indoor air pollution. As a true environment health problem, indoor air pollution stems primarily from unvented cooking and heating appliances. Pollutants from human origin such as from respiratory infection and cigarette smoking, and effluent from building materials including formaldehyde and radon gas, constitute indoor environment pollutants. The major factors in indoor air pollution result from the air conditioned system design and "passive smoking".

Food safety - This problem is largely unsolved and with the continuous growth of the modern urban civilization it will become a permanent challenge. Industrial food processing, refrigeration, chemical preservation, rapid transformation and retailed distribution are technological areas that have most contributed to avoiding food spoilage and the enhancement of food safety. Public attention should be focused on the consumption of chemically contaminated food, in the same manner it is now being alerted to "passive smoking".

On another hand the mis-use of technology will enhance errors and food will serve as a major vehicle for the spread of infection, fungal mycotoxins and contamination. The addition of chemicals either during plant growth as fertilizers or pesticides, or during food processing to enhance appearance and other characteristics, are responsible for dissemination of potentially toxic chemical substances for human health. Extensive distribution systems and modern agriculture raise the potential for massive exposure from a contaminated crop at a central point in the production and distribution of the human food chain. It must be emphasized that the etiology of most cancers is environmental and the main factor seems to be carcinogens transported through the human diet (Fishbein 1979).

Industrial disasters - Whether stemming from deficient design or inappropriate location, the massive environmental contamination from accidents and disasters is a major environmental health problem. Recent examples are the radioactive contamination produced by the nuclear power generating station of Chernobyl, the chemical release of methyl isocyanate in Bhopal, India, the release of dioxins and related compounds into the environment at Seveso, and the fire at Basel, Switzerland, that poured tons of toxic compounds in the Rhine.

These massive environmental disasters affected large populations, either killing outright or producing lifetime chronic injury or disease. They have generated a growing public concern about technology in the sense that the economic benefits of industrial development may be overshadowed by the risk to human health and environment.

Accidents as an environmental health concern - Traumatic injury and death forms an increasingly high proportion of the human health impact of technology, and in some age groups is the main cause of morbidity and death. While risks from major accidents are a major source of concern there appears to be a curious tolerance with accidental death and injury associated with human fault. Thus, the area of accidents offers the greatest opportunity for education programmes for risk reduction and progress in life saved and morbidity avoided.

The expansion of the synthetic chemical industry - This challenge is the most important threat to environmental health and will be the central theme of this paper, as large-scale ecosystems are already being affected by chemical pollution. Thousands of new chemicals in vastly increasing quantities have been introduced in every aspect of human life in the last forty years. Experience has shown that quantity and use is proportional to exposure, and therefore to a range of new pathologic effects. A vast amount of legislation has been implemented in the last thirty years to provide the basis for regulations and enforcement of standards for control of waste pollution, both in the United States and in Europe. That legislation addresses air, water, natural resources, clean water supplies, waste water and solid waste disposal, and the regulation of chemicals used in food and consumer utilities. The regulatory activities have proven effective in stemming the rate of environmental pollution. However, the chemical substances control is impaired by the lack of basic toxicologic information on thousands of these chemical substances, which is a massive challenge to environment health research. The lack of regulatory legislation and administrative infrastructure with the resultant increasing contamination of agricultural land, coastal waters, food, air and water supplies, also poses a significant challenge to environmental health with direct links to the health of large-scale ecosystems.

ENVIRONMENTAL HEALTH COMMITMENTS

There is growing awareness and concern about the impact of pollution and environmental degradation on the environmental health of the planet and all that is living on it. Major international governmental bodies such as WHO and United Nations are according high priority to topics related to environmental protection, taken in its broader sense, and numerous projects have been developed to assess the general outcomes, such as the green house effects and the ozone depletion on the atmosphere. A variety of more regional studies involving diverse terrestrial and marine ecosystems are also issues of concern. However, it has often been overlooked that "the human is the animal of interest (Ducatman 1993) and that "sustainable development and the shepherding of planetary resources are impossible without healthy humans" (Goldstein 1993).

This is well illustrated by the fact that the report "Our Common Future" (World Commission on Environment and Development 1987) produced by the United Nations Environmental Programme under the leadership of Brundtland, does not mention human health. Similarly health was nearly ignored in the preliminary agenda of the United Nations Conference on Environment and Development in Rio de Janeiro, Brazil, 1992, and only the clear sightedness of some of the Conference leaders prevented the entire issue being overlooked.

The reasons for this worrying situation are multiple, one of the most relevant being the relative lack of participation of public health scientists, (in the traditional concept of public health), in the developing area of environmental health. At a recent Symposium held at the Massachusetts Institute of Technology, the conclusion was reached that "physicians for whatever reason, denial, apathy, or the lack of formal training, have not participated in the environmental debate" (De Hart 1993).

Insufficient cooperation exists between the official departments responsible for Public Health and Environment, a situation largely responsible for the lack of involvement by health scientists in environmental issues.

From the Working Group on the Strategy for Implementing the European Charter (WHO 1990b) held in Düsseldorf, August 1990, it was recommended that, at the international level, WHO/EURO should develop and harmonize within the Region, systems for collecting and assessing of environmental health information and assemble and disseminate information regarding status and trends in health of people in the Region in association with the state of the environment. As a conclusion it was proposed "that the European Centre should develop a record linkage system for morbidity and mortality in the study of environmental occupational factors and human diseases". In the same document is outlined a proposed table of contents for the future report "Concern For Europe's Tomorrow".

The relationship between environment and human health is of fundamental importance (Amaral-Mendes 1993a). However the relation is a complex one, and it will be the indispensable role of health scientists, including medical doctors, to use all the expertise pertaining to their scientific disciplines (pathology, biochemistry, immunology, epidemiology etc.) in addressing the problem. This expertise, based on human assessments, does not exclude animal and plant based assessments; on the contrary it should be considered as part of an integrated evaluation of the health status of all the biota, including humans, composing a given ecosystem. This integrated evaluation might be an index of "ecosystem health" based upon assessments at all levels of the ecosystem, and might give relevant information about potential threats to human health. "A broad based view might be that pre-pathological or pathological changes in organisms in a given environment are indicative of the potential for humans associated with that environment to incur the same effects" (Amaral-Mendes et al. 1993b).

CONCEPTS UNDERLYING ENVIRONMENTAL MEDICINE

A definition of environmental medicine has been proposed by the Occupational and Clinical Toxicology Committee of the American College of Occupational Medicine: "The discipline of environmental medicine combines, epidemiologic and toxicologic approaches. It uniquely seeks to understand external causation and then to adopt policy, engineering, or human factor interventions to prevent or mitigate the causal outcomes..." (Ducatman et al. 1993).

Clinical environmental medicine is the study of detectable human disease or adverse health outcomes from exposure to physical, chemical and biologic environmental factors (Ducatman 1993). In the definition of Environmental Medicine, the importance of prevention is highlighted and should be considered the founding basis of environmental health.

Having defined environmental medicine and clinical environmental medicine in particular, it will be useful to consider the aspects that characterize the multidisciplinary nature of Environmental Medicine and even broadly Environmental Health.

WHO defined health as a "state of general physical, psychological and social well-being and not merely the absence of disease or infirmity" (WHO 1946). Against this background there is a dynamic situation determined by endogenous and exogenous factors (Kramers et al. 1992). The determinants for health status are shown in Figure 1.

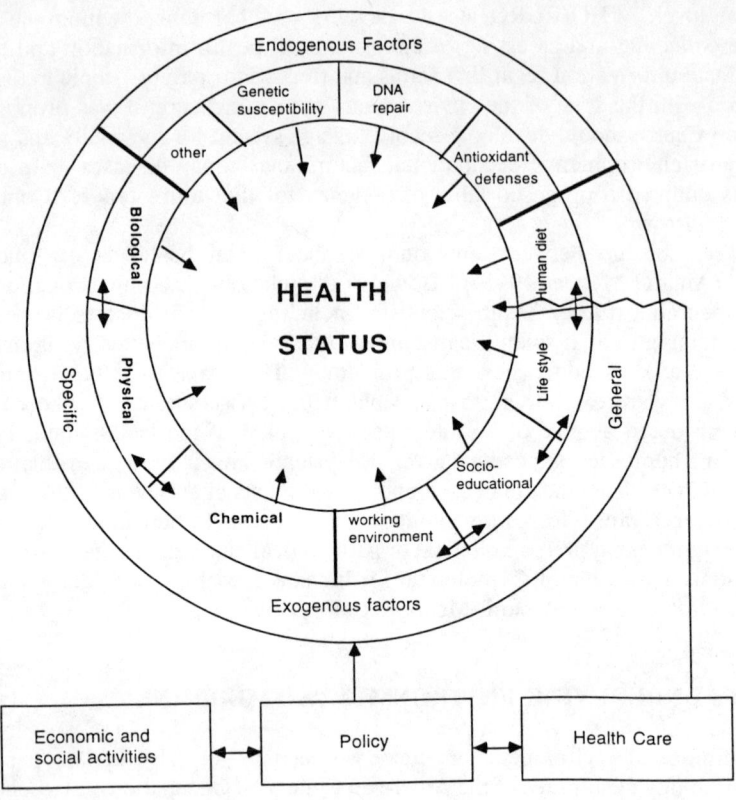

Figure 1. The determinants of the Health Status. The environment (Biophysical, physical, chemical) represents only one part of the picture.

Healthy status may be influenced positively or negatively by the exogenous factors. The environmental health, as previously defined, is influenced not only by the exogenous factors taken in narrow sense, but by the social environment as well. Two important aspects characterize the multidisciplinary nature of environmental medicine and environmental health. The first aspect is the difference marking the distinction between toxicology and ecotoxicology, and probably the best example to illustrate evolution from the orthodox concepts of traditional Public Health to the modern requirements of Environmental Health. Toxicology is the science of poisons and has a strong connotation with forensic situations. It rather addresses individuals or small communities exposed to toxic substances, mostly chemicals. Toxicology most often involves the study of one suspected toxic compound of which the effects can be readily reproduced under experimental laboratory conditions, enabling the determination of dose-effect curves, which is also true in occupational settings. Ecotoxicology considers effects of exposure to chemicals on the diverse biota composing an ecosystem, be it plants, animals or humans. It addresses populations rather individuals, and considers subacute and chronic aspects preferably to the acute ones. Ecotoxicology has to cope with complex mixtures of unidentified substances, sometimes synergetic other times antagonistic, or greatly dependent on physico-chemical or biological environmental factors (heat, cold, pH, predation chains etc.).

Thus, ecotoxicology needs to develop its ability to predict the long-term outcome of exposure to chemical pollutants, implying the development of long-term monitoring. These peculiarities make ecotoxicology a new scientific discipline, radically different from classical or forensic toxicology of which the methods are absolutely not adapted to the monitoring of populations over prolonged periods.

Toxic and hazardous wastes represent significant challenges, both now and in the future, for public systems of the U.S. and the world (The Institute of Medicine 1988). In the past twenty years, the public has become increasingly concerned about hazardous substances released into the environment. As a result, the U.S. Congress enacted and has continued to authorize the Comprehensive Environmental Response, Compensation and Liability Act, commonly known as Superfund. That act created the Agency for Toxic Substances and Diseases Registry (ATSDR) to carry out the health-related responsibilities of Superfund (ATSDR 1992). This Agency health studies programme's purpose is "to increase understanding of the relationship between exposure to hazardous substances and adverse health effects." (Bashor 1933). Bashor adds that this is accomplished primarily through epidemiological surveillance. In this context, the biomarker-based approach will now be considered, believing that is more prone to yield accurate information at a time, when successful prevention may still be considered, at each level of the ecosystem.

The biomarker-based approach is the second important aspect underlying environmental health. The term biomarker does not refer to a new concept, but is just a new name for a monitoring principle already in existence. The present emphasis put on the biomarker-based approach is due to two major developments. First, the recent scientific progress in the fields of biochemistry, cytogenetics and molecular biology have led to our improved understanding of mechanisms and metabolic processes involved in the interaction between chemicals and organisms. Second, the development of new improved analytical techniques have become much more sensitive and more selective. Biomarkers, in this modern sense, provide nowadays one of the best tools in environmental health (Depledge et al. 1991). This approach has been amply documented in the NATO Advanced Research Workshop held at Texel, Netherlands, May 1991 (Peakall and Shugart 1991).

Biomarkers were defined by the National Academy of Sciences as "xenobiotically-induced variations in cellular or biochemical components or processes, structures, or functions, measurables in a biological system or samples" (NRC 1989). Depledge (1993) proposed a slight modification, defining ecotoxicological biomarker as "A biochemical, cellular, physiological or behavioral variation that can be measured in tissue or body fluid samples or at the level of the whole organism that provides evidence to and/or effects of, one or more chemical and physical pollutants."

From a medical point of view, the new concept of biomarkers could be a breakthrough in the field of pathology, confronted with the theoretical demands of modern science (Hecht 1991). In particular, in the field of environmental health, coped with the environmental pathology of the health effects due to the increasing uses of xenobiotics in modern life.

CONCLUSION

Environment is one of the new frontiers of medicine.

(Beaupére 1991)

Concern about the degradation of the environment has become general among those who are conscious of the threats to a sustainable development of life on the surface of this planet. It is increasingly realized that damage or injury inflicted to any one species, whatever its position in the hierarchy of biota, will ultimately affect all other species, including man. Pollution of air, water and soil is widespread, other nuisances develop, species are exterminated, ecosystems are destroyed. More and more frequently, the health of ecosystems is imperiled and human environmental pathologies develop.

Physicians are long concerned with the visible health effects of environmental degradation, of which they observe the harmful effects on their patients. They have long watched the insidiously developing effects of air and drinking water pollution, of food chain contamination, of hazardous industrial processes, of ruthless waste disposal; they are insistently warning decision makers (e.g. Jurin 1991); they have developed their skills in environmental management.

Medicine, that is human medicine, has a long tradition of the use of biomarkers in assessing multiple departures from the health status; it is assumed that in ecosystem health assessment this experience could be helpfully shared with all those - from many scientific disciplines - involved in this research, crucial for the future of life on earth.

REFERENCES

Amaral-Mendes J.J. (1993) An effort to link environmental health and ecotoxicology. Special session on Human Environmental Health and ecotoxicology. First SETAC World Congress on Ecotoxicology and Environmental Chemistry: a Global Perspective, Lisbon

Amaral-Mendes J.J., Depledge M.M., Daniel B., Hal Grook R.S., Kloepper-Sams P., Moore M.N., Peakall D.B. (1993b) The conceptual basis of the Biomarker Approach. In Peakall D.B., Shugart L.K. (eds.) Biomarkers: Research and Application in the Assessment of Environmental Health. Springer-Verlag, Berlin, Heildelberg

ATSDR Health Assessment Guidance Manual (1992) Lewis Publishers

Bashor M.M. (1993) Public Health Assessments as a tool in identifying human exposure to environmental pollutants. In Travis C.C. (ed.) Use of Biomarkers in Assessing Health and Environmental Impacts of Chemical Pollutants. NATO ASI Series. Plenum Press, New York pp:63-66

Beaupére J. (1991) Santé et Environment. Le Médecin de France 664:3

Bell I.R. (1982) Clinical Ecology: A New Medical Approach to Environmental Illness. Common Knowledge Press, Bolinas, California

Calow P. (1993) Ecosystem health - unhealthy? Presentation at NATO Advanced Research Workshop on Evaluating and Monitoring the Health of Large-scale Ecosystems. Chateau Montebello, Quebec, Canada

Council of EEC (1986a) Implementation of research and development in the field of environmental protection. O.J. (Official Journal) (1986), L (Legislation) Nº 159 of June 14, 32 pp.

Council of EEC (1986b) Call for research proposals. O.J. (1986) S (Public Contracts) Nº 116 of June 19, 48pp.

Council of EEC (1986c) Call for research proposals in the field of environmental protection. O.J. (19988) C (Communication) Nº 99 of April 14, 5pp.

Council of EEC (1988) Communication of the Commission to the Council of Ministers Concerning the Efforts to Implement the Community Programme In the Field of Toxicology for Health Protection. 86/C 184/01, Council Doc. 5766/88

Council of Ministers EEC (1991) Resolution of the Council of Ministers of Health in November 11, on the basic policy on health matters. O.J. C Nº 304, November 23

Council of Ministers EEC (1993) Resolution of the Council of Ministers concerning a community programme on policy and action related to environment and sustainable development. February 1, O.J. C Nº 138, May 17

Daly H.E. (1993) The perils of free trade. Economists routinely ignore its hidden costs to the environment and the community. Scientific American, November:24-29

Depledge M.H., Amaral-Mendes J.J., Daniel B., Halbrook R., Kloepper-Sams P., Moore M.N., Peakall D.B. (1992) The conceptual basis of the biomarker approach. In Biomarkers: Research and Application in the Assessment of Environmental Health. Peakall D.B., Shugart L.R. (eds). NATO ASI Series H: Cell Biology. Vol. 68, Springer-Verlag, Berlin, Heidelberg pp:15-20

Depledge M.H. (1993) The rational basis for the use of biomarkers as ecotoxicological tools. In Fossi M.C., Leonzio C. (eds) Nondestructive Biomarkers in Vertebrates. Lewis Publications, Boca Raton pp:261-285

Ducatman A.M. (1993) Occupational physicians and environmental medicine. J.O.M 5:251-259

Ducatman A.M., Chase K.H., Farid I., LaDou J., Logan D.C., McCunney R.J., Milroy W.C., Mitchell F., Monosson I., Sunderman Jr. F.W. (1993) What is environmental medicine? J.O.M 32:1130-1132

Fishbein L. (1979) Potential Industrial Mutagens. Elsevier, Amsterdam 425 pp.

Gatien J.G. (1991) Environmental illness and clinical ecology: Issues of validity. NATO/CCMS Pilot Study on Indoor Air Quality, Oslo, Norway

Goldstein B.D. (1993) Global issues in environmental medicine. J.O.M 35:260-264

Green G.M. (1987) The challenges of the twenty-first century for environmental health. In Foá V., Emmet E.A., Maroni M., Colombi A. (eds.) Occupational and Environmental Chemical Hazards. Cellular and Biochemical Indices for Monitoring Toxicity. Ellis Horwood Ltd; Publishers, Chichester. pp:24-31

Grieco A. (1987) Present and Future of Occupational Medicine. In Foá V., Emmet E.A., Maroni M., Colombi A. (eds.) Occupational and Environmental Chemicals Hazards. Cellular and Biochemical Indices for Monitoring Toxicity. Ellis Horwood Ltd., Chichester pp:17-23

De Hart R.L. (1993) Accepting the environmental medicine challenge. J.O.M 35:265-266

EC (1986) Offical Journal, Communications N° 184 of July 27 pp:1-2

Hecht A. (1990/1) Will pathology in the future be reduced to clinical pathology only? European Pathology, Newsletter of the European Society of Pathology, The Institute of Medicine, Committee for the Study of the Future of Public Health. National Academic Press, Washington DC

Jurin J.L. (1988) Lettre ouvert à Monsieur le Préfet de la Moselle. Le Médecin de France (1991) 664:31

Kramers P.G.N, Lebret E., v.d. Mheen P.J. (1992) Effects on public health (of the Environment). In: National Environmental Outlook 2. 1990-2010. RIVM, Bilthoven pp:409-433

NRC (Nuclear Regulatory Commission). (1989) Biological markers in reproductive toxicology. National Academy Press, Washington DC 395 pp.

OECD (1981) Organization for Economic Cooperation and Development, Direction of Agriculture, Food and Fisheries C(81) N°. 30, Add. 3 and 4

Peakall D.B. and Shugart L.R. (1991) Biomarkers, Research and application in the Assessment of Environmental Health. Proceedings of the NATO Advanced Research Workshop on Biological Markers, Texel, The Netherlands. Springer Verlag, Berlin (Published in cooperation with NATO Scientific Affairs Division)

Randolph T.E. (1982) Emergence of the speciality of clinical ecology. Clinical Ecology

1: 84-90
RIVM (Rÿksinstituut voor Volksgezondheid en Milieu hygiene) (1992) National Environmental Outlook 2. 1990-2010. RIVM, Bilthoven 530 pp.
Tarkowski S. (1987) Environment and Health in Europe: A Regional Strategy. WHO/EURO, Copenhagen
World Commission on Environment and Development (1987) Our Common Future. Repoprt of the World Commission on Environment and Development, Oxford University Press, Oxford, UK
WHO (World Health Organization) (1946) Constitution. New York
WHO (1975) Early detection of health impairment in occupational exposure to health hazards. WHO Technical Report Series 571, Geneva.
WHO (1980) Health for All (HFA). WHO Regional Office for Europe, Copenhagen
WHO (1985) Targets for Health for All. WHO Regional Office for Europe,Copenhagen
WHO (1987) Thirtieth World Health Assembly Resolution. WHA 30.43
WHO (1989) European Charter of Environment and Health. WHO Regional Office for Europe, J.E. Asvall, Copenhagen
WHO (1990a) Environment and Health. The European Charter and Commentary. Conclusions of the First European Conference on Environment and Health, Frankfurt, 7-8 December, 1989. WHO/EURO Publications. European Series, n° 35, Regional Office for Europe, Copenhagen
WHO (1990b) The initial workplan of the European Centre for Environment and Health: A Discussion Document. WHO/EURO (Document ICP/CEH 210/13). Regional Office for Europe Working Group Report, Rome, Italy
WHO Commission on Health and Environment (1992) Our Planet, our Health. WHO, Geneva

RAPPORTEUR'S REPORT

E. Martinko
Environmental Monitoring and Assessment Program, U.S.E.P.A,
Office Research and Development,
401 M Street SW, Washington DC 20460
USA

SUMMARY OF PRESENTATIONS

The task of defining ecosystem health is difficult because of the complex nature of ecosystems and the multitude of problems that must be considered. The initial section, therefore, sets the stage by considering the definition of ecosystem health, its relationship to and comparison with the health services, its relationship to the delivery of ecosystem services, and its utility in communicating with the public.

Rapport suggests that ecosystem health is a concept around which a new integrative science can emerge by drawing from a number of disciplines including the biological, social, and health sciences. He cautions, however, that the health analogy cannot be followed strictly because of the multitude of problems associated with ecosystem variability and responses.

Calow argues that definitions of health boil down to the assumption of a core norm, from which deviations represent the state of health. This norm can be defined in various ways, not all of which are acceptable. He concludes that since the ecosystem approach is often used to manage ecosystems for human health and services, it is possible to define states of ecosystem health in these terms.

Ramade summarizes a number of qualitative and quantitative criteria for the abiotic and biotic components of an ecosystem that included structural and functional criteria for defining a "healthy" ecosystem. He concludes that much additional research is needed to define ecosystem health, especially in the assessment issue areas of key species identification; improved knowledge of functional bioindicators; impingement of anthropogenic perturbations; chemical cycle influence; and, impingement at regional and landscape scales.

Pratt and Cairns develop the concept of "ecosystem services," and summarize the activities or functions of ecosystems that are perceived to be beneficial to human society. They then discuss the correlation that should exist between ecosystem health and ecosystem services, a relationship that is not well documented.

Amaral-Mendes outlines some of the approaches used in toxicology to evaluate human health effects of exposure to chemicals. Also, he notes that in the field of ecotoxicology, the evaluation of the toxicological effects of human exposure to toxic compounds relies on a multimedia approach to the analysis of environmental risk and pollution control strategies. Although believing that more agreement is needed on the bases for exposure assessment and the processes involved, he suggests that the multimedia approach to the evaluation of human environmental health provides an indispensable link to ecosystem health. He then points to recent developments in the field of biomarkers and outlined their uses to determining changes in biological responses that can be related to exposure to an environmental contaminant.

DISCUSSION

Topics for the discussion included: 1) defining ecosystem health, 2) integrating science and policy, and 3) involving other groups.

Defining Ecosystem Health

Although many participants were uncomfortable with the subjective basis for using the term "ecosystem health," most agreed that it was a useful term for communicating with non-scientific audiences and politicians. Concern was expressed about the focus on system processes and ecosystems at the possible expense of biotic relationships, landscapes, ecoregions and other levels of organization. Other points of discussion are summarized as follows:

1. A healthy system embodies elements of sustainability, resilience and stability.
2. Ecosystem health is scale dependent.
3. Goals for the use of the ecosystem health concept are resource management oriented.
4. A "normal" state for ecosystem health must include humans as part of the ecosystem and recognize the influence of demographics.
5. "Zonation" or allocations of space for uses must be taken into account in maintaining ecosystem services, ecosystem sustainability, and future options.

Integrating Science and Policy

A cultural difference between scientists and policy-makers was recognized and discussed. Scientists tend to be more reductionist and long-term in their views in contrast with the more pragmatic, economic, and shorter-term perspective of policy-makers. A number of comments and suggestions were made regarding these points:

1. Scientists can be trained to understand the political process and to fill the gap between science and policy. Sufficient rewards must be provided to encourage scientists to involve themselves in policy.
2. Scientists need to recognize the boundaries between science and advocacy and respond accordingly.

3. Scientists must establish processes to develop mainstream consensus an be prepared to render best judgment in the face of diverse scientific opinions or less than adequate data on cause-effect relationships, rather than retreat from the decision-making process.
4. Better feedback mechanisms need to be developed between science and policy in the framework of adaptive management.
5. Policy-makers can develop more scientific literacy and realistic expectations for science.
6. Policy-makers must make a commitment to the integration of science into the decision-making process.
7. Scientists must encourage a more "open" process for public input and more effectively use lobbies and scientific organizations to bridge the gap between science and policy.

Involving Other Groups

The involvement of the public must be enhanced by encouraging more environmental literacy particularly through education and self-discovery. Criteria for determining if this goal is being met include increases in: 1) awareness, 2) knowledge, 3) attitudes/values, and 4) action.

Several other issues were discussed concerning the involvement of other groups: 1) The role of the media needs to be recognized and utilized more effectively through selected contacts with journalists, pre-set agendas for interviews and more critical or vocal responses to bad journalism; 2) Involvement of other groups can be effective in networking for monitoring and surveying with the recognition that caution should be exercised with respect to organization, data quality, interpretation or misinterpretation of data; 3) Involvement of other groups should not become a substitute for avoiding scientific research.

CONCLUSIONS

1. "Ecosystem health" is a useful term for communicating with politicians and the non-scientific public. The subjective basis for the term, however, presents difficulties for communication among scientists.
2. Establishing what is healthy is a difficult task, since a "norm" may not exist and the determination of a "state" of health is heavily dependent on societal values.
3. The health analogy is useful in some contexts, but should not be followed strictly.
4. Comparative studies must be conducted to provide yardsticks for measurement and criteria for defining ecosystem health.
5. Additional research is needed in the assessment issue areas of key species

identification, bioindicator development, systems and chemical cycle influences, and landscapes or ecoregion scale effects.
6. The relationship between ecosystem health and ecosystem services should be studied and documented.
7. Scientists must do more to develop mainstream consensus and take an active role in the decision-making process.
8. Feedback mechanisms need to be developed between science and policy in the framework of adaptive management.
9. Involvement of the public must be increased through the use of education, self-discovery and the media.
10. Biomarkers have a potential role in ecosystem health monitoring as early warning indicators and as a linkage to human health.

Session II

QUANTITATIVE INDICES FOR ECOSYSTEM HEALTH ASSESSMENT

INTRODUCTION

Chair: Peter Calow

Three crucial questions for this session are:

1) Can ecosystem health be defined?
2) Can general states of ecosystem health be defined?
3) Can specific states of health be defined for particular ecosystems?

My own expectation is that if general states of health can be defined they will be very general, probably very dependent upon energy dynamics and possibly hard to measure. The specific states of health are likely to be more detailed and specified in terms of both structure and function. My own feeling is that it will be hard to define these states *a priori*, and this will put emphasis on a comparative approach (What is, with what should be, defined by before/after, similar systems at same time comparisons) which present challenges in the avoidance of, for example, pseudo replication monitoring. Scale is also of crucial importance in determining what can be measured here and what results are obtained. These are issues that will be addressed to a greater or lesser extent in the papers of this session. Ecosystem health may also be judged by the health of individual organisms in ecosystems and here, signs of cancer can play an important role. But to what extent can carcinogenicity in particular and within-organism responses in general say anything about higher level -particularly more global responses? Measurement is crucial in science, so clear signals will need to come out of this session to guide discussion in sessions 3 to 5.

6. ECOLOGICAL AND ECONOMIC SYSTEM HEALTH AND SOCIAL DECISION MAKING

Robert Costanza
Director, University of Maryland International Institute for Ecological Economics
Professor, Center for Environmental and Estuarine Studies
University of Maryland
Box 38, Solomons, MD 20688
USA

HEALTH AND INTEGRITY OF COMPLEX SYSTEMS

All complex systems are, by definition, made up of a number of interacting parts. In general, these components vary in their type, structure, and function within the whole system. Thus a system's behavior cannot be summarized simply by adding up the behavior of the individual parts. Contrast a simple physical system (say, an ideal gas) with a complex biological system (say, an organism). The temperature of the gas is a simple aggregation of the kinetic energy of all the individual molecules in the gas. The temperature, pressure, and volume of the gas are related by simple relationships with little or no uncertainty. An organism, however, is composed of complex cells and organ systems. The state of an organism cannot be surmised simply by adding up the states of the individual components, since these components are themselves complex and have different, noncommensurable functions within the overall system. Indicators that might be useful for understanding heart function - pumping rate and blood pressure, for instance - are meaningless for skin or teeth.

But to understand and manage complex systems, we need some way of assessing the system's overall performance (its relative "health"). The EPA has recently begun to shift the stated goals of its monitoring and enforcement activities from protecting only "human health" to protecting overall "ecological health." Indeed, EPA's Advisory Board (SAB 1990:17) recently stated:

> *EPA should attach as much importance to reducing ecological risk as it does to reducing human health risk. These very close linkages between human health and ecological health should be reflected in national environmental policy. When EPA compares the risks posed by different environmental problems in order to set priorities for Agency action, the risks posed to ecological systems must be an important part of the equation.*

Although this statement gives the concept of ecological health importance as a primary EPA goal, it begs the question of what ecosystem health is, while tacitly defining it as analogous to human health. The dictionary definitions of health are: "1. the condition of being sound in mind, body, and spirit; 2. flourishing condition or well-being." These definitions are rather vague. In order to meet the mandate for effectively managing the environment we must construct a more rigorous and operational definition of health that is applicable to all complex systems at all levels of scale, including organisms, ecosystems, and economic systems. Past explicit or implicit definitions of ecosystem health have included:

- Health as homeostasis
- Health as the absence of disease
- Health as diversity or complexity
- Health as stability or resilience
- Health as vigor or scope for growth
- Health as balance between system components

All of these concepts represent pieces of the puzzle, but none is comprehensive enough to serve our purposes here. I wish to elaborate on the concept of ecosystem health as a comprehensive, dynamic, hierarchical measure of system resilience, organization, and vigor. These concepts are embodied in the term "sustainability," which implies the system's ability to maintain its structure (organization) and function (vigor) over time in the face of external stress (resilience). A healthy system must also be defined in light of both its context (the larger system of which it is part) and its components (the smaller systems that make it up).

In its simplest terms, then, health is a measure of the overall performance of a complex system that is built up from the behavior of its parts. Such measures of system health imply a weighted summation of a more complex operation over the component parts, where the weighting factors incorporate an assessment of the relative importance of each component to the functioning of the whole. This assessment of relative importance incorporates "values," which can range from subjective and qualitative to objective and quantitative as we gain more knowledge about the system under study. In the practice of human medicine, these weighting factors or values are contained in the body of knowledge and experience embodied in the medical practitioner.

Figure 1 shows the progression from directly measured "indicators" of a component's status, through "endpoints" that are composites of these indicators, to health with the help of "values." Measures of health are inherently more difficult, more comprehensive, require more modeling and synthesis, and involve less precision, but are more relevant than the endpoints and indicators from which they are built. It remains to determine which general approaches to developing these measures of health for ecosystems are most effective.

Health is also a scale-dependent characteristic. Figure 2 indicates this relationship by plotting a hypothetical curve of system life expectancy on the y axis vs. time and space scale on the x axis. We expect a cell in an organism to have a relatively short life span, the organism to have a longer life span, the species to have an even longer life span, and the planet to have a longer life span. But no system (even the universe itself in the extreme case) is expected to have an infinite life span. A healthy and sustainable system in this context is one that attains its full expected life span. For individual humans, this is also an often used cumulative indicator of health (at least in retrospect).

Since ecosystems experience succession as a result of changing climatic conditions and internal developmental changes, they too have a limited (albeit fairly long) life span. The key is differentiating between changes due to normal life span limits and changes that cut short the life span of the system. Things that cut short the life span of humans are obviously contributors to poor health. Smoking, AIDS, and a host of other ailments do just this. Human-induced eutrophication in aquatic ecosystems causes a radical change in the nature of the system (ending the life span of the more oligotrophic system while beginning the life span of a more eutrophic system). We would have to call this process "unhealthy" using the above definition since the life span of the first system was cut "unnaturally" short. It may have gone eutrophic eventually, but the anthropogenic stress caused this transition to occur "too soon."

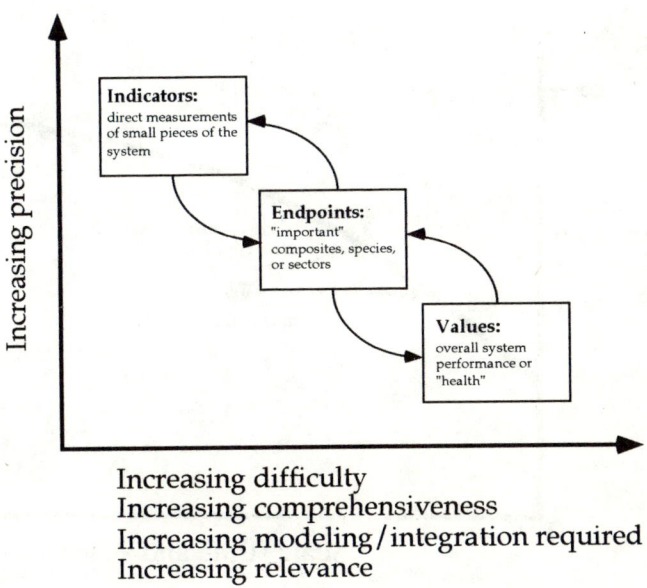

Figure 1. Relationship of indicators, endpoints, and values.

How can we create a practical definition of system health that is applicable with equal facility to complex systems at all scales? Let us first lay out the minimum characteristics of such a definition. First, an adequate definition of ecosystem health should integrate the concepts of health mentioned above. Specifically it should be a combined measure of system resilience, life expectancy, balance, organization (diversity), and vigor (metabolism). Second, the definition should be a comprehensive description of the system. Looking at only one part of the system implicitly gives the remaining parts zero weight. Third, the definition will require the use of weighting factors to compare and aggregate different components in the system. It should use weights for components that are linked to the functional dependence of the system on the components, and the weights should be able to vary as the system changes to account for "balance." And fourth, the definition should be hierarchical to account for the interdependence of various time and space scales. Costanza et al. (1992b) develop the following definition of health:

An ecological system is healthy and free from "distress syndrome" if it is stable and sustainable - that is, if it is active and maintains its organization and autonomy over time and is resilient to stress.

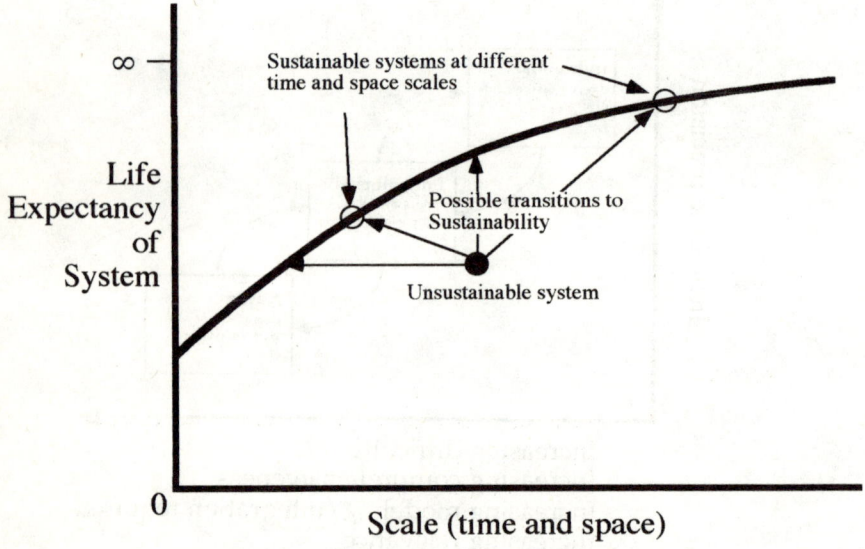

Figure 2. Health and sustainability as scale (time and space) dependent concepts.

Ecosystem health is thus closely linked to the idea of sustainability, which is seen to be a comprehensive, multiscale, dynamic measure of system resilience, organization, and vigor. This definition is applicable to all complex systems from cells to ecosystems to economic systems (hence it is comprehensive and multiscale) and allows for the fact that systems may be growing and developing as a result of both natural and cultural influences. According to this definition, a diseased or unhealthy system is one that is not sustainable and will not achieve its maximum life span. The time and space frame is obviously important in this definition. Distress syndrome (Rapport 1989) refers to the irreversible processes of system breakdown leading to the termination of the system before its normal life span. To be healthy and sustainable, a system must maintain its metabolic activity level as well as its internal structure and organization (a diversity of processes effectively linked to one another) and must be resilient to outside stresses over a time and space frame relevant to that system.

What does this mean in practice? Table 1 lays out the three main components of this proposed concept of system health (resilience, organization, and vigor) along with related concepts and measurements in various fields. What we are looking for is an index or group of indices that combine these three basic aspects of system performance and health - vigor, organization, and resilience. To operationalize these concepts (especially organization and resilience) will require a heavy dose of systems analysis, synthesis, and modeling, combined with broad-based input from the full range of stakeholders involved in the management of ecosystems.

THE HOLLING MODEL

In particular, the concept of system resilience is critical. One broad conceptual application of these ideas to ecological and economic systems, with the goal of maximum generality, is the model of Holling (1987, 1992). Holling proposes four basic functions common to all complex systems and a spiraling evolutionary path through them (Figure 3). The functions (boxes) are: 1) **Exploitation** (e.g. r-strategists, pioneers, opportunists, entrepreneurs); 2) **Conservation** (e.g. K-strategists, climax ecosystems, consolidation, rigid bureaucracies); 3) **Release** (e.g. fire, storms, pests, political upheavals); and 4) Reorganization (e.g. accessible nutrients, abundant natural resources). Within this model, systems evolve from the rapid colonization and exploitation phase, during which they capture easily accessible resources, to the conservation stage of building and storing increasingly complex structures. Examples of the exploitation phase are early successional ecosystems colonizing disturbed sites or pioneer societies colonizing new territories. Examples of the conservation phase are climax ecosystems or large, mature bureaucracies. The release or "creative destruction" (Schumpeter 1950) phase represents the breakdown and release of these mature structures via periodic events like fire, storms, pests, or political upheavals. The released structure is then available for reorganization and uptake in the exploitation phase. The amount of ongoing creative destruction that takes place in the system is critical to its behavior. The conservation phase can often build elaborate and tightly bound structures by severely limiting creative destruction (the former Soviet Union is a good example), but these structures become "brittle" and susceptible to massive and widespread destruction (i.e. the former Soviet Union). If some moderate level of release is allowed to occur on a more routine basis, the destruction is on a smaller scale and leads to a more resilient system. It could be argued that patterns of behavior with moderate levels of ongoing creative destruction evolved in those local communities and human cultures which managed to survive for thousands of years or more.

Table 1. Indices of Vigor, Organization, and Resilience in Various Fields.

Component of health	Related concepts	Related measures	Field of origin	Probable method of solution
Vigor	Function Productivity Throughput	GPP,NPP,GEP,GNP Metabolism	Ecology Economics Biology	Measurement
Organization	Structure Biodiversity	Diversity Index Average mutual information (Ulanowicz 1986) Predictability (Turner et al. 1989)	Ecology	Network analysis
Resilience		Scope for growth (Bayne 1987) Population recovery time (Pimm 1984) Disturbance absorption capacity (Holling 1987)	Ecology	Simulation modeling
Combinations		Ascendancy (Ulanowicz 1986) Index of Biotic Integrity (Karr 1981)	Ecology	

Creative destruction, in terms of shocks or surprises, seems to be crucial for system resilience and integrity. Similarly, it has been argued that episodic events, such as the Chernobyl accident, the Rhine chemical spill, the death of seals in the North Sea, are shocks to the social-cultural value system and may stimulate positive change towards more resilient ecological economic systems (Berkes and Folke 1994).

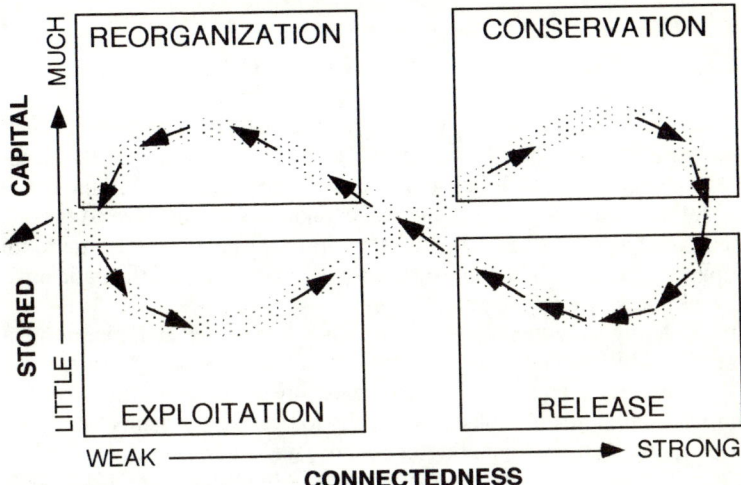

Figure 3. The four general system functions and the flow of events between them (from Holling 1987, 1992). The arrows show the speed of that flow in the ecosystem cycle, where arrows close to each other indicate a rapidly changing situation and arrows far from each other indicate a slowly changing situation. The cycle reflects changes in two attributes, i.e. (1) the y axis: the amount of accumulated capital (nutrients, carbon) stored in variables that are dominant keystone variables at the moment and (2) the x axis: degree of connectedness among variables. The exit from the cycle indicated at the left of the figure indicates the stage where a flip is most likely into a less or more productive and organized system, i.e. devolution or evolution as revolution.

Fire climax systems are a good example of the range of possibilities for creative destruction. In their unmanaged state, fire climax systems burn over extensive areas relatively often, but because of the high fire frequency the amount of fuel is insufficient to allow highly destructive fires. The more frequent, small to moderate size fires release nutrients stored in the litter and support a spurt of new growth, without destroying all the old growth. On the other hand, when fires are suppressed and controlled, fuel builds up to high levels and (because control and suppression are never perfect - remember the former Soviet Union), when the fire does come it wipes out much of the forest.

The Holling model raises some interesting questions about the relationships between diversity, stability, resilience, control, creativity, surprise, and evolution in ecological and economic systems that are ripe for further analyses, and that are central to developing indices of ecosystem health.

COMPLEX SYSTEMS, SCALE, AND HIERARCHY

Understanding and modeling the dynamics of linked ecological and economic systems, ranging in size from the biosphere as a whole to regional landscapes to local ecosystems, is critical for defining good performance and health, and for designing a sustainable future. But integrated ecological economic systems have so far received only very limited direct attention. Several current approaches may be relevant to this problem and a cooperative synthesis among ecologists, economists, mathematicians, computer scientists, and many others is essential.

In modeling complex systems, the issues of scale and hierarchy are central (O'Neill et al. 1989). Some claim that the natural world, the human species included, contains a convenient hierarchy of scales based on interaction-minimizing boundaries; scales ranging from atoms to molecules to cells to organs to organisms to populations to communities to ecosystems (including economic, and/or human dominated ecosystems) to bioregions to the global system and beyond (Allen and Starr 1982; O'Neill et al. 1986). By studying the similarities and differences between different kinds of systems at different scales and resolutions, one might develop hypotheses and test them against other systems to explore their degree of generality and predictability.

The term "scale" in this context refers to both the resolution (spatial grain size, time step, or degree of complication of the model) and extent (in time, space, and number of components modeled) of the analysis. The process of "scaling" refers to the application of information or models developed at one scale to problems at other scales. In both ecology and economics, primary information and measurements are generally collected at relatively small scales (i.e. small plots in ecology, individuals or single firms in economics) and that information is then often used to build models at radically different scales (i.e. regional, national, or global). The process of scaling is directly tied to the problem of aggregation (the process of adding or otherwise combining components), which in complex, non-linear, discontinuous systems (like ecological and economic systems) is far from a trivial problem (O'Neill and Rust 1979; Rastetter et al. 1992).

For example, in applied economics, basic data sets are usually derived from the national accounts, which contain data that are linearly aggregated over individuals, companies, or organizations. Sonnenschein (1974) and Debreu (1974) have shown that, unless one makes very strong and unrealistic assumptions about the individual units, the aggregate (large scale) relations between variables have no resemblance to the corresponding relations on the smaller scale.

Rastetter et al. (1992) describe and compare three basic methods for scaling that are applicable to complex systems. All of their methods are attempts to utilize information about the nonlinear small-scale variability in the large-scale models. They list: 1) partial transformations of the fine-scale mathematical relationships to coarse-scale using a statistical expectations operator that incorporates the fine-scale variability; 2) partitioning or subdividing the system into smaller, more homogeneous parts (i.e. spatially explicit modeling); and 3) calibration of the fine-scale relationships to coarse-scale data, when this data is available. They go on to suggest a combination of these methods as the most effective overall method of scaling in complex systems.

A primary reason for aggregation error in scaling complex systems is the non-linear variability in the fine-scale phenomenon. For example, Rastetter et al. (1992) give a detailed example of scaling a relationship for individual leaf photosynthesis as a function of radiation and leaf efficiency to estimate the productivity of the entire forest canopy. Because of non-linear variability in the way individual leaves process light energy, one cannot simply use the fine-scale relationships among photosynthesis, radiation, and efficiency along with the average values for the entire forest to get total forest productivity without introducing significant aggregation error. One must somehow understand and incorporate this nonlinear fine-scale variability into the coarse-scale equations using some combination of the three methods mentioned above. Method 1 (statistical expectations) implies deriving new coarse-scale equations that incorporate the fine-scale variability. The problem is that incorporation of this variability often leads to equations that are extremely complex and cumbersome (Rastetter et al. 1992). Method 2 (partitioning) implies subdividing the forest into many relatively more homogenous levels or zones and applying the basic fine-scale equations for each partition. This requires a method for adjusting the parameters for each partition, a choice of the number of partitions (the resolution) and an understanding of the effects of the choice of resolution and parameters on the results. Method 3 (recalibration) implies simply recalibrating the fine-scale equations to coarse-scale data. It presupposes that coarse-scale data are available (as more than simply the aggregation of fine-scale data). In many important cases, however, this coarse-scale data is either extremely limited or is not available. Thus, while a judicious application of all three aggregation methods is necessary, from the perspective of complex systems modeling, the partitioning approach (Method 2) seems to hold particular promise because it can take fullest advantage of emerging computer technologies and data bases.

From the scaling perspective, hierarchy theory is a potentially useful tool for partitioning systems in ways that minimize aggregation error. According to hierarchy theory, nature can be partitioned into "naturally occurring" levels that share similar time and space scales, and that interact with higher and lower levels in systematic ways. Each level in the hierarchy sees the higher levels as constraints and the lower levels as noise. For example, individual organisms see the ecosystem they inhabit as a slowly changing set of constraints and the operation of their component cells and organs is what matters most to them. However, Norton and Ulanowicz (1992) suggest that what appears to be "noise" at a lower level could be turned into significant perturbations on the higher level. This can happen when a critical mass of components participate in a "trend," a behavioral pattern, which affects the slower processes at the higher level. The rapid and extensive human uses of fossil fuels could be seen as such a trend, causing perturbations at the global atmospheric level, which might feed back and radically alter the framework of action at the lower level.

Shugart (1989) explains the relationship between scales: "Clearly, natural patterns in environmental constraints contribute substantially to the spatial pattern and temporal dynamics of particular ecosystems ... these patterns, especially temporal ones, may resonate with natural frequencies of plant growth forms (i.e. phenology and longevity) to amplify environmental patterns." The simplifying assumptions of hierarchy theory may ease the problem of scaling by providing a common (but somewhat generalized) set of rules that could be applied at any scale in the hierarchy.

MULTISCALE EXPERIMENTAL ECOSYSTEM RESEARCH CENTER (MEERC)

To test some of these ideas, we have recently established a Multiscale Experimental Ecosystem Research Center (MEERC) at the University of Maryland. This Center will construct a series of cosms at several time, space, and complexity scales and carry out an integrated experimental and modeling research program aimed at understanding and modeling ecosystems at each of these scales (from microcosms to mesocosms to small and large regional watersheds - macrocosms). MEERC is focused on assessing the response of these systems to nutrient and toxicant perturbations, and how (and why) these responses change with scale. The ultimate goal is to develop and test a set of performance (health) indicators for these systems, and to develop and test a set of scaling principles that will allow the extrapolation of results across scales.

It will be possible within this framework to further develop and test some of the ideas about ecosystem health mentioned earlier. Detailed, dynamic simulation models of all the experimental systems will be developed and the program's experimental design will allow the models to be rigorously tested and calibrated. The combined models and experiments will allow various indices of vigor, organization, and resilience to be calculated and tested over a range of scales. One of the results of the MEERC program will thus be a unique opportunity to test various ecosystem health indices over a range of scales. If these tests are encouraging, then the indices may be usefully applied to full-scale ecosystems, such as the Chesapeake Bay.

INTEGRATING SCIENCE AND POLICY UNDER UNCERTAINTY

A primary limiting factor in applying all of this to the problem of managing ecological and economic systems is the legalistic, adversarial system of governance we have evolved for managing ecosystems, and the lack of communication and linkage between science and policy. Ludwig et al. (1993) accurately identify many of the underlying reasons for non-sustainable resource use. They conclude by enumerating five basic principles of effective management: (1) include human motivation; (2) act before scientific consensus is reached; (3) rely on scientists to recognize problems, but not to remedy them; (4) distrust claims of sustainability; and (5) confront uncertainty.

Ecological research on the topics identified in the Sustainable Biosphere Initiative (SBI) (Lubchenco et al. 1991) can contribute to achieving ecosystem health and sustainability, if the research is appropriately linked to policy in the manner suggested by Ludwig et al. (1993). A unique feature of the SBI document was that in identifying the research needs for a sustainable biosphere, a group of ecologists pinpointed many areas of research that go well beyond the boundaries of traditional ecology and require a broad, interdisciplinary collaboration. Narrow, traditional ecological research is not relevant by itself, but the broad interdisciplinary research recommended in the SBI can be. But in order for the recommended SBI research to actually be relevant, some additional major changes in how we view science in general, and especially the linkages between science and environmental policy, are going to be needed.

As Ludwig et al. point out, one of the primary reasons for the problems with current methods of environmental management is the issue of scientific uncertainty. Not just its existence, but the radically different expectations and modes of operation that science and policy/management have developed to deal with it. If we are to solve this problem, we must understand and expose these differences and design better methods to incorporate uncertainty into the policy making and management process.

To understand the scope of the problem, it is necessary to differentiate between risk (which is an event with a known probability, sometimes referred to as statistical uncertainty) and true uncertainty (which is an event with an unknown probability, sometimes referred to as indeterminacy). Most important environmental problems suffer from true uncertainty, not merely risk.

Science treats uncertainty as a given, a characteristic of all information that must be honestly acknowledged and communicated. Over the years scientists have developed increasingly sophisticated methods to measure and communicate the uncertainty arising from various causes. It is important to note that the progress of science has, in general, uncovered more uncertainty rather than leading to the absolute precision that the lay public and some policy makers often mistakenly associate with "scientific" results.

The scientific method can only set boundaries on the limits of our knowledge. It can define the edges of the envelope of what is known, but often this envelope is very large and the shape of its interior can be a complete mystery. Science can tell us the range of uncertainty about global warming, the potential impacts of toxic chemicals, or the possible range of fish population dynamics, and maybe something about the relative probabilities of different outcomes, but in most important cases it cannot tell us which of the possible outcomes will occur with any degree of accuracy.

Our current approaches to environmental management and policy making, on the other hand, abhor uncertainty and gravitate to the edges of the scientific envelope. The reasons for this are clear. The goal of policy is making unambiguous, defensible decisions, often codified in the form of laws and regulations. While legislative language is often open to interpretation, regulations are much easier to write and enforce if they are stated in clear, black and white, absolutely certain terms.

As they are currently set up, most environmental regulations, particularly in the United States, demand certainty, and when scientists are pressured to supply this nonexistent commodity, there is not only frustration and poor communication, but mixed messages in the media as well. Because of uncertainty, environmental issues can often be manipulated by political and economic interest groups. Uncertainty about global warming is perhaps the most visible current example of this effect. In order to rationally use science to make policy, we need to deal with the whole envelope of possible futures and all their implications, and not delude ourselves that certainty is possible.

The "precautionary principle" is one way the environmental regulatory community has begun to deal with the problem of true uncertainty. The principle states that rather than await certainty, regulators should act in anticipation of any potential environmental harm in order to prevent it. The precautionary principle is so frequently invoked in international environmental resolutions that it has come to be seen by some as a basic normative principle of international environmental law (Cameron and Abouchar 1991). But the principle offers no guidance as to what precautionary measures should be taken. It "implies the commitment of resources now to safeguard against the potentially adverse future outcomes of some decision" (Perrings 1991), but does not tell us how much resources or which adverse future outcomes are most important.

This aspect of the "size of the stakes" is a primary determinant of how uncertainty is dealt with in the political arena. Methods are not currently in place to deal with high values of either stakes or uncertainty, which require a new approach, what might be called "post-normal" or "second order science" (Funtowicz and Ravetz 1991). This "new" science is really just the application of the essence of the scientific method to new territory. The scientific method does not, in its basic form, imply anything about the precision of the results achieved. It does imply a forum of open and free inquiry without preconceived answers or agendas aimed at determining the envelope of our knowledge and the magnitude of our ignorance.

Implementing this view of science requires a new approach to environmental protection that acknowledges the existence of true uncertainty rather than ignoring it, and includes mechanisms to safeguard against its potentially harmful effects, while at the same time encouraging development of lower impact technologies and the reduction of uncertainty about impacts. The precautionary principle sets the stage for this approach, but the real challenge is to develop scientific methods to determine the potential costs of uncertainty, and to adjust local incentives so that the appropriate parties pay this cost of uncertainty and have appropriate incentives to reduce its detrimental effects. Without this adjustment, the full costs of environmental damage will continue to be left out of the accounting, and the hidden subsidies from society to those who profit from environmental degradation will continue to provide strong incentives to degrade the environment beyond sustainable levels.

Ecological research (and scientific research in general) in this context should be focused on defining the edges of the knowledge envelope. This "edge-focused" research should lead to a much more effective use of science as a way to anticipate and head off problems and to link with the policy process.

For example, had this "policy-linked, edge-focused" research been the norm, we could have easily anticipated the greenhouse effect and taken steps to minimize its potential impacts. Arhaneus first described the effect and human's potential impact on it almost 100 years ago (Arhaneus 1896), but it remained a scientific curiosity until the 1980s when enough data and models had been assembled to demonstrate that the effect was, in fact, likely to cause global warming. There is still much uncertainty about the magnitude of the warming and especially about its ultimate impacts, but science can do a very good job of anticipating potential problems if we focus the effort on that function, rather than on demonstrating impacts that have already occurred or trying to predict exactly what will happen. To be relevant, ecological research should therefore focus on the edges, as well as the range of uncertainty about these impacts. It should develop better methods to communicate uncertainty and reduce its detrimental impacts, and to link more effectively with other disciplines and the policy process. How can it do this? Ludwig et al.'s principals are a good guide. We need to:

(1) Include human motivation by developing linkages with the social sciences, particularly economics, to develop a comprehensive transdisciplinary synthesis. One effort in this direction has come to be called "ecological economics" (Costanza 1991).
(2) Act before scientific consensus is reached by focusing on the edges of our knowledge and employing the precautionary principle to guide action (Perrings 1991).
(3) Rely on ecologists and other scientists to recognize the edges and worst cases, but do not rely on them to remedy the problems themselves. Research needs to be "policy-linked" and "edge-focused."
(4) Distrust claims of sustainability and confront uncertainty by shifting the burden of proof from the public to the parties that stand to gain from resources use. One mechanism for doing this is through the use of "environmental assurance bonds" that require resource users to post a bond large enough to cover the worst case damages with the potential for refund if the damages are less (Costanza and Cornwell 1992).

DATA QUALITY

The quality of scientific information in policy-relevant fields of research is difficult to assess, and quality control in these important areas is correspondingly difficult to maintain. Frequently there are insufficient high quality measurements for the presentation of the statistical uncertainty in the numerical estimates that are crucial to policy decisions. We have proposed a grading system for numerical estimates that can deal with the full range of data quality - from statistically valid estimates to informed guesses (Costanza et al. 1992). By analyzing the underlying quality of numerical estimates, summarized as spread and grade, one can provide simple rules whereby input data can be coded for quality, and these codings carried through arithmetical calculations for assessing the quality of model results. For this we use the NUSAP (numeral, unit, spread, assessment, pedigree) notational system. It allows the more quantitative and the more qualitative aspects of data uncertainty to be managed separately. By way of example, we have applied the system to an ecosystem valuation study that used several different models and data of widely varying quality to arrive at a single estimate of the economic value of wetlands (Costanza et al. 1992a). The NUSAP approach illustrates the major sources of uncertainty in this study, and can guide new research aimed at the improvement of the quality of outputs and the efficiency of the procedures.

THE MEDICAL MODEL: COMBINING SCIENCE AND JUDGMENT

How does one use this evolving (and still admittedly vague) definition of system health and sustainability in environmental management? The "medical model" broadly interpreted is an appropriate starting point. By this is meant the interactive process of system analysis, diagnosis, prognosis, and treatment. It also implies an interactive combination of scientific analysis and expert judgment.

But medical practice itself is evolving. There is a growing recognition of the efficacy of a more preventive approach to health maintaince and the need to involve the "patient" much more directly in the decision-making process. The "patient" in ecosystem management is the ecosystem itself and it is only partially and imperfectly represented by the range of stakeholders who interact with the system. Some of these stakeholders are interested in exploitative uses of the system, while others are interested in preventing all human impacts on the system. In the context of a comprehensive, dynamic measure of system resilience, organization, and vigor neither of these extremes represents a very healthy interaction. Humans are part of the system and to define a healthy ecosystem as one with no human interactions is to write off the entire planet for humans and any hope of achieving a sustainable system. On the other hand, to treat ecosystems as merely passive sources of raw materials to be exploited and destroyed is obviously not a very healthy or sustainable interaction either. What we want is a way to discover and design positive human interactions with ecosystems and to differentiate those interactions from negative ones. Human interactions that enhance ecosystem resilience, organization, and vigor are, according to the definitions developed so far, the kinds of interactions that lead to enhanced health, while interactions that degrade ecosystem resilience, organization, and vigor degrade the health of the system.

Health assessment can thus become an integral part of an adaptive ecosystem management. But first we must develop a more comprehensive understanding of the way ecological and economic systems operate and evolve, and the limits of our ability to understand and predict these systems.

A CASE STUDY: THE CHESAPEAKE BAY

The Chesapeake Bay is the largest estuary in North America, and it has been the subject of more scientific study and political wrangling than any other body of coastal water in the world. It has become clear that what happens in the Bay is in large part a function of activities in the drainage basin, but the focus of most past studies has been rather narrow. We are only now beginning to develop a comprehensive picture of the Bay and of its connections to a sprawling and often densely populated watershed - not only in ecological, but also in demographic, cultural, political, and economic terms.

Fundamental in gaining an historical and spatial perspective of human activities in the Chesapeake Bay watershed is conceptualizing the Bay system as the combination of the drainage basin and the estuary itself. We have assembled and mapped data on past and present human activities in the Chesapeake watershed in order to gain this perspective (Costanza and Greer 1994).

The Chesapeake watershed comprises an ecological system whose beauty and productivity have led to high rates of human population settlement and growth. These high population growth rates have in turn, directly and indirectly, caused a troublesome infirmity, including declining fisheries, receding wetlands, vanishing seagrasses and a devastated oyster industry. These trends have also led to a decline in the quality of human life. Traffic congestion, disappearing natural and agricultural areas, swelling landfills, and overtaxed water treatment facilities have all affected the tenor of life in the Bay region.

What can we learn from the Chesapeake example? In many ways the Chesapeake was and is a "best case scenario" for ecosystem management, but it still has a long way to go. What follows is one possible synthesis with an eye toward determining what may be extrapolated to ecosystem management problems in other areas (Costanza and Greer 1994).

1) A necessary first step for effective action is the creation of a broad consensus about both the essentials of the problem and common goals shared by all interest groups. In the Chesapeake this was relatively easy because a large percentage of the population had direct experience with the Bay and could directly perceive its decline. This decline was perceived as a loss of system organization through the decline of major biotic components of the system (sea grass beds and several fish species) and the disappearance of some species altogether, caused by a deterioration in water quality due to eutrophication and increased levels of suspended sediments. Estuarine ecosystems are particularly resilient, however (Costanza et al. 1993) and there was a broadly shared view that the system could recover quickly if the stresses were removed.

There was also a broad pre-existing consensus on the common goal of protecting the "health" of the Bay ecosystem and reversing this decline. In most ecosystem management cases, these features are missing. There may be small groups of people who are directly affected by the decline of a particular ecosystem, but not a broad cross section, or the impacts may be so subtle and distributed that they are hard to perceive without some sophisticated tools. Achieving a consensus on a common goal across interest groups in ecosystem management is even more difficult, since in many cases interest groups are in direct conflict over management goals. Consider, for example, the spotted owl controversy in the Pacific Northwest. These general characteristics often lead to severe "social traps." It is therefore necessary in most cases to take special steps to achieve a broad consensus. One of the most effective tools in this regard is the Adaptive Environmental Management (AEM) technique developed by Carl Walters and Buzz Holling (Walters 1986) or, using different terminology, the methods of Environmental Dispute Resolution developed by Gail Bingham and others (Bingham 1986). It is achieving this first step that often proves the biggest barrier, which, if not surmounted, can lead to cycles of building and destroying bureaucratic structure. The Chesapeake was able to avoid most of the lurking social traps by maintaining a relatively broad consensus and effective dialogue between interest groups around the common goal of protecting the health of the ecosystem.

2) A second step seems to be achieving broad consensus on the details of the problem and the methods of solution. In the Chesapeake this was effected by a large, relatively coordinated, EPA-funded study. In general, funding for detailed studies and the development of action plans is difficult to assemble. But if step 1 can be achieved, step 2 has a chance.

3) The third step of implementation of the remedial action plans follows directly from steps 1 and 2. A key here is holding the coalition together long enough to reach this stage and finding the resources to effect the plans. In the Chesapeake Bay (which we have already indicated is a "best case" scenario) implementation is proceeding, but there are many rough spots. Now that the real extent of the problem and the magnitude of resources necessary to fix it are known, the coalition is showing strain. This may well be the point when things fall apart. To effect a stable solution, it is important that the short-term, local incentives that drive the system away from its long-term goals are corrected (i.e. social traps are removed). Pollution taxes, environmental assurance bonds, and other incentive-based instruments are effective ways to do this, as are continuing education and dialogue on the problems of the Bay ecosystem and the goals of sustainable ecosystem management. If the Chesapeake can successfully implement this final stage, it will truly be a model of sustainable ecosystem management worthy of being emulated.

SOCIAL TRAPS

The process of short-run and local incentives getting out of sync with long-term and global goals has been well studied in the last decade under several rubrics (Hardin 1968; Axelrod 1984), but one particularly effective representation is John Platt's notion of "social traps" (Platt 1973; Cross and Guyer 1980; Teger 1980; Brockner and Rubin 1985; Costanza 1987). In all such cases the individual decision-maker may be said to be "trapped" by the local conditions into making what turns out to be a bad decision viewed from a longer or wider perspective. People go through life making decisions about which path to take based largely on "road signs," the short-run, local reinforcements that we perceive most directly. These short-run reinforcements can include monetary incentives, social acceptance or admonishment, and physical pleasure or pain. In general, this strategy of following the road signs is quite effective, unless the road signs are inaccurate or misleading. In these cases we can be trapped into following a path that is ultimately detrimental because of our reliance on the road signs. For example, overfishing is a social trap because by following the short-run economic road signs, fishermen are led to exploit the resource to the point of collapse.

The elimination of social traps requires intervention - the modification of the reinforcement system. Indeed, it can be argued that the proper role of a democratic government is to eliminate social traps (no more and no less) while maintaining as much individual freedom as possible. Cross and Guyer list four broad methods by which traps can be avoided or escaped from. These are education (about the long-term, distributed impacts); insurance; superordinate authority (i.e. legal systems, government, religion); and converting the trap to a trade-off (i.e. correcting the road signs).

Education can be used to warn people of long-term impacts that cannot be seen from the road. Examples are the warning labels now required on cigarette packages and the warnings of environmentalists about future hazardous waste problems. People can ignore warnings, however, particularly if the path seems otherwise enticing. For example, warning labels on cigarette packages have had little effect on the number of smokers.

The main problem with education as a general method of avoiding and escaping from traps is that it requires a significant time commitment on the part of individuals to learn the details of each situation. Our current society is so large and complex that we cannot expect even professionals, much less the general public, to know the details of all the extant traps. In addition, for education to be effective in avoiding traps involving many individuals, all the participants must be educated, and this is usually not possible.

Governments can, of course, forbid or regulate certain actions that have been deemed socially inappropriate. The problem with this direct, command-and-control approach is that it must be rigidly monitored and enforced, and the strong short-term incentive for individuals to try to ignore or avoid the regulations remains. A police force and legal system are very expensive to maintain, and increasing their chances of catching violators increases their costs exponentially (both the costs of maintaining a larger, better-equipped force and the cost of the loss of individual privacy and freedom).

Religion and social customs can be seen as much less expensive ways to avoid certain social traps. If a moral code of action and belief in an ultimate payment for transgressions can be deeply instilled in a person, the probability of that person's falling into the "sins" (traps) covered by the code will be greatly reduced, and with very little enforcement cost. On the other hand, the problems with religion and social customs as means to avoid social traps are: the moral code must be relatively static to allow beliefs learned early in life to remain in force later, and it requires a relatively homogeneous community of like-minded individuals to be truly effective. This system works well in culturally homogeneous societies that are changing very slowly. In modern, heterogeneous, rapidly changing societies, religion and social customs cannot handle all the newly evolving situations, nor the conflict between radically different cultures and belief systems.

Many trap theorists believe that the most effective method for avoiding and escaping from social traps is to turn the trap into a trade-off. This method does not run counter to our normal tendency to follow the road signs; it merely corrects the signs' inaccuracies by adding compensatory positive or negative reinforcements. A simple example illustrates how effective this method can be. Playing slot machines is a social trap because the long-term costs and benefits are inconsistent with the short-term costs and benefits. People play the machines because they expect a large short-term jackpot, while the machines are in fact programmed to pay off, say, $0.80 on the dollar in the long term. People may "win" hundreds of dollars playing the slots (in the short run), but if they play long enough, they will certainly lose $0.20 for every dollar played. To change this trap to a trade-off, one could simply reprogram the machines so that every time a dollar is put in $0.80 would come out. This way the short-term reinforcements ($0.80 on the dollar) are made consistent with the long-term reinforcements ($0.80 on the dollar), and only the dedicated aficionados of spinning wheels with fruit painted on them would continue to play.

MODIFYING INCENTIVES FOR IMPROVED MANAGEMENT

In the context of social traps, the most effective way to make global and long-term goals consistent with local, private, short-term goals is to somehow modify the local, private, short-term incentives (Platt 1973; Costanza 1987). These incentives are any combination of the reinforcements that are important at the local level, including economic, social, and cultural incentives (Perrings et al. 1992). We must design the social and economic instruments and institutions to bridge the gulf between the present and future, between the private and social, between the local and global, between the ecological and economic parts of the system.

One policy that has often been recommended, and which is consistent with this idea of modifying local incentives, is the "polluter pays principle." This principle would require the payment of "pollution taxes" (Pigou 1920) to account for the damages to ecological systems by private polluters or resource users. One factor limiting the adoption of this approach has been the high degree of uncertainty and unpredictability associated with ecological damages. How big should the tax be? If it is too low, the polluters are not paying the full cost to society and will continue to overpollute. If it is too high, the polluters will be subsidizing society and the cost of their products will be too high.

One way to handle this uncertainty about the true damages is the idea of a flexible environmental assurance bonding system. (Costanza and Perrings 1990; Costanza and Cornwell 1992). This variation of the deposit-refund system is designed to incorporate environmental criteria and uncertainty into the market, and to induce positive environmental technological innovation. It works in this way: in addition to charging for known environmental damages, an assurance bond equal to the current best estimate of the largest potential future environmental damages would be levied and kept in an interest-bearing escrow account for a predetermined length of time. In keeping with the precautionary principle, this system requires the commitment of resources now to offset the potentially catastrophic future effects of current activity. Portions of the bond (plus interest) would be returned if, and only if, the agent could demonstrate that the suspected worst case damages had not occurred or were less than originally assessed. If damages did occur, the portions of the bond would be used to rehabilitate or repair the environment, and possibly to compensate injured parties. By requiring the users of environmental resources to post a bond adequate to cover uncertain future environmental damages (with the possibility for refunds), the burden of proof (and the cost of the uncertainty) is shifted from the public to the resource user. At the same time, agents are not charged in any final way for uncertain future damages and can recover portions of their bond in proportion to how much better their performance actually was compared with the worst case.

Deposit-refund systems, in general, are not a new concept. They have been effectively applied to consumer policy, conservation policy, and other efficiency objectives. Deposit-refund systems can be market generated or government initiated and are often performance based. For example, deposit-refund systems are currently effectively used to encourage the proper management of beverage containers and used lubricating oils (Bohm 1981).

Strong economic incentives are provided by the bond to reduce pollution, to research the true costs of environmentally damaging activities, and to develop new innovative, cost-effective pollution control technologies. The bonding system is an extension of the "polluter pays principle" to "the polluter pays for uncertainty as well" or the "precautionary polluter pays principle" (4P) (Costanza and Cornwell 1992). It would allow a much more proactive (rather than reactive) approach to environmental problems because the bond is paid up front, before the damage is done. It would tend to foster prevention rather than clean up by unleashing the creative resources of agents on finding more environmentally benign technologies, since these technologies would also be economically attractive. Competition in the marketplace would lead to environmental improvement rather than degradation. The bonding system would deal more appropriately with scientific uncertainty and the inherent unpredictability of ecosystems. The 4P approach has several potential applications. Any situation with large true uncertainty is a likely candidate, and these situations abound in the modern world, especially in ecosystem management and especially in managing coastal and estuarine ecosystems.

TOWARD A HEALTH-BASED, SUSTAINABLE GOVERNANCE SYSTEM

Achieving a healthy and sustainable total system at multiple scales will require some fundamental shifts in how we conceive of the world and humanity's place in it. Achieving these changes within the current system of governance seems unlikely. Short- term political time frames and interest group pressure mitigate against the long-range, cooperative consensus necessary to achieve overall system health and sustainability. It also ignores the distribution issues that create unfair and ultimately politically unstable solutions.

Solutions to the problems of ecosystem health and sustainability will only be robust and effective if they are fair and equitable. Ethicist John Rawls has argued that policies that represent an overlapping consensus of the interest groups involved in a problem will most likely be fair, effective, and resilient. The normal political process tends to accentuate conflict, and majority voting often sidetracks any effort to achieve overlapping consensus since the minority is always left out. The policies resulting from majority voting often are unfair to the minority and are not resilient since the minority spends all of its time fighting the decision and trying to build a new majority to overthrow the previous majority.

How can we develop policies based on overlapping consensus that can be truly effective in combining all the diverse information and interests necessary to provide stable and resilient solutions to environmental and other social problems? The tools of Alternative Dispute Resolution (ADR) are a good place to start (Bingham 1986). ADR uses a "policy dialogue" format that includes from the beginning all the major parties affected by a policy. ADR techniques are also effective tools for institution building since they can develop a sense of common purpose among people from different institutions who would otherwise be needlessly and destructively competing. The real constraints are the lack of sustained communication between the political, public, business, and academic realms.

Achieving a healthy and sustainable system at multiple scales from local to regional to global will no doubt require a more adaptive, collaborative approach to both understanding and managing ecosystems (Walters 1986), and one that acknowledges the range of possible interactions between humans and nature and the huge uncertainty about our understanding of these interactions (Ludwig et al. 1993; Costanza 1993). We need to move beyond the sterile and polarizing confrontation between the two current extreme views of the goals of ecosystem management. These extremes can be characterized (oversimplistically, I admit) as, on the one hand, "any change induced by humans is bad", leading to the policy goal of removing humans and their effects from all ecosystems, to, on the other hand, "any change induced by humans is OK if it leads to increased private profits," leading to the policy goal of maximal exploitation of ecosystems with no regard for their sustainability.

The goal of sustainable management for ecosystem health can break this impasse by allowing the possibility of positive human interaction with ecosystems. It can do this by moving beyond the simplistic definitions of ecosystem health (like homeostasis or no change) that have dominated to ones that involve the more sophisticated (but more difficult to implement) ideas of multiscale system vigor, organization, and especially resilience. This more complex view of ecosystem structure and function on which this concept of health is based is beginning to emerge as an overlapping consensus among ecologists of several different stripes (Holling et al. 1994). It can become a basis for building a broader overlapping consensus that could allow for sustainable governance.

REFERENCES

Allen T.F.H., Starr T.B. (1982) Hierarchy. University of Chicago Press, Chicago

Arhaneus S. (1896) On the influence of carbonic acid in the air upon the temperature of the ground. Philosophical Magazine and Journal of Science 41:227-237

Axelrod R. (1984) Evolution of Cooperation. Basic Books, New York

Bayne B.L. et al. (1987) The Effects of Stress and Pollution on Marine Animals. New York, Praeger

Berkes F., Folke C. (1994) Investing in cultural capital for a sustainable use of natural capital. In Jansson A. M., Hammer M., Folke C., and Costanza R. (eds.) Investing in Natural Capital: an Ecological Economics Approach to Sustainability. Island Press, Washington DC

Bingham G. (1986) Resolving environmental disputes: a decade of experience. Conservation Foundation, Washington DC

Bohm P. (1981) Deposit-Refund Systems. Resources for the Future Inc. The John Hopkins University Press, Baltimore and London

Brockner J., Rubin J.Z. (1985) Entrapment in Escalating Conflicts: A Social Psychological Analysis. Springer-Verlag, New York 275 pp.

Cameron J., Abouchar J. (1991) The precautionary principle: a fundamental principle of law and policy for the protection of the global environment. Boston College International and Comparative Law Review 14:1-27

Costanza R. (1987) Social traps and environmental policy. BioScience 37:407-412

Costanza R. (ed.) (1991) Ecological Economics: The Science and Management of Sustainability. Columbia University Press, New York

Costanza R. (1993) Developing ecological research that is relevant for achieving sustainability. Ecological Applications 3:579-581

Costanza R., Cornwell L. (1992) The 4P approach to dealing with scientific uncertainty. Environment 34:(12)20-46

Costanza R., Funtowicz S.O., Ravetz J.R. (1992a) Assessing and communicating data quality in policy relevant research. Environmental Management 16:121-131

Costanza R., Greer J. (1994) The Chesapeake Bay and its watershed: a model for sustainable ecosystem management? In Holling C. S., Light S. (eds.) Barriers and Bridges for the Renewal of Regional Ecosystems. Columbia University Press, New York (in press)

Costanza R., Kemp M., Boynton W. (1993) Predictability, scale, and biodiversity in coastal and estuarine ecosystems: implications for management. Ambio 22:88-96.

Costanza R., Norton B., Haskell B.J. (eds.) (1992b) Ecosystem Health: New Goals for Environmental Management. Island Press, Washington DC 269 pp.

Costanza R., Perrings C. (1990) A flexible assurance bonding system for improved environmental management. Ecological Economics 2:57-76

Cross J.G., Guyer M. J. (1980) Social Traps. University of Michigan Press, Ann Arbor

Debreu G. (1974) Excess demand functions. Journal of Mathematical Economics 1:15-23

Funtowicz S.O., Ravetz J.R. (1991) A new scientific methodology for global environmental problems. In Costanza R. (ed.). Ecological Economics: The Science and Management of Sustainability. Columbia University Press, New York pp:137-152

Hardin G. (1968) The tragedy of the commons. Science 162:1243-1248

Holling C.S. (1987) Simplifying the complex: the paradigms of ecological function and structure. European Journal of Operational Research 30:139-146

Holling C.S. (1992) Cross-scale morphology, geometry and dynamics of ecosystems. Ecological Monographs 62:447-502

Holling C. S., Schindler D.W., Walker B.W., Roughgarden J. (1994) Biodiversity in the functioning of ecosystems: an ecological synthesis. In Perrings C.A., Mäler K.G., Folke C., Holling C. S., Jansson B.O. (eds.) Biodiversity Loss: Ecological and Economic Issues. Cambridge University Press, Cambridge, UK (in press)

Karr J.R. (1981) Assessment of biotic integrity using fish communities. Fisheries 6:21–27

Lubchenco J., Olson A.M., Brubaker L.B., Carpenter S.R., Holland M.M., Hubbell S.P., Levin S.A., MacMahon J.A., Matson P.A., Melillo J.M., Mooney H.A., Peterson C.H., Pulliam H.R., Real L.A., Regal P.J., Risser P.G. (1991) The sustainable biosphere initiative: an ecological research agenda. Ecology 72:371-412

Ludwig D., Hilborn R., Walters C. (1993) Uncertainty, resource exploitation, and conservation: lessons from history. Science 260:17-36

Norton, B.G., Ulanowicz R.E. (1992) Scale and biodiversity policy: a hierarchical approach. Ambio 21:244-249

O'Neill R.V., DeAngelis D.L., Waide J.B., Allen T.F.H. (1986) A Hierarchical Concept of Ecosystems. Princeton University Press, Princeton, NJ

O'Neill R.V., Johnson A.R., King A.W. (1989) A hierarchical framework for the analysis of scale. Landscape Ecology 3:193-205

O'Neill R.V., Rust B. (1979) Aggregation error in ecological models. Ecological Modelling 7:91-105

Perrings C. (1991) Reserved rationality and the precautionary principle: In R. Costanza (ed.). Ecological Economics: The Science and Management of Sustainability. Columbia University Press, New York pp:153-166

Perrings C., Folke C., Mäler K.G. (1992) The ecology and economics of biodiversity loss: the research agenda. Ambio 21:201-211

Pigou A.C. (1920) The Economics of Welfare. Macmillan, London

Pimm S. L. (1984) The complexity and stability of ecosystems. Nature 307:321–326

Platt J. (1973) Social traps. American Psychologist 28:642-651

Rapport D.J. (1989) What constitutes ecosystem health? Perspectives in Biology and Medicine 33:120–132

Rapport D.J., Regier H.A., Hutchinson T.C. (1985) Ecosystem behavior under stress. American Naturalist 125:617–640

Rastetter E.B., King A.W., Cosby B.J., Hornberger G.M., O'Neill R.V., Hobbie J.E. (1992) Aggregating fine-scale ecological knowledge to model coarser-scale attributes of ecosystems. Ecological Applications 2:55-70

Schumpeter J.A. (1950) Capitalism, Socialism and Democracy. Harper & Row, New York

Science Advisory Board (SAB) (1990) SAB-EC-90-021. EPA, Washington DC

Shugart H.H. (1989) The role of ecological models in long-term ecological studies. In Likens G.E. (ed.) Long-Term Studies in Ecology: Approaches and Alternatives. Springer- Verlag, New York pp:90-109

Sonnenschein H. (1974) Market excess demand functions. Econometrica 40:549-563

Teger A.I. (1980) Too Much Invested to Quit. Pergamon, New York

Turner M.G., Costanza R., Sklar F.H. (1989) Methods to compare spatial patterns for landscape modeling and analysis. Ecological Modelling 48:1–18

Ulanowicz R.E. (1986) Growth and Development: Ecosystems Phenomenology. Springer-Verlag, New York

Walters C.J. (1986) Adaptive Management of Renewable Resources. McGraw-Hill, New York

7. NEW APPROACHES TO THE ASSESSMENT OF MARINE ECOSYSTEM HEALTH

John S. Gray
Department of Biology
Section of Marine Zoology and Marine Chemistry
University of Oslo
P.b. 1064, 0316 Oslo
Norway

INTRODUCTION

The most pressing environmental issues that the marine environment is faced with today are climate change, habitat destruction, chemical wastes, eutrophication and altered sediment transport. These stressors lead to altered fluxes of nutrients and energy, to loss of habitat diversity and genetic diversity, altered species distributions and abundances, and on an increasing scale to extinctions. Yet over the past decade there has been a fundamental change in perceptions of man's effects on the environment. Whereas in the 1980's it was believed that in the marine environment only local scale effects were found and that only the coastal margins were affected, we now know that regional scale effects already occur (e.g. eutrophication of the Baltic Sea) and there are indications that significant biological effects can be found in the open ocean. This latter finding is the result of application of new techniques, so-called biomarkers. Biomarkers range from genetic and biochemical techniques, through physiological approaches such as measurement of scope-for-growth in bivalves and fish to population and community indicators. Likewise, in the 1980's it was believed that ecosystem models would give predictions of likely effects of pollutants, but this goal has not been achieved and in my opinion is probably not achievable.

EXPECTED TRENDS IN STRESSED ECOSYSTEMS

Odum (1985), Rapport et al. (1985), and Rapport (1989) have compiled lists of ecosystem properties that are symptomatic of stressed systems. Odum listed 18 trends that were expected in stressed ecosystems. These ranged from general system properties such as community respiration, P/R, P/B ratios, efficiency of resource usage and degree of parasitism to properties of communities which are only parts of ecosystems. In fact most of the so-called ecosystem properties are impossible to measure in the marine environment as the limits to ecosystems cannot be defined, as they can for a lake, a wood or a grassland in the terrestrial environment. Rather than trying to define an ecosystem and the characteristics that indicate wether that such a system is stressed or not I believe that it is a more pragmatic approach to examine how stress effects can be measured at local and regional scales. In such an analysis I believe, like Rapport (1990), that one needs a pluralistic approach rather than relying on one type of measure alone.

MEASUREMENT OF LOCAL-SCALE EFFECTS

In the last decade much work has concentrated on development of reliable biomarkers of both exposure and effect (Gray 1992a). Under the auspices of the Unesco/ International Oceanographic Commission/ UNEP/ International Maritime Organization's Group of Experts on Effects of Pollution (GEEP), workshops have been held in Oslo, (Bayne et al. 1990), Bermuda (Addison and Clarke 1991) and the North Sea (together with the International Council for the Exploration of the Sea) (Stebbing et al. 1992), in order to inter-calibrate newly developed techniques. The workshops have proved invaluable tools and have established that biological effects techniques are reliable, sensitive and cheap to use. Table 1 lists some of the new biomarker techniques that are now widely used in routine monitoring of the marine environment.

One of the most interesting data sets using biomarkers is that shown in Figure 1. At each station dabs (*Limanda limanda*) were sampled and analyzed for the cytochrome P-450 system, using EROD as the measured response (Renton and Addison 1992). An increase in EROD indicates increased stress in response to PAH stressors (Table 1). Acetyl cholinesterase on the other hand decreases in response to organophosphorous as a stressor (Galgani et al. 1992). Figure 1 shows a clear gradient of decreasing effects of pollution from the Elbe estuary, site 3. However, at the Dogger Bank, site 9, there was an unexpected indication of increased pollution. Subsequent chemical analyses have shown that the site studied is a sedimentation basin where pollutants accumulate. Some biological effects techniques are thus highly sensitive, cheap and reliable indicators of stress effects.

However, a word of caution is necessary since conflicting results have been obtained through application of different variants of techniques that were thought reliable. Many biomarkers vary with the physiological state of the organism, their sex and age. A proper understanding of such variability is necessary before a routine monitoring program using biomarkers is started. Too hasty and too simplistic approaches have led to problems in interpretation of biological effects data using biomarkers, (e.g. the North Sea Task Force set up to monitor the health of the North Sea).

The effects of tributyltin (TBT), on the stocks of dogwhelk, *Nucella lapillus*, around the coast of England is a salutary lesson that predictions of effects in the field cannot always be made from laboratory experiments alone. Tributyltin was used in paints to inhibit fouling on ship hulls. Toxicity tests on a variety of organisms suggested that LC_{50} concentrations were around 1 $\mu g\ l^{-1}$. Yet field studies showed that sterility, called imposex, occurred at sea water concentrations of 0.001 μg l⁻¹. Dramatic declines in populations have occurred around the coast of England (Bryan et al. 1986), and especially in harbour areas where TBT was used on large numbers of pleasure craft. Following bans in the use of TBT dogwhelk populations are now recovering.

Table 1. Categorization of biomarkers based on their diagnostic value.

Biomarker	Category	Remark
I. DNA Integrity		
a) Strand breakage	A	Potential genotoxic insult
b) Adducts	B	Exposure to specific class of chemicals (mutagens, carcinogens)
c) Photoproducts	C	Exposure to UV-B light
d) Flow cytometry	B	Indicative of chromosomal abberations
II. Protein		
a) P-450E	B	Exposure to class of chemicals (PAHs)
b) Alad and porphyrins	B/C	Exposure to class of chemicals (lead; PCBs)
c) Choline esterase	C	Exposure to class of chemicals (organophosphates)
d) GTH metabolism	A	Chemically-induced oxidate stress
e) Metallothioneins	C	Exposure to metals
III. Scope for Growth	B/C	Integrated effect of general stress
IV. Histopathology	C	Specific deleterious lesions have occurred

(A) = biomarkers of exposure only; (B) = biomarkers of exposure with uncertain eventual consequence; (C) = Biomarkers of known deleterious consequence based on mechanistic understanding.

At the community level reliable and highly sensitive techniques for assessing effects of stressors have been developed in recent years. Examples from effects of oil exploration (Gray et al. 1991) at the Ekofisk field in the North Sea illustrate this point.

The oil companies traditionally monitored benthic communities and chemistry of sediments and used changes in diversity indices and number of taxa to illustrate effects of oil exploration (mostly caused by dumping of drilling muds). The general belief was that effects were found to a 1 km radius from the point source of pollution, giving an approximate area effected of 3 km^2. Gray et al. (1991) using multivariate analysis techniques found that effects were clearly discernible to a 3 km radius. The effects were directly correlated with amounts of mud, hydrocarbons and barium in the sediment, all stemming from the discharged drilling cuttings. Thus the effect was found to 27 km^2 instead of 3 km^2. Further studies (Gray and Olsgard in prep.) on many fields in the Norwegian sector have shown that effects are similar at other fields and the area of impact may exceed 75 km^2 in some cases. Warwick and Clarke (1991) show further examples of the effectiveness of using multivariate analyses on a variety of community data from sediments to coral reefs and plankton. These studies illustrate that use of the older univariate indices such as diversity are outmoded and should no longer be applied.

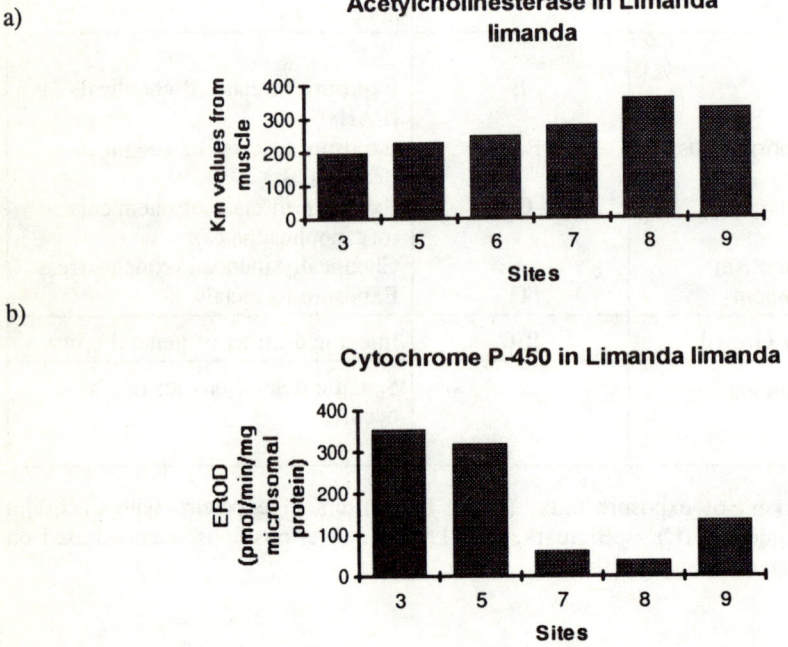

Figure 1. Application of biomarkers to a transect from the River Elbe estuary (site 3) to the Dogger Bank (site 9) North Sea. a) from Renton and Addison (1992); b) from Galgani et al. (1992).

Using the oilfield data mentioned above combining the results of the biological analyses with chemical data in newly developed multivariate techniques (Clarke and Ainsworth 1993) has given a strong indication that at a doubling of background levels of hydrocarbons and heavy metals detectable effects on the biological systems can be found. Whether or not such a "rule-of-thumb" can be used to suggest that biological effects will be found at a doubling of background levels needs thorough appraisal under a variety of environmental conditions and perturbations. Our data does however, suggest that there are a number of biological effects techniques presently applicable that are highly sensitive as well as being cheap and robust. Furthermore, it is now possible, using such techniques, to detect effects on individuals, populations and communities in areas not previously thought to be contaminated.

MEASUREMENT OF REGIONAL SCALE EFFECTS

Perhaps the most dramatic regional scale effect of man's activities in the marine environment is that of eutrophication (Gray 1992b). The Baltic Sea provides a particularly good illustration (Rosenberg et al. 1990). The scale of the effect is best illustrated by the low oxygen content and presence of hydrogen sulphide in the deep water masses. In 1969 and 1972 over 100,000 km^2 of the sea bed were anoxic. The oxygen content of the deep waters has been declining since the turn of the century. The water in the Baltic has a long residence time with replacement periods of water near Stockholm being up to 30 years and the period between replacements varies. Increased nutrient inputs are clearly linked to the eutrophication.

Other indicators of eutrophication are that fisheries landings increased dramatically and that the mean and maximum depth of macroalgae at a number of fixed sites has decreased (Kautsky et al. 1986). Yet what is needed are not measures of the end points of the change, such as oxygen depletion, but indicators of the onset of change. To my knowledge, no dose-response relationship has been established that indicates the relationship between either the quantity of nutrient inputs or the ratios of nutrients and the response of the system in terms of primary production. Wassman (1990) has suggested a relationship between the amount of primary production and export of material from the system, as envisaged by Odum (1985), but export production has not been related to nutrient input. As the system becomes progressively affected by nutrient inputs, production of benthos and fish increase, but again no quantitative relationship has been established showing where the system turns from a highly productive one to one affected by anoxia. Gray (1992b) suggests that measurements of the carbon and nitrogen content in sediments may give an indication of the likely eutrophication state.

Another regional scale problem in the marine environment is that of changes in sediment transport to coastal areas. In many areas, due to hydroelectric power development sediment has been retained and has led to large-scale changes in coastal systems. The most dramatic is the Aswan Dam which has led to dramatic sediment retention and subsequent erosion of the River Nile delta from the sea on a huge scale, with loss of agricultural areas due to the delta sinking and floods increasing, and loss of nutrients to the whole of the eastern Mediterranean with consequent loss of fisheries. Other areas however, suffer from deforestation and increased erosion as a result. This is a particularly severe problem in South-east Asia and on Pacific islands where deforestation occurs on steep-sided slopes and, with the high natural rainfall, erosion of the deforested patches is acute. As a result coastal ecosystems including coral reefs, sea-grass beds and mangroves are being destroyed on large scales (Milliman 1990; Morrison 1981).

The most pressing problem of all, however, remains that of the growth of human populations. Demographic trends (United Nations 1991) show that increasingly in both developed and third world countries the coastal zone is a preferred area for habitation. As a consequence coastal habitats are being destroyed, in the name of development, on increasing scales. Wetlands are reclaimed, mangroves removed, and sand dunes destroyed with severe damage to coastal ecosystems as a consequence. Since in many areas of the globe human populations residing in the coastal area rely on the integrity of coastal ecosystems for food, the long term consequences of habitat destruction are far-reaching.

Monitoring the scale of coastal habitat destruction over time should be relatively straight-forward using remote-sensing tools and Geographical Information Systems, yet there is a general lack of awareness of the scale of the problems and a consequent lack of effort in obtaining appropriate quantitative data. From these analyses two major problems arise that of separating natural from man-made effects and what I call the linkage problem of examining effects at different levels of organization.

SEPARATING NATURAL FROM POLLUTION-INDUCED EFFECTS

Since the 1950's Britain has run a survey of the plankton, notably the zooplankton covering a large area of the North Atlantic i.e. the Continuous Plankton Recorder programme (Colebrook and Robinson 1986). Analysis of this long-term data shows that there is a concordance in the temporal variations in zooplankton abundance in widely different geographical areas. These temporal variations can be related to long-term changes in the position of the Gulf Stream which varies irregularly but over time periods of from 3-4 to 7 or more years. Cyclonic weather patterns also influence fluctuations of planktonic organisms over periods of ten or more years (Colebrook and Robinson 1986). Few other data sets comprise such long series yet there can be no doubt that other marine systems will be affected by similar physical processes.

An even more dramatic long-term change affecting not only the marine but also terrestrial systems, almost globally, is that of the El Nino-Southern Oscillation (ENSO) phenomenon. In the marine domain effects are detected in large scale changes in planktonic systems, fisheries, sea birds and coral reefs (UNEP/GEMS 1992).

Climate change will exacerbate the problem further in that the expected scenarios for coastal systems include changed frequency of storms and frequency of episodic events such as floods in addition to raised sea level and temperatures. Thus coastal systems are to be expected to be altered by a wide range of phenomena. It is against this background of known long-term changes that one has to use indicators of ecosystem health. Change is an inherent part of marine systems and must be taken into account when assessments of ecosystem health are made. To this end a range of new strategies for monitoring marine systems have been proposed (Gray et al. 1991; Gray and Jensen 1993).

In addition the role of proper sample design has been a neglected part of health assessment. New designs for monitoring (Underwood 1991, 1993) present significant advances in allowing proper statistical evaluation of change. A further problem is that often change is not detected, not because there is no change, but because the sampling design has been inadequate. Type II statistical error is nearly always neglected. Type I error is the probability of detecting apparent change when none has occurred and is usually the basis for planning environmental surveys (e.g. the use of $p = 0.05$). Type II error is the probability of failing to detect a change when one has, in fact, occurred. Peterman (1990) and Fairweather (1991) should be consulted for reviews of these problems.

Thus the development of the range of new techniques, some of which have been described here such as biomarkers and multivariate statistical analyses techniques within the last few years, enables us to give a better appraisal of effects of man-induced changes on ecosystem health. Improved sampling designs (mentioned above), also need to be incorporated into programmes designed to assess the health of ecosystems so that monitoring programmes become more efficient and better able to detect changes in space and time than at present.

THE LINKAGE PROBLEM

Finally, those of us engaged in biological effects studies are constantly being asked to justify our findings by reference to effects at higher levels of organization than that measured. If one measures a biochemical effect at the cellular level then one is asked what is the consequence for the individual? If one detects significant effects on individuals then the question is raised but is there any detectable effect on the population? Likewise at the community level I have been asked if measured effects on benthos have any significant effects on fish or fish food i.e. almost an ecosystem level effect. This phenomenon I call "the linkage problem".

I do not believe that it is necessary for us to provide a justification for effects at higher levels. Of course scientifically it would be nice to know that reduction in scope for growth in bivalves had a consequence on subsequent fitness or that changes in benthos community structure does have an influence on fish stocks, but such links are not a necessity. Ecosystem health indicators are I believe, warnings that the system under study is being impacted. The warning can be at a variety of different levels. At the biochemical level there may be indication that there is a problem with heavy metals or polycyclic aromatic hydrocarbons (PAHs) which have induced a stress response in a fish or invertebrate. If this warning is at the same time accompanied by effects on scope for growth and/or significant changes in benthic community structure then that should be sufficient indication that the system is impacted, without the need to invoke higher order effects. Clearly the more indicators that are used the safer will be the assessment that the system under study is impacted. What is needed then is a suite of indicators at various levels of organization (see also Rapport 1990).

REFERENCES

Addison R.F., Clarke K.R. (1990) The IOC/GEEP Bermuda Workshop. J. Exp. Mar. Biol. Ecol. 138:1-182

Bayne B.L., Clarke K.R., Gray J.S. (1988) Biological effects of pollutants. Results of a practical workshop. Mar. Ecol. Progr. Ser. 46:1-278

Bryan G.W., Gibbs P.E., Hummerstone L.G., Burt G.R. (1986) The decline of the gastropod *Nucella lapiluus* around south-west England: evidence for the effect of tributyltin from antifouling paints. J. Mar. Biol. Ass. UK 66:611-640

Clarke K.R., Ainsworth M. (1993) A method for linking multivariate community structure to environmental variables. Mar. Ecol. Progr. Ser. 92:205-219

Colebrook J.M., Robinson G.A. (1986) The continuous Plankton recorder Survey. IOC Technical Series, 31:33-42

Fairweather P.G. (1991) Statistical power and design requirements for environmental monitoring. Aust. J. Mar. Freshwater Res. 42:555-568

Galgani F., Bocquené G., Cadiou Y. (1993) Evidence of variation in cholinesterase activity in fish along a pollution gradient in the North Sea. Mar. Ecol. Progr. Ser. 91: 77-82

Gray J.S., Clarke K.R., Warwick R.M., Hobbs G. (1990). Detection of initial effects of pollution on marine benthos: an example from the Ekofisk and Eldfisk oilfields N. Sea. Mar. Ecol. Progr. Ser. 66:285-299

Gray J.S. (1992a) Biological effects of marine pollutants and their detection. Mar. Poll. Bull. 25:48-50

Gray J.S. (1992) Eutrophication in the sea. In Colombo G., et al. (eds.) Eutrophication and Population Dynamics. Proceedings of the 25th EMBS. Denmark pp:3-15

Gray J.S., Calamari D., Duce R., Portmann J.E., Wells P.G., Windom H.L. (1991) Scientifically based strategies for marine environmental protection and management. Mar. Poll. Bull. 22: 432-440

Gray J.S., Jensen K. (1993) Feedback monitoring: a new way of protecting the environment. TREE 8: 267-268

Kautsky N., Kautsky H., Kautsky U., Watson M. (1986) Decreased depth penetration of *Fucus vesiculosus* (L.) since the 1940's indicates eutrophication of the Baltic Sea. Mar. Ecol. Progr. Ser. 28:1-8

Milliman J.D. (1990) Flux and fate of fluvial sediment and water in coastal seas. In Mantoura R.F.C., Martin J.M., Wollast R. (eds.) Ocean Margin Processes in

Global Change. John Wiley and Sons, New York pp:68-69

Morrison R.J. (1981) Factors determining the extent of soil erosion in Fiji. Environmental Studies Report No. 7, Institute of Natural Resources, University of the South Pacific, Suva, Fiji. 16 pp.

Odum E.P. (1985) Trends expected in stressed ecosystems. Bioscience 35: 419-422

Peterman R.M. (1990) Statistical power analysis can improve fisheries research and management. Can. J. Fish. Aquat. Sci. 47: 2-15

Rapport D.J. (1989) Symptoms of pathology in the Gulf of Bothnia (Baltic Sea): ecosystem response to stress from human activity. Biol. J. Linnean Soc. 37:33-49

Rapport D.J. (1990) Challenges in the detection and diagnosis of pathological change in aquatic ecosystems. J. Great Lakes Res. 16:609-618

Rapport D.J., Regier H.A., Hutchinson T.C. (1985) Ecosystem behaviour under stress. Am. Nat. 125:617-640

Renton K.W., Addison R.F. (1993) Hepatic microsomal mono-oxygenase activity and P4501A mRNA in North Sea dab *Limand limanda* from contaminated sites. Mar. Ecol. Progr. Ser. 91: 65-69

Rosenberg R., Elmgren R., Fleischer S., Jonsson P., Persson G., Dahlin H. (1990) Introduction-Marine eutrophication in Sweden. Ambio 19:102-108

Stebbing A.R.D., Dethlefsen V., Carr M. (1992) Biological effects of contaminants in the North Sea. Mar. Ecol. Progr. Ser. 91: 1-361

United Nations Population Division (1991) World Population Prospects 1990. U.N. New York

UNEP/GEMS (1992) The El Nino Phenomenon. UNEP, Nairobi 36pp.

Underwood A.J. (1991) Beyond BACI: experimental designs for detecting human environmental impacts on temporal variations in natural populations. Aust. J. Mar. Freshwater Res. 42:569-587

Underwood A.J. (1993) The mechanics of spatially replicated sampling programmes to detect environmental impacts in a variable world. Aust. J. Ecol. 18:99-117

Warwick R.M., Clarke K.R. (1991) A comparison of some methods for analysing change in benthic community structure. J. Mar. Biol. Ass. UK 71:225-244

Wassmann P. (1990) Relationship between primary and export production in the boreal coastal zone of the north Atlantic. Limnol. Oceanogr. 35: 464-471

8. USING BIOLOGICAL CRITERIA TO PROTECT ECOLOGICAL HEALTH

James R. Karr
Institute for Environmental Studies
University of Washington FM-12
Seattle, WA 98195
USA

INTRODUCTION

Among organisms that inhabit Earth, the human species is unique for three reasons. First, our population size and the extent of our geographic distribution gives us an impact that is unprecedented in the history of life on Earth. Five times in the last 600 million years, cataclysmic events, driven by major geological or astronomical forces such as meteorite strikes or climate change, have set back the evolutionary process with a spasm of extinction that was global in scope (Wilson 1992). As we move into the 21st century, the human species, a biological agent, is likely to be responsible for the next massive extinction event.

Second, the evolution of culture and of technology have been instrumental in changing the relationship between humans and their environment, largely by altering the carrying capacity of Earth for humans. Human society operates, especially since the dawn of the industrial revolution, as if free of the environmental constraints that control populations of all other species. We have been lulled into believing that our ability to alter carrying capacity through technological innovation is unlimited. But we do not operate without the constraints that influence all biology because rapidly evolving human cultures are imbedded in and dependent on slowly changing ecological systems.

Third, only humans are capable of recognizing the threat posed by their activities and capable of planning to minimize or avoid those threats. Before the 1960s, environmental protection was largely the province of conservationists, individuals and groups, dedicated to protecting habitats from destruction, especially overharvest by commodity groups (timber harvest, agriculture, excessive sport or commercial hunting and fishing).

By 1970, recognition of the pervasive and complex nature of environmental issues was growing. The 1972 United Nations Conference on the Human Environment-Stockholm Conference-led to creation of the United Nations Environment Program (UNEP). In the United States, many legislative initiatives sought to reverse the trend. Those laws typically dealt with narrowly defined issues such as chemical contamination of single media (Clean Water and Clean Air Acts), threats to populations of species or groups of species (Endangered Species and Marine Mammal Acts), use of specific classes of land (National Forest Management Act), hazardous waste management (Resources Conservation and Recovery Act), and cleanup of contaminated waste disposal sites ("Superfund").

Only the National Environmental Policy Act (NEPA), passed in 1969 by Congress and signed into law on January 1, 1970 by President Richard Nixon, attempted to articulate a broad societal vision. That law sought to establish policies that would encourage harmony between humans and their environment (Hildebrand and Cannon 1993). Unfortunately, that broad vision was lost as government and nongovernment bureaucracies were established to prepare environmental impact statements. Those reports became ends rather than means and often were not scientifically sophisticated (Schindler 1976; Fairfax 1978).

By the 1980s, special interest constituencies ranging from the wise-use movement to Earth First fostered an attitude of confrontation. At the same time, a growing number of voices tried to focus attention on the need to protect the health or integrity of local and global life-support systems (Karr and Dudley 1981; Rapport 1989). As we enter the 1990s, a wave of initiatives directed toward sustainability have developed, where sustainability may be defined in ecological (Ludwig et al. 1993; Levin 1993), economic (World Comm. on Env. and Dev. 1987), social (Yale Law School 1993), and security (Myers 1993) terms.

Although some individuals deny that environmental topics deserve societal attention (Ray and Guzzo 1993; Simon and Kahn 1984), beyond these fringes exists a growing recognition that environmental issues are the principle challenge facing human society in the next millennium. This issue like no other in the history of human society is developing a consensus that includes societal leaders on a global scale: citizens (Dunlap et al. 1993); scholars (NRC 1993ab); universities (Talloires Declaration); governments (Rio Summit); business leaders (Schmidheiny 1992); labor (United Steelworkers of America 1990); and world religions (Briggs 1993).

At long last the battleground is shifting from whether or not society faces serious environmental challenges to how best to respond to those challenges. Recognition is widespread that our approach to the natural world is not only a disaster but a crime. It is conceptually and morally wrong like slavery, child labor, and murder. We should think of environmental problems as another type of failed relationship for human society.

BIOTIC IMPOVERISHMENT

Human influence is not only pervasive, it is also expanding. Massive biotic impoverishment, the systematic reduction in earth's ability to support living systems, is the major consequence of that expansion. Biotic impoverishment occurs in a number of forms (Table 1). Unfortunately, societal recognition of the extent to which Earth's life support systems are being degraded is limited; we know even less about the long-term effects of that impoverishment on human society. As a direct result of our relative ignorance, the right to a quality environment is threatened for a substantial and growing number of humans. That threat is exhibited in many forms of social injustice but rarely is the connection to environmental rights understood for those living today and for future generations. Global, regional, and local environmental challenges, then, are tied to the interactions of three factors: (1) human reproductive behavior and resource consumption causes (2) biotic impoverishment and (3) social injustice. In short, human reproductive and consumptive behavior are changing the face of the earth.

Table 1. Global forms of biotic impoverishment that result from actions of human society.

1. Soil depletion/desertification/salinization
2. Depletion of renewable natural resources
3. Depletion and contamination of water supplies
4. Extinction of species
5. Habitat destruction and fragmentation
6. Epidemics and pest outbreaks, introduction of exotics
7. Chemical contaminantion
8. Global climate change, ozone depletion
9. Reduction in human cultural diversity
10. Reduced quality of human life, economic deprivation

SEEKING SOLUTIONS - MEASURING SOCIETAL WELL-BEING

Recognition of a problem, however, does not guarantee a solution. If the premise that biotic impoverishment is the problem has merit, then the solution must come from efforts to prevent that impoverishment, to protect the integrity of living systems. Individual health as measured by conventional medicine and economic health as assessed by macroeconomics, the conventional measures of well-being, are not adequate to protect societal interests. Widespread recognition of that need has stimulated philosophers, ecologists, and economists to search for innovative approaches to define societal health based on ecological integrity (Karr 1992, 1993) or ecosystem health (Rapport 1989; Costanza et al. 1992). Properly selected measures of ecological health can be used to determine if life-support systems are degraded and identify the factors responsible for degradation. They may even be used to track the success of restoration programs.

IMPEDIMENTS

The major impediments to integrating the concept of ecological health into the conscience of human society include the tendency of western society to 1) dissociate human welfare from the life-support systems upon which we depend, 2) evaluate human-environment interactions on short time scales and over local areas, and 3) isolate individual problems in efforts to solve them without reference to the larger matrix within which they exist. Societal ecological literacy (Orr 1992) is not sufficient to recognize the clear distinctions between the advocates of "technological sustainability" with faith in technological answers and market solutions and advocates of "ecological sustainability" who recognize that we cannot escape our creaturehood. Our economies are wrapped in and ultimately constrained by physical, chemical, and biological processes on an isolated, finite planet.

Technological extravagance, endorsed by many as a replacement of critical components of life support systems, is especially shortsighted because the secondary and tertiary influences of technology bring unanticipated threats to human welfare. The first humans to use wheels did not, for example, anticipate the need to bury productive agricultural land under pavement to provide roads for those wheels. Similarly, the inventor of the internal combustion engine did not anticipate the effects of that engine on the global atmosphere. The ozone depleting characteristics of CFCs were not anticipated when refrigeration dependent on use of CFCs spread.

In these and numerous other examples, human society sought to resolve environmental challenges with a narrow conceptual analysis followed by implementation of a solution involving a new technology. Typically, technology was designed to replace or speed a natural process upon which humans and other members of Earth's biota depend. Further, most efforts to correct ecological imbalance treated the symptom rather than the root cause of the problem, as Meffe (1992) notes in discussing the inadequacies of salmon hatcheries to maintain salmon stocks in the Northwest.

ESTABLISHING STANDARDS

Society's approach to protecting its long-term and short-term interests is to establish technical or behavioral standards to which individuals, corporations, or governments must adhere. Standards may be design criteria used in construction of high-rise buildings, highways, or bridges; ethical standards for lawyers or psychiatrists; or approved procedures for disposal of toxic materials. Problems arise when these standards are abused (e.g. poorly constructed buildings collapse) or when design criteria are not adequate to the task. Many cities in the western United States are, for example, retrofitting highway bridges constructed without adequate regard for the occurrence of earthquakes.

Enforcement of the Clean Water Act (CWA) in the United States provides an excellent example of standards, criteria, and enforcement processes that fail to protect societal interests. Twenty years ago the Congress recognized that the quality of water resources continued to decline. Signs of degradation ranged from threats to human health and declining fish and shellfish populations. In 1969 the Cuyahoga River near Clevelend, Ohio, caught fire.

Under the leadership of Senator Edmund Muskie of Maine, the Congress moved to extend the Nation's commitment to high quality water resources. Those efforts resulted in passage of a revised Clean Water Act (Public Law 92-500) in 1972. The CWA contained several revolutionary approaches as it shifted away from a narrow focus on navigable waters, organic and point-source pollution, and construction of wastewater treatment plants. For the first time, for example, substantial attention was given to the need to control non-point sources of pollution. Other visionary goals included a mandate to "restore and maintain the chemical, physical, and biological integrity of the nation's waters." The CWA, thus, clearly established a legal foundation for protection of the aquatic biota. Unfortunately, that vision was not reflected in the implementing regulations developed in support of the CWA (Karr 1990).

Although the 1972 CWA did not explicitly call for improvements in the ecological health of the nation's water resources, that was, I believe, the intent of the Congress when it used the biological integrity phrase (Karr and Dudley 1981, Karr 1991, Adler et al. 1993). That broad vision has remained clear as the CWA and its subsequent reauthorizations (PL 95-127, PL 100-4) set targets such as fishable and swimmable, antidegradation, no net loss, zero discharge (of pollutants), and protection of ecological integrity.

Integrity implies an unimpaired condition or the quality or state of being complete or undivided; biological integrity for streams implies the presence of an adaptive assemblage of organisms having a species composition, species richness, and functional organization comparable to that of natural habitat in the region. Biological integrity is the condition of aquatic communities inhabiting unimpaired water bodies of a specified region. Edwards and Ryder (1990) recently used the phrase "harmonic community" in a similar context in a discussion of the goal of restoring ecological health to the Laurentian Great Lakes. Healthy ecological systems are more likely to withstand perturbations imposed by natural environmental processes as well as the many disruptions induced by human society. At the same time, they require minimal support for management.

The United States Environmental Protection Agency established water-quality standards to accomplish these goals. In the U. S. water quality standards consist of two parts: designated uses and criteria. Designated uses are the purposes or benefits to be derived from a water body (e.g., aquatic life or drinking water). Criteria are the conditions presumed to support or protect the designated uses. Historically, numeric chemical criteria were the foundation of water quality standards. For example, dissolved oxygen may not fall below 5 mg/l if the designated use is a coldwater fishery. Some individuals consider the antidegradation goal, to prevent further decline in resource condition, to be a third component of water quality standards.

Unfortunately, water resource managers implemented the CWA with regulations directed toward reducing the release of chemical contaminants (conventional "pollution"-- SAB 1990; Karr 1990, 1991). This approach implicitly but incorrectly assumed that crystal clear, distilled water flowing down concrete channels was the goal of the CWA.

But despite massive expenditures based on implementation of chemical criteria, water resources, especially their biological components, are in steep decline. Evidence that the biota of aquatic systems continues a downward spiral is widespread (Karr et al. 1985; Moyle and Leidy 1992; Williams and Neves 1992; Allen and Flecker 1993). A disproportionate number of aquatic organisms as compared to terrestrial vertebrates are classed as rare to extinct (Master 1990). But the water resource crisis extends beyond degradation in water quality (chemical contamination) and loss of species to include homogenization of the biota (regional extinction and introduction of exotics--Culotta 1991), declines in commercial and sport harvests of aquatic resources, and fish consumption advisories to protect human health.

Failure to protect the health of aquatic ecosystems is due to two major problems. First, chemical criteria based on toxicity tests do not incorporate complex ecological effects (Levin et al. 1989). Second, degradation is caused by a broader array of factors than chemical contamination, the primary focus of conventional water quality programs (Karr 1991). The number of human activities that degrade healthy water resources is huge and their cumulative impacts create even greater complexity. Degradation often begins in upland areas of the watershed or catchment as a result of human actions that alter the vegetative cover of the land surface. These changes, combined with modification of stream corridors, alters the quality of water delivered to the stream channel as well as the structure and dynamics of those channels and their adjacent riparian environments. Impact from human activities upon five primary classes of variables may result in degradation of water resources (Karr 1991; NRC 1992).

1. **Water Quality**: Temperature; turbidity; dissolved oxygen; acidity; alkalinity; organic and inorganic chemicals; heavy metals; toxic substances.

2. **Habitat Structure**: Substrate type; water depth and current velocity; spatial and temporal complexity of physical habitat.

3. **Flow Regime**: Water volume; temporal distribution of flows.

4. **Energy Source**: Type, amount, and particle size of organic material entering stream; seasonal pattern of energy availability.

5. **Biotic Interactions**: Competition; predation; disease; parasitism; mutualism.

I recently discovered that George Perkins Marsh (1857) recognized this structure nearly 150 years age when he noted the effects of human actions on flow ("The general character of our water courses has become in fact more torrential"), habitat structure ("continually changes the beds and banks of the streams"), energy source ("changes…in the character of our waters involve great changes in the nutriment which nature supplies to the fish"), and water quality ("annual mean temperature has been raised or lowered" and "fill the water with mud and other impurities"). Of the five factors, only biotic interactions were not explicitly mentioned in Marsh's analysis of how human actions have "produced an almost total change in all the external conditions of piscatorial life..."

USING BIOLOGICAL CRITERIA TO DETECT AND DEFINE DEGRADATION

Long-term success in protecting water resources requires careful thought about assessment endpoints, including development of biological criteria. Biocriteria are narrative expressions or numerical values that describe the acceptable biological condition of aquatic communities inhabiting streams and rivers of a given designated aquatic life use. Biocriteria should be reflective of the biological integrity of the particular region under study. Because ecological health ultimately depends on the presence of a healthy biological system, measures of biological condition provide the proper endpoint for most analyses. Ecological health cannot be protected by narrow dependence on chemical criteria, measures of habitat condition, or analyses of instream flow.

The basic premise, that the biota provide a sensitive screening tool to measure the health of a water resource, depends on the assumption that the greater the anthropogenic impact in a watershed the greater the impairment of that biota. A corollary is that streams and rivers not subject to anthropogenic impact contain natural communities of aquatic organisms that exhibit unimpaired conditions.

These assumptions provide the scientific foundation for formulating hypotheses regarding the impact of human activity on the health of the biota of running waters. That is, impairment reflects a departure from the natural condition due to human disturbances. Natural disturbances such as floods or drought may also affect the aquatic biota as part of normal ecological processes. Such natural disturbances must be considered when interpreting data but should not be considered impairment.

The development of biocriteria requires definition of the attributes of healthy systems, especially those attributes that society seeks to protect, and characterization of the behavior of those attributes under varying human influence. Numerous attributes of the biota have been used to assess the condition of water resources. Some are difficult and expensive to measure while others are not. Some provide reliable evaluations of biological conditions while others, perhaps because they are highly variable, are more difficult to interpret. Thus, the choice of attributes to be measured and assessed is critical to the success of any program to monitor biological conditions.

Toxicologists have long recognized the importance of individual health in evaluating the extent to which human actions have degraded the water resource. Ecologists proposed use of information theoretic measures of diversity to assess if species richness (the number of species) or relative abundances of species have been altered. Fish biologists have used a variety of measures of population structure to assess the health of populations of target species (e.g. sport fish).

Historically, most biological evaluations were designed to detect a narrow range of factors degrading water resources. For example, the biotic index (Chutter 1972; Hilsenhoff 1987) is designed to detect the influence of oxygen demanding wastes ("organic pollution") or sedimentation as is the saprobic index developed early in this century (Kolkwitz and Marsson 1908).

With increased understanding of the complexity of biological systems and the complex influences of human society on those systems, more integrative approaches to assessing biological integrity have been developed. Some (Ulanowicz 1990; Kay 1990; Kay and Schneider 1992) advocate use of thermodynamics while others concentrate on richness or diversity (Wilhm and Dorris 1968). The best approach is an integrative combination; that is, efforts to protect biotic integrity should include evaluation of a broad diversity of biological attributes. The most obvious elements (units) are the species of the biota but additional critical elements include the genetic diversity within those species and the assemblages (communities and landscapes) upon which those species depend. Processes (or the functional relationships) span the hierarchy of biological organization from individuals (growth, metabolism) to populations (mutation and recombination rates, reproduction, recruitment, demography, dispersal, speciation) and assemblage-communities-ecosystems (nutrient cycling, competition, predation, energy flow, metapopulation dynamics). For example, an important process in streams is the interaction that involves larval stages of mussels attaching to fish gills and, thus, being widely dispersed.

Because biocriteria are used in bioassessment programs to evaluate the impact of stress placed by human action on the health or integrity of the system, the selection of biological monitoring approaches is critical. When known stressors exist, it may be prudent to narrow the range of attributes measured to those likely to be influenced by those stressors. In more complicated water resource problems or for long-term status and trends monitoring, more attributes should be measured. Finally, programs to monitor for effects of environmental actions should have especially broad perspectives to insure sensitivity to all forms of degradation. At least four loosely defined components should be evaluated:

- **Structure**: Species richness (total and selected groups); relative abundances, including the extent to which one or a few species dominate.

- **Composition**: The identity of the species that make up the biota, including but not limited to rare and endangered species; native vs. exotic species.

- **Individual Health**: Proportion of diseased or deformed individuals; contaminant levels in tissues; metabolic processes; individual reproductive success.

- **Ecological Processes**: Trophic structure of the biota, including abundance of top carnivores; relative abundances of trophic specialists and generalists; population demography; recruitment rates; predation rates.

Comprehensive assessments that include all four are more likely to protect ecological health. For each component, one or more metrics or attributes should be assessed. Successful metrics represent the expression of a known influence of human activities on the characteristics of the resident biota; for example, with increased human disturbance, the total species richness, the number of intolerant species, and the number of trophic specialists declines, while the number of trophic generalists or dominance by a tolerant species increases.

ASSESSING BIOLOGICAL INTEGRITY

The complexity of biological systems and the varied impacts of human society require a broadly based, multimetric approach (Karr 1991) that integrates diverse information. Attributes evaluated include species richness, indicator taxa (both intolerant and tolerant), relative abundances of species, trophic, or reproductive guilds, and the incidence of hybrids, disease, and anomalies, such as lesions, tumors, or fin erosion in fish and deformed head capsules in invertebrates.

To implement biological criteria, biologists need formal, standard methods to sample the biota, evaluate the resulting data, and clearly communicate the condition of that biota. A sound biological monitoring program that uses biocriteria to assess biological integrity should have several attributes: firm conceptual foundation based in ecological principles; metrics that span individual, population, assemblage, and landscape levels; metrics that include elements and processes of biological systems; defined reference condition, the condition of an unimpaired site(s) against which a study site(s) is evaluated; collection and tabulation of high quality data; and creative analysis and synthesis of information about relevant biological attributes. Finally, the resident biota is the central focus because that biota integrates conditions across the watershed and over the time scale of the resident biota.

Each biological attribute used as a metric in a multimetric assessment tool is selected because a quantitative association exists between that attribute and the magnitude of human disturbance. That is, each metric is a working hypothesis based in extensive knowledge of natural history and the behavior of biological systems under stress from human actions.

The Index of Biotic Integrity (IBI) (Karr 1981, 1991; Karr et al. 1986) or its conceptual clones typically incorporates a dozen biological attributes into a single index (Table 2). No multimetric index can include all attributes because sampling to directly measure all components is expensive and time consuming. Thus, the challenge is to extract adequate biological information from comprehensive field samples of the resident biota. IBI was developed for use with fishes in the streams of Illinois and Indiana. Its success stimulated others to adopt IBI in water resource evaluations over a wider geographic area (Ohio EPA 1987; Lyons 1992; Oberdorff and Hughes 1992) and for taxa other than fish (e.g. benthic invertebrate communities - Ohio EPA 1988; Plafkin et al. 1989, Kerans and Karr in press; riparian birds - Brooks et al. 1991). At least one state (Nebraska) combines fish and invertebrate metrics into a single index. Ohio uses separate indexes for fish and invertebrates.

Table 2. Biological attributes used in assessment of fish and benthic invertebrates. (From Karr 1981 and Kerans and Karr in press.)

Category	Metric

FISH INDEX OF BIOTIC INTEGRITY
  Species richness and composition
      1.   Total number of native fish species
      2.   Number and identity of darter species
      3.   Number and identity of sunfish species
      4.   Number and identity of sucker species
      5.   Number and identity of intolerant species
      6.   Proportion of individuals as green sunfish
  Trophic composition
      7.   Proportion of individuals as omnivores
      8.   Proportion of individuals as insectivorous cyprinids
      9.   Proportion of individuals as top carnivores
  Fish abundance and condition
      10.   Number of individuals in sample
      11.   Proportion of individuals as hybrids
      12.   Proportion of individuals with tumors, fin erosion, and skeletal anomalies

BENTHIC INDEX OF BIOTIC INTEGRITY
  Taxa richness
      1.   Total taxa richness
      2.   Number of intolerant snail and mussel species
      3.   Mayfly taxa richness
      4.   Caddisfly taxa richness
      5.   Stonefly taxa richness
  Proportion of individuals
      6.   Proportion of individuals as *Corbicula*
      7.   Proportion of individuals as oligochaetes
      8.   Proportion of individuals in the two most abundant species
      9.   Proportion of individuals as omnivores and scavengers
      10.   Proportion of individuals as collectors-filterers
      11.   Proportion of individuals as grazers-scrapers
      12.   Proportion of individuals as strict predators (excluding chironomids and flatworms)
  Total Abundances
      13.   Total abundance

Metric values are compared to values expected for a relatively undisturbed stream of similar size and geographic region. Each metric is rated 5, 3, or 1 depending on whether its condition is comparable to, deviates somewhat from, or deviates strongly from the expected value.

Metric scores for a twelve metric index like the original fish IBI are summed with a potential range from 12 in areas without fish to 60 in areas with fish faunas equivalent to those in pristine or relatively undisturbed areas.

Regional modifications of the IBI have been very successful as long as the metrics included are selected to retain the general ecological structure of the original IBI (Miller et al. 1988). Many state agencies have adopted the biotic integrity approach for use in water management programs. Ohio EPA, for example, uses IBI and two other biological indexes to establish and maintain use designations for water bodies, and in support of their Section 319 CWA non-point source program, Section 305(b) CWA water quality inventory reports, and National Pollution Discharge Elimination System (NPDES) discharge permits (Dudley 1991). The conceptual approach of IBI has been used to evaluate a variety of aquatic environments (large rivers, lakes, estuaries and reservoirs - Dionne and Karr 1991; Jordan 1991; Miller et al. 1988) and with two major taxa (fish - Karr et al. 1986; Lyons 1992; invertebrates - Ohio EPA 1988; Lenat 1988; Kerans and Karr in press). The success of efforts to evaluate biological integrity depend less on the taxon selected for sampling and analysis than on the ecological sophistication brought to bear on the development of the analytical approach. Further, sampling and evaluation of several major taxa is likely to improve inferences about conditions at a site.

Because the biota of rivers reflects the status and quality of their landscape, use of biological criteria provides a more integrative conceptual framework for evaluating the condition of landscapes and the life support systems upon which human societies depend (Table 3). An integrity perspective forces us to recognize the interactions of surface and ground waters, the connections between water quality and water quantity, and the dependence of the aquatic biota on water and the landscape. The most direct and effective assessment of the status of rivers and the landscapes or watersheds that they drain should be based in biology.

The success of biocriteria to define and protect the quality of water resources has stimulated efforts to apply the approach to terrestrial environments and complex landscapes (Hunsaker and Carpenter 1990). Assessment of biological integrity is essential to chart the course of federal and state programs to protect the economic and ecological interests of society (Costanza et al. 1992).

Under Section 305(b) of the CWA, states are required to report the status of water resources within their boundaries. Because those reports focus on chemical conditions, they chronically underreport the extent of degradation. In Ohio, for example, conventional chemical criteria failed to detect 50% of the impairment of surface waters when compared to the more comprehensive, sensitive, and objective assessment provided by the use of biological criteria (Yoder 1991).

RIVERS, WATERSHEDS, AND LANDSCAPES

Rivers are in many ways the lifeblood of human society. They provide water for domestic and industrial uses, they serve as transportation corridors, and they provide food, recreation and scenic beauty. Those rivers and their landscapes provide the resources upon which humans depend for their societal health. In addition to their direct importance to human society, rivers are indicative of the health of the surrounding landscape just as blood samples provide important insight about the health of humans.

Table 3. Advantages of biotic integrity assessment for protection of the quality of water resources.

---

1. Broad ecological foundation
2. Provides biological meaningful evaluations
3. Flexible for special needs
4. Sensitive to broad range of degradation
5. Integrates cumulative impacts of all forms of pollution (e.g., point and non-point sources of chemical contamination, habitat or flow alteration)
6. Direct evaluation of resource condition
7. Easy to relate to general public
8. Overcomes weaknesses of single parameter chemical approaches
9. Assesses degree of degradation, not just threshold condition
10. Can be used in both space and time dimensions

Riverine ecosystems encompass an interactive mosaic of environments that extend from headwater streams and wet meadows at the upper limit of drainage basins to the ocean. This mosaic includes small streams and large rivers, riparian zones, terrestrial watersheds, and groundwater (including the hyporheic zone). Water, plants, animals, and the by-products of human society such as toxic chemicals, nutrients, and persistent debris such as plastics move among these environments without reference to county, state, or national boundaries. Individual human actions may produce small local effects, but when these impacts are combined, they spawn more complex problems that cascade across the landscape and down the river, resulting in considerable loss of natural resource values. The cumulative affect is often ignored because most conventional aquatic resource evaluations consider only local water quality parameters. Use of biocriteria provides an integrative endpoint to evaluate the conditions in a study watershed.

Many of the problems created by human actions could be resolved, or at least minimized, by a planning scale that is both tractable and logical. The regional watershed perspective provides a logical planning level. An approach based on the concept of the watershed permits, even imposes, the need for a holistic perspective with concern for both human and ecological health, a broadly defined societal well being. The implementation of watershed specific plans requires treatment of all stressors from all sources as well as the need to consider all chemical, physical, and biological effects. A watershed protection approach is also more likely than past efforts to require coordination among all interested parties, including numerous governmental bureacracies with conflicting programs and policies. Above all, an integrative watershed protection approach provides mechanisms to assess progress and develop and improve tools and programs.

CONCLUSIONS

Water's mobility, integrative qualities, and importance to all biological systems makes examination of the status of water resources a practical approach for assessing the condition of the landscapes upon which human societies depend for their long-term "health." Over the past decade, major advances have been made in the development and use of ambient biological monitoring and biological criteria to define the health of landscapes. Existing definitions and approaches for measuring the quality of water resources through biocriteria provide a template to guide the development of procedures to assess sustainability across diverse natural and human landscapes. Critical components of successful biocriteria include evaluations relative to regional expectations and use indexes that reflect the multivariate nature of biological systems, including individual, population, assemblage, and landscape perspectives.

REFERENCES

Adler R.W., Landman J.C., Cameron D.M. (1993) The Clean Water Act 20 Years Later. Island Press, Washington, DC

Allen J.D., Flecker A.S. (1993) Biodiversity conservation in running waters. BioScience 43:32-43

Briggs D. (1993) World's clerics draft global ethic: violence, sexism, environmental abuse are all targeted. September 1, 1993 Seattle Times

Brooks B.P., Croonquist M.J., D'Silva E.T., Gallagher J.E., Arnold D.E. (1991) Selection of biological indicators for integrating assessments of wetland, stream, and riparian habitats. In Biological Criteria: Research and Regulation. EPA-440/5-91-005. Office of Water, U. S. Environmental Protection Agency, Washington, DC pp:81-89

Chutter F.M. (1972) An empirical biotic index of the quality of water in South African streams and rivers. Water Research 6:19-20

Costanza R., Norton B.G., Haskell B.D. (1992) Ecosystem Health: New Goals for

Environmental Management. Island Press Washington DC
Culotta E. (1991) Biological immigrants under fire. Science 254:1444-1447
Dionne M., Karr J.R. (1991) Designing surveys to assess biological integrity in lakes and reservoirs. In Biological Criteria: Research and Regulation. Office of Water, U.S. Environmental Protection Agency, Washington, DC. EPA-440/5-91-005 pp:62-72
Dudley D.R. (1991) A state perspective on biological criteria in regulation. In Biological Criteria: Research and Regulation. U.S. Environmental Protection Agency, Washington, DC. EPA-440/5-91-005 pp:15-18
Dunlap R.E,. Gallup Jr. G.H., Gallup A.M. (1993) Of global concern: results of the health of the planet survey. Environment 35(9):7-15,33-39
Edwards C..J, Ryder R.A. (1990) Biological surrogates of mesotrophic ecosystem health in the Laurentian Great Lakes. Report to the Great Lakes Science Advisory Board, International Joint Commission, Windsor Ontario
Hildebrand J.G., Cannon J.B. (1993) Environmental Analysis: The NEPA Experience. Lewis Publishers, Boca Raton, Florida
Hilsenhoff W.L (1987) An improved biotic index of organic stream pollution. Great Lakes Entomologist 20:31-39
Hunsaker C.T., Carpenter D.E. (1990) Environmental Monitoring and Assessment Program: Ecological Indicators. United States Environmental Protection Agency, Washington, DC. EPA/600/3-90/060.
Fairfax S.K. (1978) A disaster in the environmental movement. Science 199:743-747
Jordan S.J. (1991) Fish assemblages as indicators of environmental quality in Chesapeake Bay. In Biological Criteria: Research and Regulation. U.S. Environmental Protection Agency, Washington, DC. EPA-440/5-91-005 pp:73-80
Karr J.R. (1981) Assessment of biotic integrity using fish communities. Fisheries(Bethesda) 6(6):21-27.
Karr J.R. (1990) Biological integrity and the goal of environmental legislation: lessons for conservation biology. Conservation Biology 4:244-250.
Karr J.R. (1991) Biological integrity: a long-neglected aspect of water resource management. Ecological Applications 1:66-84
Karr J.R. (1992) Ecological integrity: protecting earth's life support systems. In Costanza R., Norton B.G., Haskell B.D. (eds.) Ecosystem Health: New Goals for Environmental Management. Island Press Washington DC pp: 223-238
Karr J.R. (1993) Protecting ecological integrity: an urgent societal goal. Yale Journal of International Law 18(1):297-306
Karr J.R., Dudley D.R. (1981) Ecological perspective on water quality goals. Environmental Management 5:55-68
Karr J.R., Fausch K.D., Angermeier P.L., Yant P.R., Schlosser I.J. (1986) Assessing biological integrity in running waters: a method and its rationale. Special Publication 5 Illinois Natural History Survey. Urbana, Illinois
Karr J.R., Toth L.A., and Dudley D.R. (1985) Fish communities of midwestern rivers: a history of degradation. BioScience 35:90-95

Kay J.J. (1990) A non-equilibrium thermodynamic framework for discussing ecosystem integrity. In Edwards C.J., Regier H.A. (eds.) An Ecosystem Approach to the Integrity of the Great Lakes in Turbulent Times Spec. Publ. 90-4, Great Lakes Fish. Comm. Ann Arbor, Michigan pp:209-237

Kay J.J., Schneider E.D. (1992) Thermodynamics and measures of ecological integrity. In McKenzie D.H., Hyatt D.E., McDonald V.S.(eds.) Ecological Indicators. Vol. 1, Elsevier Applied Science, London pp:159-182

Kerans B.L., Karr J.R. (in press) Development and testing of a benthic index of biotic integrity (B-IBI) for rivers of the Tennessee Valley. Ecological Applications

Kolkwitz R., Marsson M. (1908) Okologie der pflanzlichen saprobien. Berich. Deutsch. Bot. Gesellsch. 26:505-519

Lenat D.R. (1988) Water quality assessment of streams using a qualitative collection method for benthic macroinvertebrates. Journal North American Benthological Society 7:222-233

Levin S.A. (ed.) (1993) Science and sustainability: 17 responses to Ludwig et al. (1993). Ecological Applications 3:545-589

Levin S.A., Harwell M.A., Kelly J.R., Kimball K.D. (1989) Ecotoxicology: Problems and Approaches. Springer-Verlag New York

Ludwig D., Hilborn R., Walters C. (1993) Uncertainty, resource exploitation, and conservation: lessons from history. Science 260:17-18

Lyons J. (1992) Using the index of biotic integrity (IBI) to measure environmental quality in warmwater streams of Wisconsin. United States Forest Service, Minneapolis, Minnesota. General Technical Report NC-149

Marsh G.P. (1857) Report made under authority of the Legislature of Vermont on the artificial propagation of fish. Free Press Print Burlington Vermont

Master L. (1990) The imperiled status of North American aquatic animals. Biodiversity Network News 3(3):1-2,7-8

Meffe G.L. (1992) Techno-arrogance and halfway technologies: salmon hatcheries on the Pacific coast of North America. Conservation Biology 6:350-354

Miller D.L. et al. (1988) Regional applications of an index of biotic integrity for use in water resource management. Fisheries (Bethesda) 13(5):12-20

Moyle P.B., Leidy R.A. (1992) Biodiversity in aquatic ecosystems: evidence from fish faunas. In Fiedler P.L., Jain S.K. (eds.) Conservation Biology: The Theory and Practice of Nature Conservation, Preservation, and Management. Chapman and Hall New York pp:127-169

National Research Council (1992) Restoration of Aquatic Ecosystems: Science, Technology, and Public Policy. National Academy Press Washington

National Research Council (1993a) Population Summit of the World's Scientific Academies: A Joint Statement by Fifty-eight of the World's Scientific Academies. National Academy Press Washington

National Research Council (1993b) Research to Protect, Restore, and Manage the Environment. National Academy Press Washington

Oberdorff T., Hughes R.M. (1992) Modification of an index of biotic integrity based on fish assemblages to characterize rivers of the Seine-Normandie Basin, France. Hydrobiologia 228:117-130

Ohio Environmental Protection Agency (1988) Biological criteria for the protection of aquatic life. Ohio EPA Division of Water Quality Monitoring and Assessment, Surface Water Section, Columbus, Ohio

Orr D.W. (1992) Ecological Literacy: Education and the Transition to a Postmodern World. State University of New York Press, Albany, New York

Plafkin J., Barbour M.T,. Porter K.D., Gross S.K., Hughes R.M. (1989) Rapid bioassessment protocols for use in streams and rivers: benthic macroinvertebrates and fish. United States Environmental Protection Agency, Washington

Rapport D.J. (1989) What constitutes ecosystem health? Perspectives in Biology and Medicine 33(1):120-132

Ray D.L., Guzzo L. (1992) Environmental Overkill: Whatever Happened to Common Sense. Regnery Gateway, Washington, DC

Schindler D.W. (1976) The impact statement boondoggle. Science 192:50

Schmidheiny S. (1992) Changing Course: A Global Business Perspective on Development and the Environment. MIT Press, Cambridge, Massachusetts

Science Advisory Board (SAB) (1990) Reducing risk: setting priorities and strategies for environmental protection. U.S. Environmental Protection Agency, Washington, DC. SAB-EC-90-021

Simon J., Kahn H. (1984) The Resourceful Earth: A Response to "Global 2000." Blackwell New York

United Steelworkers of America (1990) Our Children's World: Steelworkers and the Environment. Pittsburgh, Pennsylvania

Vitousek P.M., Ehrlich P.R., Ehrlich A.H., Matson P. (1986) Human appropriation of the products of photosynthesis. BioScience 36:368-373

Wilhm J., Dorris T.C. (1968) Biological parameters of water quality criteria. BioScience 18:477-481

Williams J.E., Neves R.J. (eds) (1992) Biological diversity in aquatic management. Trans. NA Wildl. Nat. Res. Conf. 57:343-43.

World Commission on Environment and Development (1987) Our Common Future. Oxford University Press, Oxford

Yale Law School (1993) Earth Rights and Responsibilities: Human Rights and Environmental Protection. Symposium. Yale Journal of International Law 18:215-411

Yoder C. (1991) The integrated biosurvey as a tool for evaluation of aquatic life use attainment in Ohio surface waters. In Biological Criteria: Research and Regulation. U. S. Environmental Protection Agency, Washington, DC. EPA-440/5-91-005 pp:110-122

9. BIOLOGICAL CHANGES IN THE GERMAN BIGHT OF THE NORTH SEA AS INDICATORS OF ECOSYSTEM HEALTH

Volkert Dethlefsen
Bundesforschungsanstalt für Fischerei
Institut für Fischereiökologie
Außenstelle Cuxhaven
Deichstr. 12, D-27472 Cuxhaven
Germany

INTRODUCTION

The pollution of the North Sea and its impact on components of ecosystems have attracted public and scientific attention since the early 1970s. Reviews reflecting the development of knowledge on anthropogenically caused ecological perturbations are to be found in *"Umweltprobleme der Nordsee - Sondergutachten Juni 1980 des Sachverständigenrates für Umweltfragen"* (Anonymous 1980), *"Pollution of the North Sea - an assessment"* (Salomons et al. 1988), *"Environmental Protection of the North Sea"* (Newman and Agg 1988), and *"Nordsee - ein Lebensraum ohne Zukunft"* (Buchwald 1990), *"Warnsignale aus der Nordsee"* (Lozán et al. 1990). In October 1987 the German parliament discussed the Protection of the North Sea in a public hearing (Schutz der Nordsee 1987).

A number of substantial scientific projects concentrated on pollution problems in this area. Results of two major German activities are published in *"Biogeochemistry and distribution of suspended matter in the North Sea and implications to fishery biology"* (Kempe et al. 1988) and *"Zirkulation und Schadstoffumsatz in der Nordsee"* (Sündermann and Beddig 1988*)*. In a number of reviews I have listed some of the biological and chemical details of the ecological changes in some of the North Sea areas (Dethlefen, 1988ab, 1989a, 1989; Dethlefsen et al. 1989). Currently a massive report on the status of the North Sea is published as part of the North Sea Task Force, an intergovernmental organization. The Task Force reports are based on data collected over many years and reviewed by groups of scientists, usually from different countries, within each section of the North Sea (North Sea Assessment Reports 1993; Subregions 1, 4a, 6, 7a, 8 and 10 - The Wadden Sea).

In this paper a brief summary will be given of some of the ecological changes observed in the North Sea in the last decade with special reference to the German Bight. The next part will be on the long and medium term variability of two biological deviations from normal - diseases of fishes and malformations of fish embryos - including regional changes of the prevalence encountered. This will be followed by a discussion of ecological changes in the North Sea in relation to anthropogenic activities.

ECOLOGICAL CHANGES IN THE NORTH SEA

The changes encountered in the North Sea range from cellular, biochemical, physiological, morphological, gross pathological deviations on individuals, to variability on the population level. They include early developmental stages, juveniles and adults and include all major groups of biota from algae to mammals including sessile and mobile benthic organisms and pelagic species.

Oxygen Deficiency and its Effects

Although first reports on the occurrence of low oxygen concentrations in bottom water in the central North Sea date back to 1902 (Gerke 1916) it was an unexpected finding in 1982 to detect the same phenomenon in the German Bight (Rachor and Albrecht 1983). In the summer of 1982 in two thirds of an area of 15,000 km^2, contents of dissolved oxygen were less than 5 ml/l corresponding to 60% saturation. Lowest values were near 1 ml/l. In areas with oxygen saturation around 10% catches of fishes were extremely low and underwater photographs revealed mortalities of ophiurids and bivalves (Dethlefsen and von Westernhagen 1983). Dyer et al. (1983) found extremely low catches including dead fish in the summer of 1982 in Danish coastal waters. Kröncke (1985) studied benthic communities in areas with recurrent low or medium dissolved oxygen concentrations. She concluded that in areas with lowest O_2 levels, three benthic species dominated known for their ability to survive low O_2 concentrations. In 1984 when oxygen concentrations were higher than during the two preceding summers, benthic species occurred which were considered to be re-immigrants. Also in the period between 1984 and 1986 oxygen concentrations were found to be reduced in the area studied. This was despite the fact that weather conditions were favourable for good oxygen supply with low temperatures and high wind velocities. It was concluded that even under favourable hydrographic conditions the high load of biologically degradable organic matter in the Southern North Sea is responsible for O_2 reductions. Mellergaard and Nielsen (1987) reported on a correlation between low oxygen in certain areas in the Eastern North Sea and increased prevalence of external diseases of dab (*Limanda limanda*). Dethlefsen (1990) had similar results with prevalence of lymphocystis and epidermal papilloma elevated in summers with low dissolved oxygen in bottom waters in the German Bight and decreasing prevalence in summers with better oxygen conditions.

Nutrients and Algae

In the period between 1962 and 1984 surface temperatures, salinity, concentrations of phosphate, nitrate and nitrite increased in the vicinity of the island of Helgoland (Radach and Berg 1986). Concentrations of nitrate tripled within a few years after 1980 and decreased in the period after 1990. Concentrations of phosphate doubled in the period between 1962 and 1972, remained constant until 1981, and decreased thereafter (Hickel 1993). The total phytoplankton biomass increased in the period studied (between March and September) by a factor of 3. Biomass of flagellates fluctuated with extremely high population densities in certain years. Direct correlations with concentrations of nutrients in seawater were not detectable. Gilbricht (1983) pointed out that increased phytoplankton stock is not necessarily a consequence of increased nutrients in the water column. Phytoplankton blooms can be expected to occur when the respective water column is stratified due to temperature or salinity, thus reducing turbulent mixing. These conditions prevail in the German Bight during calm weather periods in the summer and are aggravated by the inflow of high quantities of freshwater from rivers (Hickel 1993). Bätje and Michaelis (1986) reported increased intensities of blooms of *Phaeocystis pouchetii* in the past decade. Annually recurrent blooms of *Phaeocystis* were also reported for Dutch and French coastal waters with an increasing intensity over the last 20 years (Lancelot et al. 1987).

Changes in Benthic Communities

For many of the European estuaries changes in the benthic and fish communities have been documented as early as the beginning of the industrial revolution, i.e. around 1870 to 1890. In the Elbe, for example, fishermen complained that the fishery had deteriorated as early as 1798 (Riedel-Lorjé and Gaumert 1982). Information is available on similar changes in British estuaries (Rees and Eleftheriou 1989). German investigators found (Dörjes 1986) when comparing present results with those of Hagmeier (1925), Hagmeier and Kändler (1927) and Caspers (1938) that there has been a general decrease of many species of benthic organisms for the Helgoland area (in the inner German Bight of the Southern North Sea). The changes in macrozoobenthos communities can be subdivided into:

Species that have disappeared - Prior to the sixties the echinoderms *Astropecten irregularis*, *Solaster papposus*, *Thyone fusus*, and *Cucumaria elongata*, and the crustaceans *Upogebia deltaura* and *Callianassa subterranea*, were very abundant in the Helgoland area. They are no longer present. Also molluscan species, *Ostrea edulis* and *Helcion pellucidus*, present in the 30s and 60s in substantial populations, are no longer present along the German coast (Dörjes 1986).

Species with strong population decreases - Molluscs like *Venus ovata*, *Cultellus pellucilus*, *Crepidula fornicata*, and *Cardium fasciatum* which were considered to be most abundant in the area of Helgoland (Heincke 1894), are not abundant in this area today. Polychaetes *Glycera alba*, *Lubrinereis tetraura* (*impatiens*) and *Sabellaria spinulosa*; the echinoderm *Echinocyamus pusellis*; the sponge *Halochondria panicea;* and the crustaceans *Porcellana longicornis* and *Pisidia longicornis* also showed significant reductions in their populations.

Species with strong fluctuations - In this group Dörjes (1986) includes species with reduced numbers of individuals for prolonged periods, which after developing into rich populations are regularly reduced to their former level. Species with strong population fluctuations are the polychaetes *Pectinaria koreni* and *Pholoe minuta*, molluscs *Mysella bidentata*, *Montacuta ferruginosa*, *Abra alba*, and *Donax vitatus*, the echiurid *Echiurus echiurus*, the decapod crustacean *Corystes cassivellaunus* and the fish *Callionymus lyra*.

Further benthic studies were carried out by Rachor (1985), who investigated long term benthic changes in two areas in the German Bight, one located in a former sewage sludge dumping ground - within the period of 1961 to 1980 it received 350,000 m^3 sewage sludge annually (Dethlefsen 1986a) - and the other located in the Central German Bight - the dumping area for wastes from titanium dioxide production, receiving between 450,000t of H_2SO_4 (25%) and 750,000t H_2SO_4 (12%) annually until the end of 1989 (Dethlefsen 1986b, 1990). Rachor (1982) found benthic decreases in the sewage sludge dumping area. Mühlenhardt-Siegel (1981, 1985) was able to show clear effects of the dumping of sewage sludge on macrobenthos resulting in impoverished communities. Caspers (1979, 1987) attributed changes in the composition of macrobenthos in the sewage sludge area to natural variability and claimed that long term studies are necessary to understand causes. For the dumping area for wastes from titanium dioxide production no clear decrease of the benthic community was demonstrated (Rachor 1982). Riesen and Reise (1982) came to the conclusion that the disappearance of oysters and the polychaete *Sabellaria* have to be interpreted as effects of intensive fisheries, while the extension of banks of blue mussels (*Mytilus edulis*) could be due to eutrophication (Reise 1986).

Significant changes in the composition of macrobenthic communities were found to have occurred in the Jade Busen area (German Wadden Sea). In the period between 1939 and 1986 about 40% of the species have changed with current species number being higher than that of the 1930s (Michaelis 1987). When considering possible causes of these changes no clear answers are available. Changes in fishery intensity and structure do have a significant impact on benthos through heavy fishing gear (Rauck 1985), but Dörjes (1986) considers that pollution is amongst the most likely factors responsible for these changes. Changes in macrobenthic communities are not restricted to onshore areas. Kröncke (1988a) revisited stations on the Dogger Bank studied by Ursin in 1951/52. Increases in abundance of certain opportunistic polychaetes, like *Spiophanes bombyx*, *Magelona spp.* and *Chaetozone setosa*, were registered. *Amphiura filiformis* has increased in population density. The most obvious changes according to Kröncke (1988a) were the absence of bivalve patches consisting of *Spisula subtruncta*, *S. elliptica* and *Mactra corralina cinera*, formerly abundant species in the area. In her attempt to interpret these changes Kröncke discussed whether organic enrichment or heavy metals (Pb and Cd) might be amongst the causative factors (Kröncke 1988b). The heavy metal contamination of certain Dogger Bank species was found to be higher than that of onshore species. It is the conclusion of Rees and Eleftheriou (1989) that effects on benthos were invariably confined to areas of limited water exchange; where changes occurred they were of little significance to the North Sea area as a whole. In a study on the benthic environment of the Northern North Sea, Basford and Eleftheriou (1988) stated that there was no evidence of any important large-scale contamination of the North Sea as a whole.

FISHES

Stock size of fishes is, amongst other factors, controlled by the variability of recruitment. The most significant impact adding to this often natural variability is through fisheries (Ehrich 1993). Further anthropogenic influences like pollution can be expected to be most significant in estuaries, for example of the North Sea, which are known for their high contamination. Those fish species which either inhabit estuaries or migrate through them during their life cycles, especially those which are commercially exploited, can be expected to be the most endangered. Lozán (1990) reviewed the long term development of stocks of fish species in the river Elbe, which he states stands as an example for many European estuaries with similar development. He lists nine species for which decreases of abundance have been recorded since 1918.

Table 1. Comparison of the fish fauna in the river Elbe 1918 and today (Lozán 1990).

	1918	Today
Anguilla anguilla	133,000	significantly decreased, heavily contaminated
Platichthys flesus	537,000	decreased
Coregonus oxyrynchus	9,000	natural population extinct
Salmo salar	170	natural population extinct
Salmo trutta	400	natural population extinct
Acipenser sturio	few	extinct
Petromyzon marinus and Lampetra fluviatilis	2,300	single specimen
Osmerus eperlanus	990,000	frequent, heavily contaminated
Alosa fallax	15,000	frequent, heavily contaminated

Lozán (1990) considered low dissolved oxygen, various pollution factors and overfishing to be responsible for these decreases. The debate whether pollution or overfishing, or both, are responsible for the deterioration of fish stocks in the river Elbe dates back to 1789 (Riedel-Lorjé and Gaumert 1982) and is still unresolved. A brief summary of this debate is given in Dethlefsen (1988a).

Fish Population in the Wadden Sea

In a study of catch data of German shrimpers from 1954 to 1981 the occurrence of fish and crustacean species was analyzed (Tiews 1983). Seven inhabitants of the Wadden Sea have decreased in abundance: shore crab, butterfish, sea snails, gobies, eels, little sole and gurnard, while three of the facultative inhabitants: dab, sprat and cod, have increased in numbers to equal the former total biomass. Fifteen further species remained constant in numbers and biomass. For two of the species that decreased, correlations exist between population levels and meteorological factors such as temperature. Since fishing intensity and strategy in the Wadden Sea has remained unchanged over the period of the survey it can be questioned whether contamination might be amongst the causative factors. To the circumstantial evidence noted above, can be added the fact that contamination of water and sediments was higher in areas of clearest population decreases. In a continuation of the study Tiews (1990) found that two of the seven species declining until 1981 - shore crab (*Carcinus maenus*) and butterfish (*Pholius gunellus*) - have reversed the downward trend and increased in abundance. The other five species are still declining with the possible exception of gobies which might also show a tendency for recovery.

Mortality of Seals

Seals (*Phoca vitulina*) on the sands of the Schleswig-Holstein Wadden Sea coast have increased from a population of 1,540 specimens in 1974 to 3,800 in 1987. Reasons for this increase were the prohibition of hunting, the installation of seal protection areas, and the formation of national parks in the years of 1985 and 1986 (Heidemann and Schwarz 1990). Despite these protection measures the water quality of certain areas of the North Sea deteriorated. Seals have long been known to carry high contaminant loads. Concentrations of PCB for example reached maxima of 564 ppm in blubber (Drescher et al. 1977), and mercury reached levels of 269 ppm in livers. However direct correlations between contamination of seals and physiological deficiencies have not been demonstrated for Schleswig-Holstein specimens.

This is in contrast to the work of Reijnders (1986) and Helle et al. (1976ab) on seals in the Swedish Baltic. In February 1988 elevated mortality rates of seals were detected in the area of the Danish Kategatt island Anholt. In March 1988 the first indications of unusually high mortality rates in seal populations were detected in the Schleswig-Holstein Wadden Sea areas. Until April 1988 more than 5,840 dead seals were found along the Schleswig-Holstein west coast. Together with the Kategatt/Skagerrak area (approximately 6,700 dead seals) the Schleswig-Holstein Wadden Sea represented a hot spot which altogether included 17,000 dead seals. Most of the seals found dead were in good condition with a normal sized blubber (Vogel 1989), but they all revealed marked indications of acute disease. Pathological alterations of the lung included emphysema and oedema. In the Netherlands two types of viruses were isolated (Osterhaus and Vedder 1988), and viruses (morbillivirus, phocid distemper virus, PDV) could be detected in a large number of dead seals (Visser et al. 1991). It is difficult to decide what role contaminants played in predisposing these organisms to the attack of viruses. It was not excluded that pollution of the North Sea played a role in the high seal mortality (Heidemann and Schwarz 1990).

Impairment of Reproduction

Two types of studies have been carried out in the North Sea and in the Baltic: correlations between residues in gonads and viability of hatch, and malformations of pelagic fish embryos. Studies of the possible correlations between residues and hatch combined the advantage of *in situ* long term exposure under natural conditions and subsequent controlled experiments to quantify the hatch. Malformations of fish embryos *in situ* have to be interpreted as a result of combined impact during gonadal development and acute effects in the water column. Results of the first type of studies are published in von Westernhagen et al. (1981, 1988ab, 1989).

Malformations of pelagic fish embryos - Despite the fact that studies on the status of fish embryos, especially of commercial species in the Northeast Atlantic, have a long history, almost nothing is known about gross malformations of these early developmental stages. This is astonishing because population biologists use embryo and larval density and survival in their quantitative models. Normally fish embryos are obtained for scientific purposes using high speed gear and physical pressure developed by those gears is such that it causes mortality or gross deformities leading to death. Furthermore, most samples were preserved in formalin turning the yolk opaque and precluding the possibility to detect slight deviations from normal development.

Studies to link contamination of surface water with the occurrence of gross cytologic and cytogenetic conditions in fish embryos were first carried out in 1975 in the New York Bight by Longwell and co-workers (Longwell and Hughes 1981). They sampled embryos of Atlantic mackerel in the New York Bight and grouped them by means of differences in egg moribundity based on cell state, mitotic and chromosome irregularities and deviations, and arrest at gastrulation. Their results indicated a lower egg viability in areas of the Bight, which were more contaminated. Longwell and Hughes also showed that Atlantic mackerel was adversely impacted by hydrocarbon and heavy metal pollution in the New York Bight.

Grauman and Sukhorukova (1982) reported on mass occurrence of abnormal cod and sprat embryos in the Baltic from 1979 to 1981. Malformation rates were 12% of the embryos deformed in 1979 and 50% in 1981. The authors refer to long term data of Baltic sprat eggs beginning in 1954 with elevated abnormalities not occurring prior to 1979. In their discussion they state that sprat eggs and larvae have a high ecological plasticity being able to tolerate wide ranges of temperatures and salinity. They therefore exclude these factors as being responsible for high malformation rates and discuss whether atmospheric precipitation or river discharges might be the responsible factors.

These studies initiated a German program in the Baltic and North Sea coastal waters to investigate prevalence of grossly visible malformations in pelagic fish embryos. Some of the results are published in Dethlefsen et al. (1989), Cameron et al. (1990) and Cameron and Berg (1993). In the Western Baltic (von Westernhagen et al. 1988b) in 1983 18% of cod, 22% of flounder, and 24% of plaice embryos were defective. In 1984 28% of plaice, 32% of cod, and 44% of flounder embryos were abnormally developed.

Methods - In order to avoid mechanical damage caused by fishing gear and procedure a normal plankton net with 300 μm mesh size was towed at minimum speed close to the surface, hauls not being longer than 15 minutes. Immediately after capture, fish embryos were transferred into glass dishes and investigated unpreserved under the microscope.

Results - The predominating North Sea pelagic fish embryos in German, Dutch and Danish waters in the early spring were dab (*Limanda limanda*), flounder (*Platichthys flesus*), whiting (*Merlangius merlangus*), cod (*Gadus morhua*) and plaice (*Pleuronectes platessa*) (Cameron et al. 1992a). Developmental abnormalities were characterized by irregular cell divisions in early development stages. In later embryos circular enclosures and irregular development of peri- and blastopore were found as well as irregularities of the yolk surface. Examples for the types of malformations encountered during these earlier studies are given by Cameron et al. (1990). Results to be referred to were obtained in 3/84, 2/85, 2/86, 3/87, 3/93. For the sampling carried out after March 1984 a constant station grid was covered. Frequencies of malformation of all stages of the species investigated during these studies are depicted in Tables 2 and 3, for March 1987 and March 1993.

In Part a) of these tables, the total malformation rates not differentiated according to developmental stages for the respective species are given. Beginning in 1986 malformations were divided into slight, medium and severe deviations from normal. In part b) of the tables malformation rates are given according to developmental stages as defined by von Westernhagen (1970). From these tables it can be observed that the overall malformation rates decreased with increasing developmental time with highest prevalence of malformed embryos encountered at early developmental stage Ia and lowest at the late developmental stage prior to hatch (IV) with differences between developmental stage Ib and II being less pronounced than others.

During all cruises whiting (*Merlangius merlangus*) was the species displaying highest malformation rates of the five most frequently occurring species dab, flounder, whiting, cod and plaice. Plaice showed low malformation rates during these studies (Figure 1). In the period from 1984 to 1986 prevalence of malformed embryos of all species investigated increased and in 1993 the malformation rates were lower than in 1986/87 (see also Tables 2 and 3).

Results on the regional distribution of prevalence of malformations of embryos of dab (*Limanda limanda*), all developmental stages, are shown in Figure 2. In March 1987 between 7.7% and 56.7% of the embryos investigated were malformed with elevated levels in the central and northern part of the German Bight and off the German and Dutch Eastfrisian Islands. Elevated levels were also found in the southern part of the areas studied off the river Rhine. In March 1993 lower prevalence were found in the area studied. Only at one station located in the northeastern corner of the station grid were malformation rates higher than 15% encountered. Figure 3 contains long term data on prevalence of malformations of embryos of dab including all developmental stages and all stations covered during the respective studies. Data for 1990 are from Cameron and Berg (1992b) and data for the years 1991 and 1992 were recalculated from Cameron and Berg (1993).

Table 2a. Prevalence of malformed fish embryos and numbers of embryos investigated per species March 1987.

Spec	n all	NE	% malf	% ms	% m	% sev
Dab	21,093	366	21.6	9.9	9.7	2.0
Flo	4,670	51	20.5	9.1	9.8	1.6
Pla	3,401	8	11.2	3.8	6.0	1.4
Cod	1,014	6	15.8	8.6	6.0	1.2
Whi	992	13	54.1	18.4	26.8	8.9

Table 2b. Prevalence of malformations of different developmental stages March 1987.

Spec	n all	NE	n Ia	%	n IB	%	n II	%	nIII	%	n IV	%
Dab	21,093	366	4,488	44	694	21	2,944	32	10,062	13	2,539	5
Flo	4,670	51	1,007	41	345	22	606	29	1,819	12	842	7
Pla	3,401	8	204	26	155	12	653	25	1,376	10	1,005	2
Cod	1,014	6	207	27	38	11	157	22	413	14	193	4
Whi	992	13	256	64	19	21	203	63	420	52	81	21

Spec = species; n all = numbers investigated; NE = not developed; % malf = % malformed; % ms = slightly malformed; % m = medium malformed; % sev = severe malformed; Dab = Dab (*Limanda limanda*); Flo = Flounder (*Platichtys flesus*); Whi = Whiting (*Merlangius merlangus*); Cod = Cod (*Gadus morhua*); Pla = Plaice (*Pleuronectes platessa*)

Table 3a. Prevalence of malformed fish embryos and numbers of embryos investigated per species March 1993. Legend see Table 2.

Spec	n all	NE	% malf	% ms	% m	% sev
Dab	32,742	99	4.9	2.8	1.6	0.6
Flo	2,124	21	5.6	3.1	1.9	0.6
Pla	1,709	4	5.5	2.4	1.8	1.3
Whi	1,413	3	11.1	4.0	4.6	2.4

Table 3b. Prevalence of malformations of different developmental stages March 1993. Legend see Table 2.

Spec	n all	NE	n Ia	%	n Ib	%	n II	%	n III	%	n IV	%
Dab	3,2742	99	4,840	11	3,223	6	3,546	7	17,541	3	3,545	1
Flo	2,124	21	335	7	384	8	200	7	956	5	228	0
Pla	1,709	4	137	12	126	13	129	14	691	5	622	1
Whi	1,413	3	212	14	116	17	132	27	711	9	239	3

In the period between 1984 and 1987 prevalence of malformed embryos of dab increased reaching a total level of 22% in 1987. No data are available for 1988 and 1989. Data for the years 1990 to 1992 are not strictly comparable because a different network of stations were used in 1991 and 1992 and only few of the 1990 sampling stations were located in the German Bight. Levels of prevalence encountered for 1990 and 1992 were comparable to those found during the sampling in 1984, while the prevalence encountered in 1991 were similar to those found in 1985. Lowest malformation rates in the period studied were found in 1993. For the years 1986, 1987 and 1993 a classification of malformations into the categories slight, medium and severe (following criteria developed by Cameron et al. 1990) is available, which is depicted in Figure 4. It can be seen that the relation between slight and medium/severe deviations from normal have changed over time, the relation of slight deviations to normal being highest for sampling in 1993. In conclusion, it can be stated that prevalence of malformed embryos were lower in the period between 1990 and 1994 than in the period between 1984 - 1987 with lowest malformation rates found in 1993, thus indicating an overall downward trend over time.

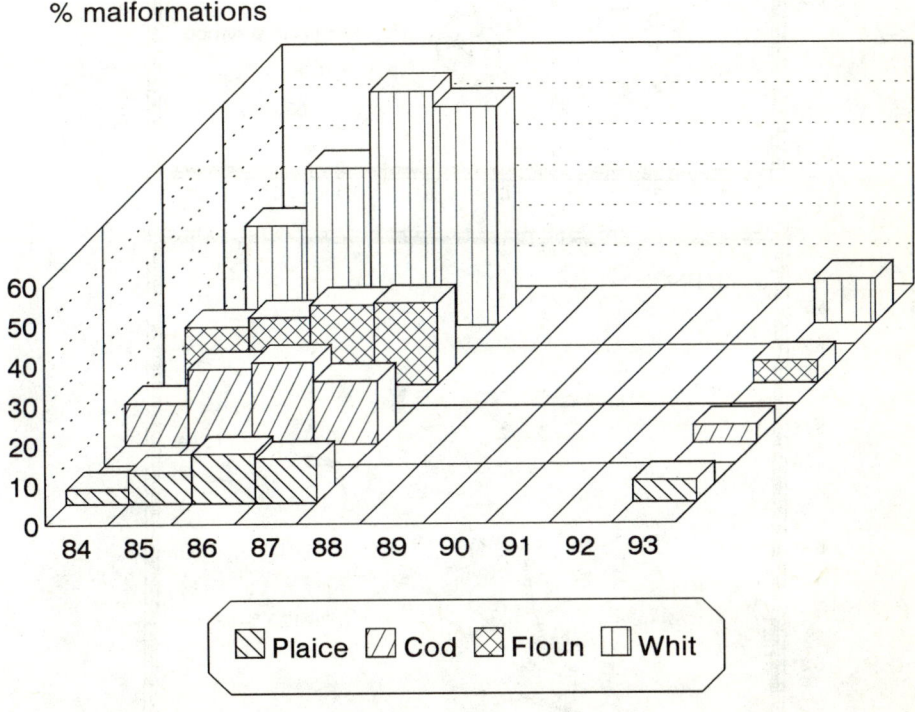

Figure 1. Malformations of pelagic fish embryos of flounder (*Platichthys flesus*), whiting (*Merlangius merlangus*), cod (*Gadus morhua*) and plaice (*Pleuronectes platessa*). Data is summarized for all developmental stages and all areas.

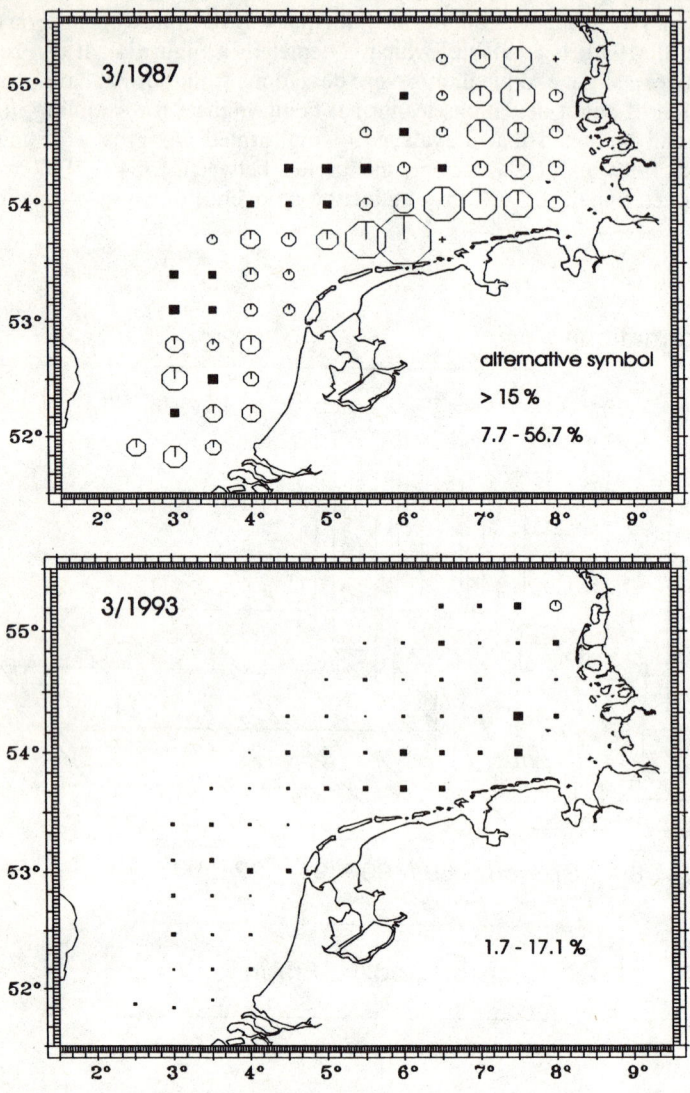

Figure 2. Malformations of dab (*Limanda limanda*) embryos - March 1987 (above) and March 1993 (below).

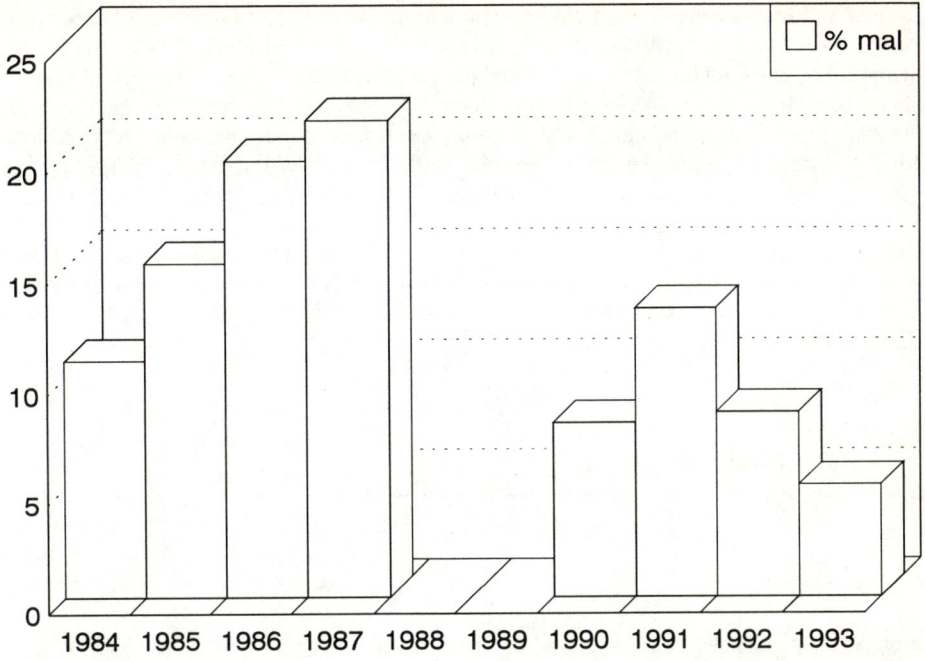

Figure 3. Malformations of dab (*Limanda limanda*) embryos - all developmental stages all stations. Values from 1990 (Cameron and Berg 1992b), values for 1991/92 (Cameron and Berg 1993).

Fish Diseases

Studies on fish diseases in the German Bight were initiated in 1977 (Möller 1979, 1981; Dethlefsen 1980, 1984a, 1985; Dethlefsen et al. 1984; Watermann 1982; Watermann and Dethlefsen 1982; Watermann 1984; Watermann and Dethlefsen 1985). Cod (*Gadus morhua*) and dab (*Limanda limanda*) were found to be the two species most frequently afflicted with external diseases. Major diseases of cod were ulcerations, skeletal deformities and fin rot. Major diseases of dab were lymphocystis, epidermal papilloma and ulcerations. Comprehensive data on epidemiology of diseases of dab in the German Bight and the North Sea were provided by Dethlefsen et al. (1987), including information on the contamination of fish and the possible contribution of biological factors to the outbreak of disease. Amongst the biological factors possibly related to the outbreak of the disease, migration of healthy and diseased specimens, population density, condition-factor, food and net injuries were discussed. Dethlefsen (1990) summarized findings of ten years of fish disease studies in the North Sea. Fisheries impact resulting in net injuries and low dissolved oxygen were amongst the non-contamination factors found to possibly contribute to high prevalence of external diseases in dab. Vethaak and ap Rheinallt (1992) provide a comprehensive review on fish diseases in the North Sea in relation to marine pollution. They conclude that only certain diseases including neoplastic growth in livers of dab are likely related to pollution.

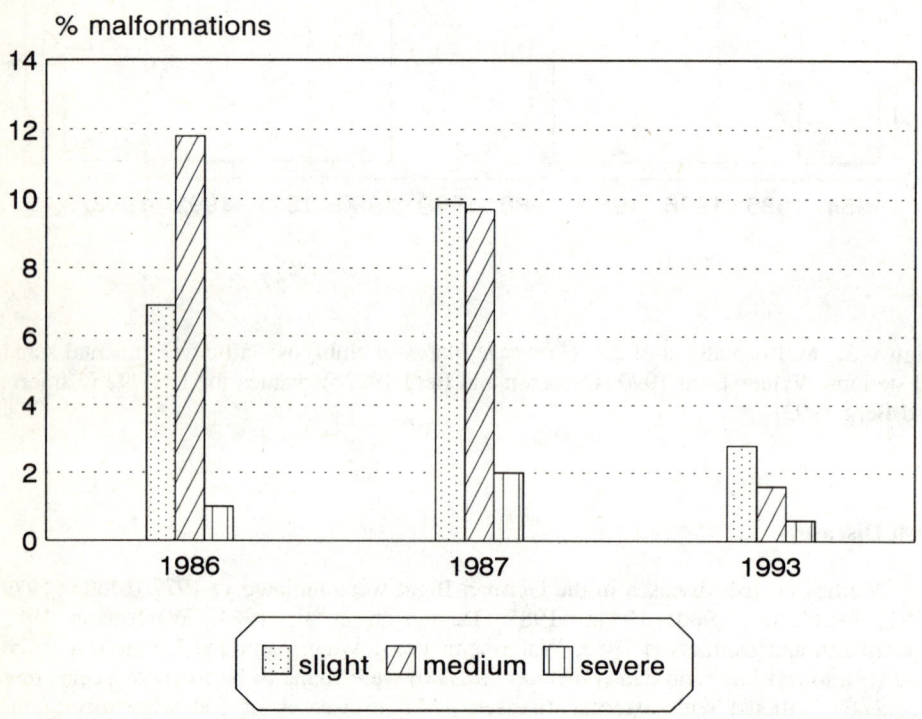

Figure 4. Slight, medium and severe malformations of dab (*Limanda limanda*) embryos - all developmental stages.

In the following discussion, newer data on the long term development of disease prevalence will be provided for two externally visible diseases of dab (*Limanda limanda*) and three representative areas in the North Sea. *Epidermal papilloma* are epithelial tumours of unknown etiology. Some authors suspect that a virus might be involved, but no definite proof has been provided. Lymphocystis is a widespread well-described viral disease known to occur in many fish species. At the beginning of the systematic studies in the North Sea both phenomena have been detected to occur frequently, especially on dab in various areas of the North Sea. The long term studies in the central part of the German Bight started in 1979 but it was not before 1982 that a standardized methodology was applied so that only data from that time on will be presented. Sampling was carried out twice a year (first two weeks in January and two weeks in May, June or July) in the former dumping area for wastes from titanium dioxide production (dumping was terminated in January 1989) northwest of the island of Helgoland (54°15N - 54°25'N / 07°26.5'E - 07°26.5'E - 07°38.5'E), the Dogger Bank (54°25'N - 54°50'N / 02°00'E - 02°31'E) and beginning in summer 1988 in the Firth of Forth (56°15'N - 56°25'N / 01°50'W - 02°10'W).

In the period between summer 1982 and summer 1986 prevalence of dab from the former dumping area afflicted with epidermal papilloma increased for each successive year from values slightly below 2% to values around 10%. In the period between 1986 and 1989 a level above 9% was maintained, from 1990 to 1992 prevalence were around 6%, and in the summer of 1993 a prevalence of 3.2 % was measured (Figure 5a).

For lymphocystis of dab in the former dumping area for wastes from titanium dioxide production, frequencies fluctuated between 7% and 16.5% in the period between 1982 and 1987 with no clear trend detectable. A maximum value was found in summer 1988 with 26.2% of the dab afflicted with *lymphocystis*. From that time until 1993 these prevalence decreased stepwise to 6.3% (Figure 5b).

Winter data for these diseases (samples were taken in January of the respective years), seem to follow a completely different pattern with lowest values in January 1983, 1986 and 1992. Winter minima and maxima occurred in the same years for both diseases. Prevalence of *epidermal papilloma* in Dogger Bank dab increased from 1982 (1.3%) to 1986 (13.5%), followed by 2.2% in the summer of 1987 to 7% in the summer of 1991, with values between 5% and 6% thereafter (Figure 5a). For *lymphocystis* of the Dogger Bank dab, again a maximum value was found for the summer of 1986 (20.6%) with a subsequent continued decrease of disease prevalence until the summer of 1993 when it reached 6.3%. For winter values of *epidermal papilloma* on the Dogger Bank, a slight but continuous increase can be observed in the period between 1981 and 1993.

The winter data for *lymphocystis* of dab on the Dogger Bank seem to more closely follow the summer trends (Figure 5b). Data for the Firth of Forth are available beginning in summer 1988. Neither for *epidermal papilloma* nor for *lymphocystis* summer or winter data are clear trends detectable (Figure 5). Summer prevalence of *epidermal papilloma* for the Firth of Forth ranged from 3.2% in 1988 to 5.6% in 1992 reaching 5.4% in 1993. Prevalence of *lymphocystis* were high during winter and summer and ranged from 22.5% in summer 1988 to 25.5% in summer 1990. Prevalence in 1993 was 22.2%.

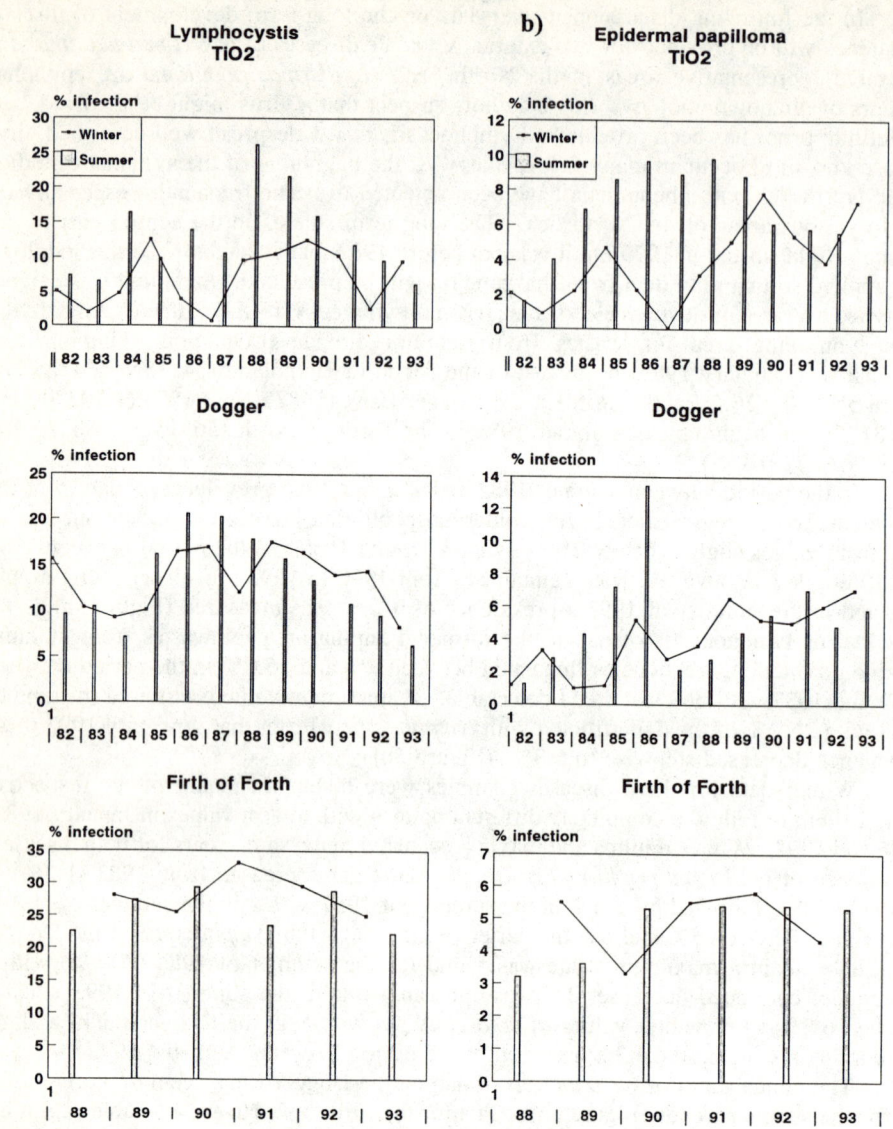

Figure 5. a) Dab (*Limanda limanda*) epidermal papilloma in the former dumping area for wastes from TiO_2-production - the Dogger and the Firth of Forth. **b)** Dab (*Limanda limanda*) lymphocystis in the former dumping area for wastes from TiO_2-production - the Dogger and the Firth of Forth.

CONCLUSIONS

The question whether the biological variability described above is an expression of ecological disease is extremely difficult to answer for a number of reasons. Population densities and composition change over time (long or short term cycles are well described for many biological systems). Disease is a concomitant factor of life and its mere presence in natural populations does not necessarily indicate an ecological imbalance. All deviations listed above may have natural causes. The reaction of organisms to internal or external stressors, natural or anthropogenic including various contaminants, involves the same physiological mechanisms and is therefore unspecific as to causes (Selye 1950). A result of stress in the phase of exhaustion is amongst others decreased resistance to disease, or impact on reproduction (Sindermann 1984). The debate on causes of elevated levels of fish diseases in wide areas of the North Sea started more than 10 years ago. It polarized the scientific community and is still unresolved (Vethaak and ap Rheinallt 1992).

Experiments

The obvious inability of scientists to provide a proof for a cause and effect relationship between pollution and diseases in the North Sea prompted the start of large scale experiments in order to reduce the *in situ* complexity. Preliminary results of Dutch studies suggest that there was little difference between the reference and the polluted mesocosms in terms of disease development (Vethaak 1992). This is in spite of the fact that concentrations of PCB and PAH in the mesocosms span a considerably wider range than the pollution levels to be found in adjacent Dutch coastal waters. The author did not discuss the fact that reference or control fish and those fish exposed to contaminated dredge spoil in the experiments are generally extremely stressed by handling or by the simple fact that they have been extracted from their natural environment and forced to live under highly stressful conditions in mesocosms. This certainly predisposes fish and renders them vulnerable to the attack of infectious diseases. When the immune system of the test animals has already been impaired prior to the studies, it is to be expected that both control and exposed fish will develop disease symptoms. The impact of handling of fish health is described in Peters (1988). According to Vethaak (1993) large basins for experiments cannot be replicated, at least in sufficient numbers, because of their associated costs. In practice potential variability between mesocosms is ignored. Apparently identical mesocosms will diverge from one another even when subjected to seemingly identical physicochemical conditions. The assumption that observed differences are due to differences between treatments rather than to variability between mesocosms may be unjustified. Nevertheless in Vethaak's experiments hepatocellular adenoma occurred after 2.5 years of exposure and other liver lesions developed before tumour formation was apparent. High prevalence of epidermal disease particularly observed during the early part of the experiments, indicated that conditions did not exactly duplicate those in the field, but there appeared to be a pollution associated risk for lymphocystis.

Since the concentrations of contaminants used in Vethaak's experiments were well within the range occurring in enclosed Dutch waters, it was clear to this author that the occurrence of diseases in wild fish populations is at least in part due to the toxic action of these pollutants (Vethaak 1993).

Natural Causes versus Pollution

It is known that many contaminants including organochlorines, heavy metals, PAHs and others have the potential to cause diseases in aquatic organisms (Meyers and Hendricks 1982). The contamination of the North Sea biota, seawater and sediments with various substances is elevated in certain estuaries, especially Elbe, Weser, Rhine but also Thames, Humber (Murray and Norton 1982). Biological effects were demonstrated to have occurred in these estuaries since the turn of the century (Rees and Eleftheriou 1989). It was recently demonstrated that elevated levels of contaminants also occurred in biota, waters and sediments of offshore North Sea areas (Sündermann and Beddig 1990, 1992). Concentrations of certain heavy metals and organochlorine compounds were shown to be elevated in plankton and benthos organisms even in regions of the central Northern North Sea. The Dogger, a shallow bank in the central Southern North Sea, has been shown to display elevated levels of certain heavy metals in surface sediments and benthic organisms. (Everaarts and Fisher pers. comm.; Kröncke 1988b). Elevated levels of DNA-adducts were demonstrated to occur in dab *(Limanda limanda)* from the outer German Bight and the eastern part of the Dogger (Tail End) (Fisser and Blömeke 1992) indicating the presence of mutagenic or carcinogenic potential in these areas. Lange (1992) measured elevated EROD- and P450- activities in livers of dab not only in the German Bight but also in fish from the central North Sea. In 1990, a suite of biological and chemical techniques has been applied on a gradient leading from the inner German Bight to the northeastern end of the Dogger. Previous work was to some extent corroborated by the data which showed that where elevated levels of some contaminants occurred at the Dogger station, biological effects were also noted in dab (Stebbing et al. 1992). Thus it seems there is ample evidence that contamination and biological effects are not restricted to estuaries and onshore areas.

The hypothesis that in certain offshore areas of the North Sea biological deviations from normal might be related to contamination therefore cannot simply be rejected. On the other hand there are no studies available that prove to the satisfaction of everyone that clear cause and effect relationships exist between contamination and biological effects in offshore areas. In all cases simple models were tested. The analysis of correlations between diseases and contaminant levels often included single substances and it was assumed that the correlations could be linear. It is not surprising that the results were negative. Even in those cases where a cause and effect relationship was established between specific diseases and contaminants, it is possible that these correlations are accidental. Although we have models available which can account for the complexity of the multiple combined action of contaminants, further anthropogenic factors, and general ecosystem interdependencies, these have not yet been applied in this context.

This may be due to the overwhelming complexity of marine ecosystems. Reasons for this complexity are to be seen in the variability of the target organism on one hand and the plasticity of the pollution factors on the other. Predisposing factors, not included in the analysis published in the North Sea context, could be of overriding importance and the extreme variability of contamination levels have just been demonstrated (Sündermann and Beddig 1990). These studies showed that meteorological and atmospheric conditions are determining flushing times in the North Sea; westerly winds have the tendency to effectively drive water bodies containing contaminants out of the North Sea into the Atlantic. Less favourable meteorological conditions, for example prolonged easterly winds that can prevail for weeks, block the water exchange with the ocean, leading to very high levels of contaminants during these periods. The range of concentrations of contaminants can be extremely variable related to the variability of weather. This means that the circulation, and its consequences for physical and chemical factors is largely driven by meteorological factors, influencing biological systems and cannot precisely be predicted (Sündermann and Beddig 1990). In the face of this complexity, how can we expect that linear models as presently applied could provide the answers we are looking for?

From what has been said above it can be concluded that even if there is a certain plausibility that some of the biological fluctuations in the North Sea might be due to pollution this is far from being unanimously accepted. Thus it seems that at present the formulation of ecological quality objectives is premature. Far better knowledge is needed which can only be achieved with an enormous increase in research effort. This requires resources which are more and more difficult to solicit for marine research.

The absence of a firm numerical basis for marine pollution management on the one hand and the failure of the assimilative capacity approach to protect the marine environment (Chadwick and Nilsson 1993) led to the formulation of the precautionary principle (a 'better-safe-than-sorry' principle), which advocates the reduction of inputs into the environment of substances with a known hazardous or toxic nature, especially where there is reason to believe that harmful effects are likely to occur (Dethlefsen et al. 1993). It seems that at least for the foreseeable future, the application of this concept will provide a better degree of safety for the protection of the North Sea environment than ill-based ecological quality objectives.

REFERENCES

Anonymous (1980) Umweltprobleme der Nordsee. Sondergutachten 1980. Der Rat von Sachverständigen für Umweltfragen. W. Kohlhammer, Stuttgart, Mainz 503 pp.
Bätje M., Michaelis H. (1986) *Phaeocystis poucheti* blooms in the Eastfrisian coastal waters (German Bight, North Sea). Mar. Biol. 93: 21-27
Basford D., Eleftheriou A. (1988) The benthic environment of the North Sea (56° to 61° N). J. Ma. Biol. Ass. UK 68:125-141
Buchwald K. (1990) Nordsee - Ein Lebensraum ohne Zukunft? Die Werkstatt Göttingen 544 pp.
Cameron P., Berg J., von Westernhagen H, Dethlefsen, V (1990) Mißbildungen bei

Fischembryonen der südlichen Nordsee. In Lozán J.L., Lenz W., Rachor E., Watermann B., von Westernhagen H. (eds.) Warnsignale aus der Nordsee: wissenschaftliche Fakten. Berlin Hamburg: Parey pp:281-293

Cameron P., Berg J., Dethlefsen V., von Westernhagen H. (1992a) Developmental defects in pelagic embryos of several flatfish species in the southern North Sea. Netherl Journ. Sea Res. 29 (1-3) 239-256

Cameron P., Berg J. (1992b) Morphological and chromosomal aberrations during embryonic development in dab (*Limanda limanda*). Mar. Ecol. Prog. Ser. 91:163-169

Cameron P., Berg J. (1993) Fortpflanzungsfähigkeit der Fische. In Geht es der Nordsee besser? Schriftenreihe SDN 1:120-129

Caspers H. (1938) Die Bodenfauna der Helgoländer Tiefen Rinne. Helgoländer wiss Meeresunters (2):1-112

Caspers H. (1979) Die Entwicklung der Bodenfauna im Klärschlamm-Verklappungsgebiet vor der Elbe-Mündung. Arb Dtsch Fischerei-Verb 27:109-134

Caspers H. (1987) Changes in the Benthos at a sewage-sludge dumpsite in the Elbe Estuary. In Capuzzo J.M., Kester J.M. (eds.) Oceanic Processes in Marine Pollution 1: Biological Processes and Wastes in the Ocean 1:201-230

Chadwick M.J., Nilsson J. (1993) Environmental quality objectives - assimilative capacity and critical load concepts in environmental management. In Jackson T. (ed) Clean Production Strategies: Developing Preventive Environmental Mangement in the Industrial Economy. Lewis Publishers pp:29-39

Dethlefsen V. (1980) Observations on fish diseases in the German Bight and their possible relation to pollution. Rapp. P-v Réun Cons. Int. Explor. Mer. 179:110-117

Dethlefsen V. (1984a) Untersuchungen zur Erkennung subletaler Schadstoffeffekte an Fischen in der Deutschen Bucht: Krankheiten Biochemie Physiologie Rückstände. Forschungsbericht BMFT 1-427

Dethlefsen V. (1984b) Diseases in North Sea fishes. Helgoländer Meeresunters 37:353-374

Dethlefsen V. (1985) Krankheiten von Nordseefischen als Ausdruck der Gewässerbelastung. Abh. Naturw. Verein Bremen 40 (3):233-252

Dethlefsen V. (1986a) Experiences of the Federal Republic of Germany with Dumping of Sewage Sludge. In Kullenberg G. (ed.) The Role of the Oceans as a Waste Disposal Option. pp:233-242

Dethlefsen V. (1986b) Überblick über Auswirkungen der Verklappung von Abfällen aus der Titandioxidproduktion in der Deutschen Bucht. Veröff Inst für Küsten- und Binnenfisch 95:1-42

Dethlefsen V. (1988a) Status report on aquatic pollution problems in Europe. Aquat. Toxicol. 11:259-286

Dethlefsen V. (1988b) Biological effects of pollution in the North Sea. In Lokke H., Tyle H., Bro-Rasmussen F. (eds.) First European Conference on Ecotoxicology - Conference Proceedings - October 17-19 1988 Copenhagen, Denmark pp:362-374

Dethlefsen V. (1989a) Ecosystem Changes in the German Bight as a Result of

Contamination A Review. ICES CM E:8
Dethlefsen V. (1989b) Fish in the polluted North Sea. Dana 8:109-129
Dethlefsen V. (1990) Ten years of fish disease studies of the Institut für Küsten-und Binnenfischerei. Arch FischWiss 40 (1/2):119-132
Dethlefsen V., von Westernhagen H. (1983) Oxygen deficiency and effects on bottom fauna in the eastern German Bight 1982. Meeresforsch. 30:45-53
Dethlefsen V., Watermann B., Hoppenheit M. (1984) Sources of variance in data from fish disease surveys. Arch FischWiss 34:155-173
Dethlefsen V., Watermann B, Hoppenheit M (1987) Diseases of North sea dab (*Limanda limanda* L.) in relation to biological and chemical parameters. Arch FischWiss 37 3:107-237
Dethlefsen V., Berg J., Cameron P., Söffker K. (1989) Untersuchungen zur Entwicklung und Anwendung neuer Methoden zur Darstellung schadstoffbedingter biologischer Effekte am Beispiel der Fortpflanzung von Fischen. Abschlußbericht BMFT Project 525-3891-MFE 0525 59 pp.
Dethlefsen V., Jackson T., Taylor P. (1993) The precautionary principle - towards anticipatory environmental management. In Jackson T.(ed.) Clean Production Strategies: Developing Preventive Environmental Mangement in the Industrial Economy. Lewis Publishers pp:41-62
Dörjes J. (1986) Langfristige Entwicklungstendenzen des Makrobenthos der Deutschen Bucht. Submitted by the Federal Republic of Germany to the Scientific and Technical Working Group International North Sea Conference 1987
Drescher H.E., Harms U., Huschenbeth U. (1977) Further Data on Heavy Metals and Organochlorines in Marine Mammals from German Coastal Waters. Reports Mar. Res. 26:153-161
Dyer M.F., Pope J.G., Fry P.D., Law R.J., Portmann J.E. (1983) Changes in fish and benthos catches off the Danish coast in September 1981. J Mar. Biol. Ass. UK 63:767-775
Ehrich S. (1993) Fischbestände. In Geht es der Nordsee besser? Schriftenreihe SDN 1:150-156
Fisser B., Blömeke B. (1992) DNA Schäden in Fischen aus der Nordsee: 32P Postlabelling Analyse von DNA-Addukten in Lebern von Klieschen aus der Deutschen Bucht und der Doggerbank. Berichte Zentr. Meeres- u. Klimaforsch. Hamburg Nr. 24:43-44
Gehrke J. (1916) Über die Sauerstoffverhältnsise der Nordsee. Ann. Hydr. Berl. 44:177-193
Gilbricht M. (1983) Eine "red tide" in der südlichen Nordsee und ihre Beziehung zur Umwelt. Helgol Meeresunters 63:393-426
Graumann G., Sukhorukova L. (1982) On the emergence of sprat and cod abnormal embryos in the open Baltic. ICES CM/J:7
Hagmeier A. (1925) Vorläufiger Bericht über die vorbereitenden Untersuchungen der Bodenfauna der Deutschen Bucht mit dem Petersen-Bodengreifer. Ber Dt wiss Kommn Meeresforsch 1:247-272

Hagmeier A., Kändler R. (1927) Neue Untersuchungen im nordfriesischen Wattenmeer und auf den fiskalischen Austernbänken. Wiss Meeresunter 16:1-90

Heidemann G., Schwarz J. (1990) Das Seehundsterben im schleswig-holsteinischen Wattenmeer 1988/89. In Lozán J.L., Lenz W., Rachor E., Watermann B., von Westernhagen H. (eds.) Warnsignale aus der Nordsee: wissenschaftliche Fakten. Berlin Hamburg: Parey pp:325-330

Heincke F. (1894) Die Mollusken Helgolands. Wiss Meeresunters NF 1:1-92

Helle E., Olsson N., Jensen S. (1976a) DDT and PCB levels and reproduction in ringed seal from the Bothnian Bay. Ambio 5:188-189

Helle E., Olsson M., Jensen S. (1976b) PCB levels correlated with pathological changes in seal uteri. Ambio 5:261-263

Hickel W. (1993) Nährstoffe, Phytoplankton, Algenblüten. Schriftenreihe SDN 1:44-58

Kempe S., Liebezeit G., Dethlefsen V., Harms U. (eds.) (1988) Biogeochemistry and distribution of suspended matter in the North Sea and Implications to Fisheries Biology. Mitt Geol-Paläontol Inst Univ Hamburg 65:XI-XXIV

Kröncke I. (1985) Makrofaunahäufigkeit in Abhängigkeit von der Sauerstoffkonzentration im Bodenwasser der östlichen Nordsee. MSc Thesis Univ. Hamburg 127 pp.

Kröncke I. (1988a) Macrofauna standing stock of the Dogger Bank. A comparison 1951-1952 versus 1985. Mitt Geol-Paläontol Inst. Univ. Hamburg 65:439-454

Kröncke I. (1988b) Heavy metals in North Sea fauna. Biogeochemistry and distribution of suspended matter in the North Sea and implications to Fisheries Biology. Mitt Geol-Paläontol. Univ. Hamburg 65:455-45.

Lancelot C., Billen G., Sourmia A., Weisse T., Coligen F., Veldhuis M.J.W, Davies A., Wassmann .P (1987) Phaeocystis blooms and nutrient enrichment in the continental coastal zones of the North Sea. Ambio 16 1:38-46

Lange U. (1992) Induktion von Enzymen des Entgiftungssystems in der Leber von Fischen: Ethoxyresorufin O-Deethylase in Klieschen (*Limanda limanda*) aus der südlichen und zentralen Nordsee. Berichte Zentr. Meeres-und Klimaforsch. Hamburg Nr. 24:37-42

Longwell A.C., Hughes J.B. (1981) Cytologic, cytogenetic and developmental state of atlantic mackerel eggs from sea surface water of the New York Bight, and prospects for biological effects monitoring with ichthyoplankton. Rapp. P-v Réun Cons. Int. Explor. Mer. 179:275-291

Lozán J. (1990) Zur Gefährdung der Fischfauna - Das Beispiel der diadromen Fischarten und Bemerkungen über andere Spezies. In Lozán J.L., Lenz W., Rachor E., Watermann B., von Westernhagen H. (eds.) Warnsignale aus der Nordsee wissenschaftliche Fakten. Berlin Hamburg: Parey pp:231-249.

Lozán J., Lenz W., Rachor E., Watermann B., von Westernhagen H. (eds.) (1990) Warnsignale aus der Nordsee: wissenschaftliche Fakten. Berlin, Hamburg, Parey pp: 1-433.

Mellergaard S., Nielsen E. (1987) The influence of oxygen deficiency of the dab population in the Eastern North Sea and the Kattegat. ICES CM/E:6

Meyers T.R., Hendricks J.D. (1982) A summary of tissue lesions in aquatic animals induced by controlled exposures to environmental contaminants, chemotherapeutic agents, and potential carcinogens. Mar. Fish. Rev. 44 (12):1-17

Michaelis H. (1987) Bestandsaufnahme des eulitoralen Makrobenthos im Jadebusen in Verbindung mit einer Luftbildanalyse. Jber 1986 Forschungsstelle Küste 38:13-97

Möller H. (1979) Geographical distribution of fish diseases in the Atlantic. Meeresforsch 27:217-235

Möller H .(1981) Fish Diseases in German and Danish coastal waters in summer 1980. Meeresforsch 29:1-16

Mühlenhardt-Siegel U. (1981) Die Biomasse mariner Makrobenthos-Gesellschaften im Einflußbereich der Klärschlammverklappung vor der Elbemündung. Helgoländer Meeresunters 34:427-437

Mühlenhardt-Siegel U. (1985) Die Weichbodengemeinschaft vor der Elbemündung unter dem Einfluß der Klärschlammverklappung. Diss. Univ. Hamburg 177 pp.

Murray A.J., Norton M.G. (1982) The field assessment of the effects of dumping of wastes at sea: 10 analysis of chemical residues in fish and shellfish from selected coastal regions around England and Wales. Fisheries Research Technical Report No. 69, Ministry of Agriculture Fisheries and Food Directorate of Fisheries Research ISSN 0308 - 5589. 42 pp.

Newmann P.J., Agg A.R (eds.) (1988) Environmental protection of the North Sea. Heinemann Professional Publ. Oxford, London 886 pp.

North Sea Assessment Reports 1, 4, 6, 7a, 8 , 10, (1993) NorthSea Task Force

Osterhaus A.D.M.E., Vedder E.J. (1988) Identification of a virus causing recent seal death. Nature 335:20

Peters G. (1988) Streß macht auch Fische krank. Naturwissenschaftl Rundschau 41 (8):303-309

Rachor E. (1982) Indikatorarten für Umweltbelastungen im Meer. Decheniana Beihefte (Bonn) 26:128-137

Rachor E. (1985) Eutrophierung in der Nordsee - Bedrohung durch Sauerstoffmangel. Abh Naturw Verein Bremen 40:283-292

Rachor E., Albrecht H. (1983) Sauerstoff-Mangel im Bodenwasser der Deutschen Bucht. Veröff Inst Meeresforsch Bremerh 19:209-227

Radach G., Berg J. (1986) Trends in den Konzentrationen der Nährstoffe und des Phytoplanktons in der Helgoländer Bucht (Helgoland Reede Daten). Ber Biol Anst Helgoland 2:1-63

Rauck G. (1985) Wie schädlich ist die Seezungenbaumkurre für Bodentiere? Infn Fischwirtsch 32:165-167

Rees H., Eleftheriou A. (1989) North Sea benthos: A review of field investigations into the biological effects of man's activities. J. Cons. Int. Explor. Mer. 45:284-305

Reijnders P.I.H. (1986) Reproductive failure in common seals feeding on fish from polluted coastal waters. Nature 324:456-457

Reise K. (1986) Gütezustand der Nordsee: Teilbereich Benthos. International North Sea Conference 1987

Riedel-Lorjé J.C., Gaumert T. (1982) 100 Jahre Elbe-Forschung. Hydrobiologische

Situation und Fischbestand 1842 - 1943 unter dem Einfluß von Stromverbau und Sieleinleitungen. Arch Hydrobiol/Suppl 61 (Unters Elbe-Aestuar 5) 3:317-376

Riesen W., Reise K. (1982) Macrobenthos of the subtidal Wadden Sea: revisited after 55 years. Helgoländer Meeresunters 35:409-423

Salomons W., Bayne B.L., Duursma E.K., Förstner U. (eds.) (1988) Pollution of the North Sea - An assessment. Springer Berlin 687 pp.

Selye H. (1950) Stress and the general adaptation syndrome. Br. Med. J. 1:1383-1392

Sindermann C.J. (1984) Fish and environmental impacts. Fourth Congress of European Ichthyologists, Hamburg, September 1982. Arch. Fisch Wiss. 35(1):125-173

Schutz der Nordsee (1987) Öffentl. Anhörung d. Bundestagsausschusses für Umwelt, Naturschutz u. Reaktorsicherheit am 5. Oktober 1987/(Hrsg Dt Bundestag Referat Öffentlichkeitsarbeit 663 pp.

Sündermann J., Beddig S. (eds.) (1988) Zirkulation und Schadstoffumsatz in der Nordsee (ZISCH). BMFT-Projekt MFU 0545 Abschlußbericht Univ. Hamburg 323 pp.

Sündermann J., Beddig S. (eds.) (1990) Zirkulation und Schadstoffumsatz in der Nordsee (ZISCH II). BMFT-Projekt MFU 0576 5 Abschlußbericht. Univ. Hamburg 193 pp.

Sündermann J., Beddig, S. (eds) 1992 Prozesse im Schadstoffkreislauf Merr-Atmosphäre: Ökosystem Deutsche Bucht (PRISMA), BMFT-Projekt MFU 0620/6, 2. Zwischenbericht Universität Hamburg 204 pp.

Stebbing A.R.D., Dethlefsen V., Carr M. (eds) (1992) Biological effects of contaminants in the North Sea. Mar. Ecol. Prog. Ser. 91. 362 pp.

Tiews K. (1983) Über die Veränderungen im Auftreten von Fischen und Krebsen im Beifang der deutschen Garnelenfischerei während der Jahre 1954-1981. Ein Beitrag zur Ökologie des deutschen Wattenmeeres und zum biologischen Monitoring von Ökosystemen im Meer. Arch FischWiss 34:1-156.

Tiews K. (1990) 35-Jahrestrends (1954-1988) der Häufigkeit von 25 Fisch- und Krebstierbeständen an der deutschen Nordseeküste. Arch FischWiss 40 (1/2):3-38

Vethaak A.D. (1992) Large-scale mesocosm study of the effects of marine pollution on the health status of fish: General methods on interim report on epidemiology. ICES CM/E:11

Vethaak A.D., ap Rheinallt T. (1992) Fish disease as a monitor for marine pollution: the case of the North Sea. Reviews in Fish Biology and Fisheries 2:1-32

Vethaak A.D. (1993) Fish diseases and marine pollution. Thesis, University of Amsterdam 162 pp.

Visser I.K.G., Teppema J.S., Osterhaus A.D.M.E. (1991) Virus infections of seals and other pinnipeds. Rev. Med. Microbiol. 2:1105-114

Vogel S. (1989) Speckdicke als Konditionsparameter bei 1988 an der Küste Schleswig-Holsteins tot aufgefundenen Seehunden (*Phoca vitulina* L.). In Bohlken H. (ed.) Zoologische und Ethologische Untersuchungen zum Robbensterben. Vorläufiger Endbericht FE-Vorhaben BMU pp:39-61

Watermann B. (1982) An unidentified cell type associated with an inflammatory condition of the subcutaneous conncective tissue in dab, *Limanda limanda* L. Short communication. J. Fish Dis. 5:257-261

Watermann B. (1984) Untersuchungen zur Histologie und Pathogenese von Hautwucherungen der Kliesche (*Limanda limanda* L.) aus der Nordsee. Diss. Univ. Hamburg 79 pp.

Watermann B., Dethlefsen V. (1982) Histology of pseudobranchial tumours in Atlantic cod (*Gadus morhua*) from the North Sea and the Baltic Sea. Helgoländer Meeresunters 35:231-242

Watermann B., Dethlefsen V. (1985) Epidermal hyperplasia and dermal degenerative changes as cell damage effects in gadoid skin. Arch. Fisch Wiss. 35:205-221

Westernhagen von H. (1970) Erbrütung der Eier von Dorsch (*Gadus morhua*), Flunder (*Platichthys flesus*) und Scholle (*Pleuronectes platessa*) unter kombinierten Temperatur- und Salzgehaltsbedingungen. Helgoländer Meeresunters 21:21-102

Westernhagen von H., Rosenthal H., Dethlefsen V., Ernst W., Harms U., Hansen P.D. (1981) Bioaccumulating substances and reproductive success in Baltic flounder *Platichthys flesus*. Aquatic Toxicol. 1:85-99

Westernhagen von H., Dethlefsen V., Cameron P., Janssen D. (1988a) Chlorinated hydrocarbon residue in gonads of marine fish and effects on reproduction. Sarsia 72:419-422

Westernhagen von H., Dethlefsen V., Cameron P., Berg J., Fürstenberg G. (1988b) Developmental defect on pelagic fish embryos from the Western Baltic. Helgoländer Meeresunters 42:13-36

Westernhagen von H., Cameron P., Dethlefsen V., Janssen D. (1989) Chlorinated hydrocarbons in North Sea whiting (*Merlangius merlangus* L.) and effects on reproduction. I. Tissue burden and hatching success. Helgoländer Meeresunters 43:45-60

10. INDICES FOR CARCINOGENICITY IN AQUATIC ECOSYSTEMS: SIGNIFICANCE AND DEVELOPMENT

Paule Vasseur
Fabrice Godet, Halima Bessi, Lucie Lambolez
Centre des Sciences de l'Environnement, BP 4025
57040 Metz Cedex 1
France

INTRODUCTION

Cancer is not particular to modern life; bone tumours were noted in specimens of dinosaurs by paleobiologists (Zimmerman 1977). It is now known that exogenous as well as endogenous factors, nonanthropogenic as well as anthropogenic compounds can cause cancer, which a large part of the public still considers "unnatural". Cancer can arise from natural causes, such as viral or parasitic infections, fungal or marine toxins, ionizing or UV radiations, hormones, but compounds belonging to specific classes of synthetic chemicals are also known to be potent carcinogens. Therefore, pollution has often been incriminated as a factor perturbing natural populations (Chu and Chiu 1990) and producing cancer epizootics in aquatic species. The questions environmentalists have been dealing with for decades are: "to what extent are anthropogenic activities responsible for carcinogenicity in ecosystems? Which etiological factors must be controlled and within what safety limits must their level be maintained in order to protect wildlife from cancer?"

This paper deals with our knowledge of the etiology of environmental carcinogenesis and with the approaches being developed at present in ecoepidemiology in order to more accurately elucidate the causal factors and to ensure prevention. The use of genetic biomarkers reflecting the exposure to carcinogens, for instance DNA adducts, which are early events in the multiple process of cancer and thus are interesting tools for prevention in environmental biomonitoring, will be discussed in terms of their significance as indices of carcinogenicity. We shall focus on aquatic ecosystems, taking some examples from studies on fish. A great deal of attention has been paid to these species in the past due to their commercial and recreational interest, so that available information on the problem of cancer relates almost exclusively to fish.

ECOEPIDEMIOLOGY

The evaluation of cancer risk in the environment requires a realistic approach taking into account the complexity of natural ecosystems. In this perspective field studies are more appropriate than lab experiments whose conditions are too restrictive compared with the multifactorial nature of the environment. Ecoepidemiological approaches are useful: they integrate epidemiological surveys of populations in the exposed environment and knowledge of their ecology including their biology and, where fish are concerned, their migrating behaviour.

A number of isolated as well as epizootic neoplasms in fish populations have been reported since the mid-19th century. These tumours may affect all organs, but more than half of documented epizootic tumours involve liver neoplasms (Sarokin and Schulkin 1992). Though skin neoplasms have also been described (Black 1983; Baumann and Harshbarger 1985), liver tumours appeared to be associated with pollution more than any other tumour type (Pierce et al. 1978; Murchelano and Wolke 1985; Malins et al. 1985; Malins et al. 1988). The first incidence of liver cancer in wild fish populations was reported by Dawe et al. (1964) from a survey of fish populations in Maryland: liver neoplasms were found with a relatively high incidence in white suckers (*Catostomus commersoni*) and multiple nodular lesions resembling hepatoma were observed in brown bullhead (*Ictalarus nebulosus*). For the first time, pollutants were evoked as a possible etiologic factor, but the role of protozoan parasiticism in the occurrence of this disease could not be excluded.

Later, hepatoma were again noted in several bottom-dwelling marine fishes from the U.S. West and East coasts: English sole (*Parophrys vetulus*) from Puget Sound and its tributaries (McCain et al. 1977), tomcod (*Microgadus tomcod*) from the Hudson River estuary (Smith et al. 1979), winterflounder (*Pseudopleuronectes americanus*) and windowpane flounder (*Scophthalmus aquosus*) from Connecticut, Rhode Island and Boston Harbor (Murchelano and Wolke 1985), white croaker (*Genyonemus lineatus*) from the Los Angeles area (Malins et al. 1987). In the North Sea, besides epidermal lesions frequently reported in flatfish (Dethlefsen 1980, 1988, 1990; Vethaak et al. 1992), hepatic neoplasia were also revealed after investigation of internal organs in dab (*Limanda limanda*) and flounder (*Platichthys flesus*) (Bucke et al. 1984; Verthaak 1993). Hepatocellular carcinomas were also noted, though less frequently, in bottom-feeding fish species from the river Rhine and its branches: bream (*Abramis brama*) and roach (*Rutilus rutilus*) (Slooff 1983; van Kreijl and Slooff 1985). The prevalence of preneoplastic and neoplastic lesions could reach 10% in specimens collected at marine polluted sites, whilst no neoplasms were found in any of the hundred fishes from unpolluted sites. The hypothesis that liver neoplasms and other liver lesions could be the result of exposure to sediment-associated chemical contaminants, was confirmed by the comparison of polluted and unpolluted areas; bottom-feeding fish species in contact with contaminated sediments were reliable indicators from this point of view. This hypothesis was supported in other respects, by experiments in fish with pure compounds, with identified field contaminants or contaminated sediment extracts, which resulted in tumours with the largest occurrence in the liver and a pattern of lesions resembling those observed in the field.

The temporal histogenesis of hepatic neoplasia in wild fish involves, at first, initial degenerative lesions, also called "megalocytic hepatosis", resulting from the cytotoxic effects of the carcinogens; it is the most common idiopatic lesion, detected in juveniles under one year of age. Lesions co-occurring generally with megalocytic hepatosis are (i) hepatocellular regeneration, which represents the compensatory proliferative response to the necrosis and (ii) foci of cellular alterations. Basophilic vacuolated foci are considered as preneoplastic lesions and represent an obligatory precursor in the induction of hepatic neoplasms. Tumorigenesis may affect the hepatic parenchyma or/and the intrahepatic biliary system, leading respectively to adenoma and cholangioma; these benign tumours may progress to malignancy and lead to hepatocarcinoma, which can be observed in animals of at least two years of age.

The geographic pattern of lesion distribution was similar in different bottom-dwelling species inhabiting the same area; only the occurrence of the lesions may differ between species (Myers and Rhodes 1988). This sustained the hypothesis of an environmental risk factor, which the authors related to exposure to contaminants. Polycyclic aromatic hydrocarbons (PAHs) appeared the most likely chemical etiologic agents from the identifiable sediment-associated chemicals. Positive statistical associations of liver lesions and hepatoma in fish with exposure to contaminants were demonstrated by comparing lesion prevalence with actual concentrations of identified pollutants in sediment; significant correlations were found with levels of PAHs in sediment (Malins et al. 1985; Myers et al. 1991). However, pollutants identified as PAHs and organochlorinated compounds (DDT and its congeners, PCBs), accounted for only a small part of the factors and the mechanisms operating in the process of hepatic neoplasia in fish. A study of Myers et al. (1991) on the etiology of hepatic neoplasm in English sole inhabiting polluted waterways and embayments of Puget Sound illustrates this point; these authors modelized neoplasm prevalence taking into account sediment contaminant levels, interaction among chemicals and fish size (an indirect measure of age). PAHs in sediments accounted for about 12% of the variation in the prevalence of neoplasms, more than any other contaminant group analyzed. Nevertheless, no more than 35% of the variation in neoplasm prevalence could be explained by the model, suggesting that a number of etiological factors are yet to be identified. In other cases, the etiology of the observed lesions remains an open issue: evidence linking effects to exposure to chemical contaminants is inconclusive, although the level of contamination of sediments may be high at some sites. A study of Malins et al. (1987) on the etiology of hepatic lesions in white croaker from several sites in marine waters adjacent to the Los Angeles area illustrates this: no strong associations between exposure to PAHs and the occurrence of liver disease could be established, though hepatic lesions were generally found at sites with highly contaminated sediments. However, the range of contaminant concentrations between sites was large: 54-2800 ng/g dry weight of sediment (d.w.s.) for PAHs, 2-1300 ng/g d.w.s. for DDT, 6-500 ng/g d.w.s. for PCBs. No more correlations could be found between the concentrations of DDT and PCBs in the liver or the metabolites of PAHs in the bile.

These examples underline some of the difficulties encountered in field studies, as far as a reliable identification of causes of disturbance are concerned due to (i) confounding variables masking true causal associations and creating false correlations, (ii) the multiplicity of potentially causal agents, some of them, if not most of them, being unknown, (iii) the unidentified chemical factors which are not currently analyzed but may act as carcinogens, cocarcinogens or tumour promoters, (iv) the interference between physiological, ecological and chemical factors, which may result in unexpected effects.

It is worthwhile to underline that physicochemical analyses of environmental media, which are essential in identifying existing pollutants and in determining the level of contamination of ecosystems, suffer from some limits. Indeed, the analytical screening carried out is not exhaustive and some classes of pollutants may escape detection because they were not searched for. Moreover, reliable and sensitive analytical techniques are not available for some compounds. In fact, only a fraction of the chemical contaminants present in sediments are currently and reliably identified: these are most often aromatic and chlorinated hydrocarbons; in contrast, chemical groups such as N-nitroso derivatives, aromatic and heterocyclic amines, which are typical genotoxic carcinogens, are rarely analyzed. Failure in physicochemical analyses of environmental media may partly explain the difficulties in identifying the key factors responsible for the lesions. Other explanations can be found in intrinsic factors such as fish movement, fish physiology and metabolism, explaining the specificities of species and individual susceptibility, which may not be reflected by analyses of the level of environmental contamination.

Attempts to express exposure in a manner more representative of the actual uptake of the contaminant by the exposed organisms, led scientists to consider (i) tissue levels of residues corresponding to persistent contaminants; (ii) metabolites excreted in the bile, for xenobiotics which are rapidly metabolized, and led finally to research biomarkers of exposure which could integrate pharmacokinetic factors such as tissue disposition, detoxication, and excretion. Trends in this sense have been noted in the last few years in approaches in ecoepidemiology which have aimed at a better prediction and an improved refinement in the explanation of risk factors, by means of:

- early reliable biological indicators of cancer risk in natural populations, which could be a useful tool in the survey of ecosystems: chromosomal abnormalities, largely used in short term tests to evaluate the genotoxicity of chemicals and of contaminated environmental media, were applied *in situ* and studied in wild species for their interest.

- indices of exposure to contaminants which could express the effective dose reaching the target tissue: biomarkers such as DNA damage which could be responsible for genetic changes, necessary steps in cancer development, were studied in this context.

CHROMOSOMAL ABNORMALITIES IN WILD SPECIES

Neoplasms and tumours detected among natural populations are classical endpoints expressing the carcinogenicity of pollutants. Nevertheless, many years are necessary for the tumorigenic process to develop. Thus, this endpoint is inappropriate to protect wild species from cancer risk. In contrast, alterations in the structure and function of DNA, which are commonly used in short-term tests as early indicators of genotoxicity, could be useful as *in situ* bioindicators for exposure to carcinogens in the environment. Cytogenetic techniques such as chromosomal aberrations and sister chromatid exchanges, though they were successfully used to detect the genotoxicity of aquatic contaminants in experiments, were rarely applied *in situ* due to the relatively large numbers of chromosomes of most fish species. Only Al Sabti and Kurelec (1985) studied aberrant metaphases in gill cells of marine mussels and concluded in the relevance of this criterium as an *in situ* bioindicator of pollution (Table 1). The "micronucleus" test was preferentially applied to *in situ* studies, because it was shown to be sensitive, simple and cost effective in lab experiments. The presence of micronuclei (MN) was studied in mussels and fishes sampled from unpolluted and contaminated sites; gill cells and hemocytes in marine bivalves and erythrocytes in fishes were investigated (Table 1). These studies showed that the background MN frequency was normally low, in the range 1-3°/$_{oo}$ cells and that this frequency increased with pollution as shown by Hose et al. (1987); however, Brunetti et al. (1988), Carrasco et al. (1990) and Wrisberg et al. (1992) failed to observe a dose-effect relationship, the level of micronuclei in samples from highly polluted sites being sometimes similar to that from reference sites. The explanation given by the authors is that the presence of toxic substances increased mortality and that the collected specimens were the most resistant organisms to polluting agents.

We believe more generally that a high level of cytotoxic contaminants may impair growth and cell division. If the rate of cell division decreases, the probability that micronuclei are formed during mitosis, decreases in parallel. Thus the sensitivity of the micronucleus indicator is associated with the use of juveniles and the survey of dividing cells. It is worthwhile noting that these conditions were not examined and fulfilled in previous studies: for instance, Carrasco et al. (1990) mentioned that the fish they analyzed were 5 - 12 years old. More research seems necessary in order to improve the use of chromosomal abnormalities as *in situ* indicators for genotoxic chemical exposure in environmental biomonitoring.

DNA ADDUCTS AS BIOMARKERS OF EXPOSURE TO CARCINOGENS

Permanent changes in DNA structure resulting from the covalent binding of electrophilic compounds or metabolites may be factors of risk in the development of cancer. Thus DNA adducts, by reflecting the amount of reactive substances that reached the target tissue, were postulated to be relevant biomarkers of exposure to chemical carcinogens. The relevance of this marker was extensively studied in human cancer to detect and control populations at risk. It was also studied experimentally in aquatic species exposed to known carcinogens (McCarthy et al. 1989; Shugart et al. 1992) and for environmental purposes in wild aquatic species collected from polluted areas. ^{32}P-post labelling analysis of DNA adducts was shown by Dunn et al. (1987), Varanasi et al. (1989), and Kurelec et al. (1989a) to provide useful information on exposure of fishes and mussels to genotoxic substances: higher amounts of liver DNA adducts could be found in polluted areas compared to reference sites or control media. The "fingerprint" pattern of adducts is nevertheless often complex in species collected from polluted areas; some spots on autoradiogramms of TLC maps of ^{32}P-labelled adducts could be identified as due to PAH metabolites, but a lot of adducts remained to be identified. This identification could be useful to get benefit from the fingerprint pattern of adducts, which seemed specific of a geographical area, in other terms, of the type of pollutants, but independent of the fish species analyzed (Varanasi et al. 1989).

Table 1. Chromosomal abnormalities studied in wild aquatic species sampled from unpolluted and polluted sites.

Species	Criteria	Cells	Range of values unpolluted	polluted	Authors
BIVALVES					
Mytilus galloprovincialis	% aberrant metaphases	gill cells	2.90 ± 2.80	4.10 ± 2.30 7.70 ± 2.70	Al Sabti & Kurelec 1985
	Micronuclei °/°° cells	gill cells	3.20 ± 0.26 3.60 ± 0.20	4.70 ± 0.25 2.20 ± 0.25 2.50 ± 0.23	Brunetti et al. 1988
Mytilus edulis	Micronuclei °/°° cells	hemocytes	1.10 ± 0.60	0.90 ± 0.50 2.87 ± 0.80 2.32 ± 1[a] 0.90 ± 0.6[b]	Wrisberg et al. 1992
FISH					
Genyonemus lineatus	Micronuclei °/°° cells	erythrocytes	0.80 ± 1.10	3.40 ± 2.70	Hose et al. 1987
Paralabrax clathratus		erythrocytes	0.60 ± 0.60	6.80 ± 5.1	
Genyonemus lineatus	Micronuclei °/°° cells	erythrocytes	0.13 ± 0.29 0.75 ± 1.44	0.03 ± 0.13 0.17 ± 0.31 0.27 ± 0.32 0.46 ± 0.59	Carrasco et al. 1990

a,b: mussels sampled at the same site on rock (a) or sediment (b).

Varanasi et al. (1989) analyzing hepatic DNA adducts of English sole and winter flounder in the area of Puget Sound and Boston Harbor, showed that there was a general agreement between levels of fluorescent metabolites of aromatic hydrocarbons in bile (FACs), DNA damage and total pollutants in sediments when fish from contaminated and reference sites were compared. But DNA adducts and FACs in bile did not correlate well, when individual fish from contaminated sites were analyzed. This is not surprising because analysis of DNA adducts gives a more direct measure of wild fish exposure to genotoxic compounds than either measurements of the level of FACs in bile, which involves both active and inactive metabolites, or the concentrations of chemicals in sediments.

It must be noted that the amount of DNA adducts cannot be used intrinsically to evaluate the degree of exposure to carcinogens, but only with reference to controls of the same species, because some adducts can also be detected in organisms from unpolluted areas (Kurelec et al. 1989b), or grown in lab (Dunn et al. 1987), as well as in control cultures of fish cells (Masfaraud 1992) or embryonic cells (Bessi 1993); an endogenous origin is probable, but an exogenous origin, involving for example food in lab species or an unknown past contamination in reference sites, cannot be excluded.

The interest of DNA adducts as biomarkers of exposure to chemical carcinogens, led to the concept of molecular dosimetry of DNA adducts as a tool in cancer risk assessment. As the probability of cancer was related to the degree of DNA damage, it was assumed that the dose of the active chemical bound to the target tissue, could be a predictive tool of tumour incidence. The relevance of this concept is being extensively studied in research at this time and it will be discussed below.

DNA ADDUCT DOSIMETRY IN RISK ASSESSMENT

Disease risk (the proportion of adversely affected organisms) is generally estimated by means of a dose-response model, which is a function of the dose of active chemical that reaches the target tissue. For instance, most models of carcinogenesis assume a mutagenic change of a normal cell to an initiated cell, followed by the proliferation of initiated cells and finally the transition from an initiated to a malignant state. Thus the incidence of cancer could be considered directly correlated with the intensity of DNA damage. To day, many authors explore the quantitative relationships between DNA adducts and cancer risk, and examine their application to intra and interspecies extrapolation. These studies, up to now, have been carried out with controlled experiments with known carcinogens; the steady state or cumulative levels of DNA adducts occurring during exposure must be considered in order to perform a valid stoichiometric analysis of adduct formation.

A study by Dashwood et al. (1988) on tumour incidence in trouts of aflatoxin B_1, was very interesting in this context. Liver tumour incidence was measured one year after a two week exposure to doses from 10 to 320 ppb aflatoxin B_1 in the diet. A strong correlation was found between liver tumour incidence and hepatic AFB_1-DNA adducts

(HADA). From these results it was further demonstrated by Bechtel (1989) and Dashwood et al. (1990) that rainbow trout and male Fisher rat exhibited similar linear correlation of hepatic tumour risk with HADA concentrations (Figure 1): the DNA adduct levels associated with a 50% tumour incidence were quantitatively similar in the two species and corresponded to about 89 HADA per 10^8 nucleotides.

Though these results are promising, adduct-risk correlations need to be validated with other carcinogenic chemicals in other species and in organs displaying various susceptibilities, before considering DNA adducts as an indicator of cancer risk across species. Gaylor et al. (1992) showed that DNA adduct concentrations corresponding to 50 % tumour incidence could differ according to the carcinogens and the animals (Table 2): the 100-fold range of DNA adduct concentrations associated with the same tumour incidence raises the question of whether we can predict tumour induction potential on the basis of DNA adduct levels. Differences may relate to DNA repair rates, the distribution of adducts on the genome and the effects of some carcinogens on cell proliferation in addition to mutagenic effects.

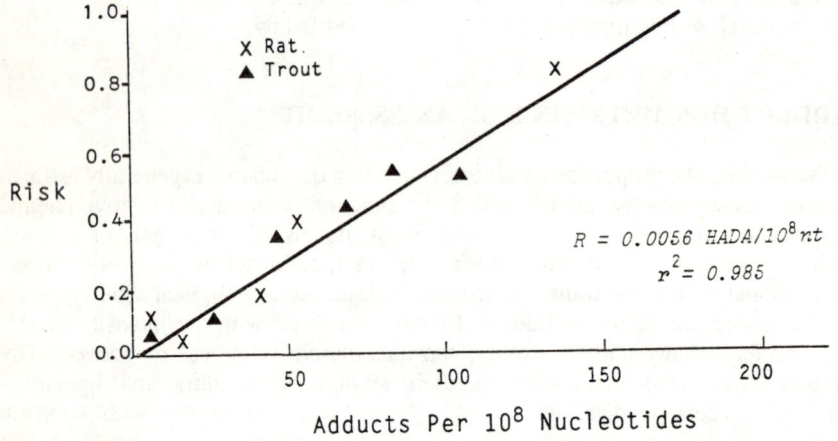

Figure 1. Chronic hepatic tumour (incidence) plotted as a function of hepatic AFB_1 DNA adducts (HADA) concentration for the male Fischer rat and rainbow trout (Bechtel 1989)

Table 2. DNA adduct levels in experimental animals associated with 50% tumour incidence (Gaylor et al. 1992)

Carcinogen	Species	Tissues	DNA adduct	Adduct level pmole/mg
MNU	Mice	Bone, thymus	O^6-Me-dG	220
ENU	Mice	Bone, thymus	O^6-Et-dG	25
NDEA	Rats	Liver	O^4-Et-dT	2
AFB_1	Rats	Liver	$N7$-AFB_1-dG	3
	Trout	Liver	$N8$-AFB_1-dG	3
2-AAF	Mice	Liver	C8-AF-dG	110
	Mice	Urinary bladder	C8-AF-dG	250
4-ABP	Mice	Liver	C8-ABP-dG	250
	Mice	Urinary bladder	C8-ABP-dG	150
	Dogs	Urinary bladder	C8-ABP-dG	110

Abbreviations: MNU, N-methyl-N-nitrosourea
ENU, N-ethyl-N nitrosourea
NDEA, N-nitrosodiethylamine
AFB_1, aflatoxin B_1
2-AAF, 2-acetylaminofluorene
4-ABP, 4-aminobiphenyl
O^6-Me-dG, O^6-methyldeoxyguanosine
O^6-Et-dG, O^6-ethyldeoxyguanosine
O^4-Et-dT· O^4-ethyldeoxythymidine
$N7$-AFB_1-dG, *trans*-8,9-dihydro-8-(deoxyguanosin-7-yl)-9-hydroxyaflatoxin B_1
C8-AF-dG, N-(deoxyguanosin-8-yl)-2-aminofluorene
C8-ABP-dG, N-(deoxyguanosin-8-yl)-4-aminobiphenyl.

The estimation of cancer risk on the basis of adduct dosimetry entails the assumption that adducts are randomly distributed on the genome and have the same effects regardless of tissue and cell specificities. In fact it was shown recently that (i) genomic DNA is not repaired equally in all parts on the genome, but preferentially in active genes, (ii) the mutability of genes present in the genome of mammalian cells is not uniform and (iii) mutational "hot spots" are present in the genome of somatic cells *in vivo* (Bohr et al. 1985; Hanawalt 1991). As a result, the notion of specific altered loci on the chromatin and the role played in cell function by the mutated genes, would need to be considered.

Genes involved in the control of cellular differentiation, quiescence of stem cells and programmed death on the one hand, and cell proliferation on the other, are factors of importance in the development of cancer. Nowadays, research emphasizes two categories of genes: (i) oncogenes, which in an altered or overexpressed form act to promote proliferation; and (ii) tumour suppressor genes, in which mutational change may result in a negation of their normal inhibitory action on proliferation. The fact that one mutation may activate *ras* oncogenes and promote tumourigenesis, provided the mutation occurs at one specific codon, 12, 13, 61, 117 or 146 (Barbacid 1987; Sloan et al. 1990), underlines the crucial role of these protooncogenes. This explains that at the present time a number of studies focus on mutations in *ras* oncogenes in mammals but also in fish (McMahon et al. 1990). Fortunately, many lines of evidence indicate that *ras* mutations in themselves are not sufficient to produce fully developed malignant tumours; additional genetic damage, perhaps mutations of tumour suppressor genes, is required to disrupt the control of cell growth and to make the tumour invasive and metastatic.

The conclusion is that an evaluation of the risk following exposure to carcinogens on the basis of total DNA damage and repair, may not be sufficient; better correlations may be obtained by investigating adduct clearance from specific genetic loci. The relevance of adduct dosimetry in cancer risk assessment has been studied up to the present time through experiments carried out with one specific carcinogen. The problems encountered will be even more complicated with mixtures of carcinogens leading to the formation of a myriad of adducts, such as in environmental conditions characterized by multiple exposure. In the environment, interactions must also be considered and may modify the dose-response relationship between initial damage on DNA and cancer risk.

Substances activating cell proliferation will potentialize the effects of genotoxic carcinogens. These substances, called tumour promoters or nongenotoxic carcinogens, may be anthropogenic contaminants, but may also have a natural origin: some strong tumour promoters, such as okadaic acid obtained from a black sponge, are found among marine toxins produced by fungi, blue-green algae, dinoflagellates...though the synergistic effects of these nongenotoxic carcinogens are well established experimentally, their implications in environmental carcinogenesis remain unknown. Little attention was devoted up to now to nongenotoxic carcinogens, compared with the genotoxic ones; yet their relation to cancer may be important and deserves to be considered.

Another area which it would be worthwhile studying relates to DNA repairing systems, whose activity may be impaired by natural or synthetic substances. The effects of contaminants on the integrity of these enzymatic systems is also little investigated, in spite of their essential role in the metabolism of DNA. Though cross-species comparisons of aflatoxin B_1 are encouraging, too little data exists at this time to judge the capacity of DNA adduct levels in predicting cancer incidence and to permit extrapolations across species. More knowledge must be collected about the mechanisms of carcinogenesis, beyond the formation of DNA lesions, before a rational approach can be evolved to identify specific or groups of key DNA lesions. Key DNA lesions are assumed to explain organ- or tissue-specificities; unfortunately, there is evidence that such lesions may be different for different chemicals. What are the repairing capacities of the cells? Which genes are concerned in the genomic alterations? Which allele is concerned by the mutation? An answer to these questions proves necessary to assess cancer risk from DNA damage.

AREA OF FUTURE RESEARCH

Notwithstanding rapid progress in fundamental cancer research, the elucidation of mechanisms in environmental carcinogenesis and the finding of relevant key genomic lesions will take time. Nevertheless, environmentalists have to deal with the problem of cancer in ecosystems and to take decisions with prevention and remediation in view. Prevention means not only controlling the quality of complex effluents before their release into the environment, but also determining the maximum limit values of toxic contaminants in wastewaters and quality objectives in the environment. Besides the choice of the appropriate bioassay among the genotoxicity tests proposed in biomonitoring (Godet et al. 1993), the real problem is to define the level of acceptable contamination. Zero pollution is an utopia, but what could "safety" levels of contamination in natural ecosystems be? Difficulties are linked to the multiplicity of contaminants, synergistic/antagonistic interactions and the gaps in our knowledge of cause-effect relationships in the environment.

Both field studies and controlled experiments deserve to be pursued and developed to provide us with a better understanding of mechanisms of lesions and to establish more realistic dose-effect relationships in an environmental context.

Field studies are essential to discern the anthropogenic factors that could be responsible for the observed effects and the levels of contamination associated with deleterious effects. An examination of the range of contaminant levels found in "healthy" sites, where neither preneoplastic nor neoplastic lesions are observed, will also be very fruitful with the establishment of safety levels of concentration in the environment in view. From our point of view, level of contamination would be considered as a whole and not pollutant by pollutant, because of interactions among the mixture of contaminants. For instance, inducers of metabolizing enzymes, such as PCBs, may enhance the effects of potent carcinogens by activating the formation of electrophilic metabolites.

Hypotheses arising from field studies on direct and indirect factors of risk need to be validated. Controlled experiments are indispensable and complementary to *in situ* studies for this purpose. It would be necessary for them to be conducted with native species in conditions as close as possible to the environmental ones. Enclosures and mesocosms could provide a high potential resolution. We think that artificial rivers or a series of experimental basins connected to the aquatic environment could be very useful in allowing us to perform experiments in flowthrough conditions, with the natural water derived from the ecosystem and controlled native species at different stages of life. Biochemical and cytogenetic studies on genitors, embryos, juveniles and adults would allow us to correlate contamination with the temporal development of early DNA lesions and karyotypic abnormalities, and with effects on reproduction, carcinogenesis and development of populations.

At this time, genomic damage, such as DNA adducts and chromosomal abnormalities, can only be interpreted as biomarkers of exposure to potent carcinogens.

The extent to which these anomalies in somatic cells reflect a threat for populations remains to be established. There is no doubt that there is a probability that gametes and embryos can be altered when mutations are detected in somatic tissues. Chromosomal mutations are generally lethal for embryos: then consequences on the reproduction and the size of populations would be rapid. Point mutations, in contrast, which may be present as recessive mutations on heterozygote genes, will be silent but inheritable. As they may be expressed after damage or deletion of the functionning allele, recessive mutations may have consequences in the long term, by rendering the descendants more sensitive to the effects of potent carcinogens.

CONCLUSION

We need research and time to fill the gap in our present knowledge. *In situ* studies carried out to survey contaminated ecosystems would be of major importance in this concern; but confounding factors, the multiplicity of potentially causal agents which may remain unidentified, the interactions between physiological, ecological and chemical factors explain that field studies are often inconclusive. The failure of ecoepidemiology in identifying the factors of cancer risk explains the fact that experimental studies carried out in controlled conditions are needed to better appreciate the carcinogenic potential of suspected contaminants. In order to partly overcome the drawbacks of the traditional experimental approach, improvements may be attempted by performing experiments in a series of different size enclosures, mesocosms, or basins in flowthrough conditions; this could provide a high potential resolution, and offer the possibility of testing controlled native species in conditions close to the aquatic environment. Extensive interdisciplinary and often interinstitutional collaborations are a necessity and a condition for successful research in this area; collaborative interdisciplinary research and communication are not the least difficult part of the problem and must be greatly encouraged.

REFERENCES

Al-Sabti K., Kurelec B. (1985) Induction of chromosomal aberrations in the mussel, *Mytilus galloprovincialis* Watch. Bull. Environ. Contam. Toxicol. 35:660-665

Barbacid M. (1987) *ras* Genes. Ann. Rev. Biochem. 56:780-813

Baumann P.C., Harshbarger J.C. (1985) Frequencies of liver neoplasia in a feral fish population and associated carcinogens. Mar. Environ. Res. 17:324-327

Bechtel D.H. (1989) Molecular dosimetry of hepatic aflatoxin B_1-DNA adducts: linear correlation with hepatic cancer risk. Regul. Toxicol. Pharmacol. 10:74-81

Bessi H. (1993) Transformation morphologique des cellules SHE et communication intercellulaire des V79 appliquées à la détection de cancérogènes non génotoxiques. Thèse Doctorat Univ. Metz

Black J.J. (1983) Field and laboratory studies of environmental carcinogenesis in Niagara river fish. J. Great Lakes Res. 9:326-334

Bohr V.A., Smith C.A., Okumoto D.S., Hanawalt P.C. (1985) DNA repair in an active gene: removal of pyrimidine dimers from the DHFR gene of CHO cells is much more efficient than in the genome overall. Cell 40:359-369

Brunetti R., Majone F., Gola I., Beltrame C. (1988) The micronucleus test: examples of application of marine ecology. Mar. Ecol. Prog. Ser. 44:65-68

Bucke D., Watermann B., Feist S. (1984) Histological variations of hepato-splenic organs from the North Sea dab, *Limanda limanda* L. J. Fish. Dis. 7:255-268

Carrasco K.R., Tilbury K.L., Myers M.S. (1990) Assessment of the piscine micronucleus test as an *in situ* biological indicator of chemical contaminant effects. Can. J. Fish. Aquat. Sci. 47(11):2123-2136

Chu M.M.L., Chiu A (1990) Environmental carcinogenesis and biotechnology. J. Biotech. 16:17-36

Dashwood R.H., Arbogast D.N., Fong A.T., Hendricks J.D., Bailey G.S. (1988) Mechanisms of anti-carcinogenesis by indole-3-carbinol: detailed *in vivo* DNA binding dose-response studies after dietary administration with Aflatoxin B1. Carcinogenesis 9:427-432

Dashwood R.H., Loveland P.M., Fong A.T., Hendricks J.D., Bailey G.S. (1990) Combined *in vivo* DNA binding and tumour dose-response studies to investigate the molecular dosimetry concept. Mutation and the Environment, Part D:335-344

Dawe C.J., Stanton M.F., Schwartz F.J. (1964) Hepatic neoplasms in native bottom-feeding fish of deep creek lake, Maryland. Cancer Res. 24:1194-1201

Dethlefsen V. (1980) Observations on fish diseases in the German Bight and their possible relation to pollution. Rapp P-v Réun. Cons. Perm. Int. Explor. Mer. 179:110-117

Dethlefsen V. (1988) Assessment of data on fish disease. In Newman P.J., Agg A.R. (eds.) Environmental Protection of the North Sea. pp:276-285

Dethlefsen V. (1990) Ten years fish disease studies of the Federal Research Board Fisheries Hamburg. Arch. Fish. Wiss. 40:119-132

Dunn B.P., Black J.J., Maccubbin A. (1987) ^{32}P-Postlabeling analysis of aromatic DNA adducts in fish from polluted areas. Cancer Res. 47: 6543-6548

Gaylor D.W., Kadlubar F.F., Beland F.A. (1992) Application of biomarkers to risk assessment. Environ. Health Perspect. 98:139-141

Godet F., Vasseur P., Babut M. (1993) Essais de génotoxicité *in vitro* et *in vivo* applicables à l'environnement hydrique. Rev. Sc. Eau 6:285-314

Hanawalt P.C. (1991) Heterogeneity of DNA repair at the gene level. Mutat. Res. 247:203-211

Hose J.E., Cross J.N., Smith S.G., Diehl D. (1987) Elevated circulating erythrocyte micronuclei in fishes from contaminated sites off Southern California. Marine Environ. Res. 22:167-176

van Kreijl C.F., Sloof W. (1985) Mutagenic activity in Dutch river water and its biological significance for fish. In Zimmermann F.K., Taylor-Mayer R.E. (eds.) Mutagenicity Testing in Environmental Pollution Control. pp: 86-104

Kurelec B., Garg A., Krca S., Chacko M., Gupta R.C. (1989a) Natural environment surpasses polluted environment in inducing DNA damage in fish. Carcinogenesis 10(7): 1337-1339

Kurelec B., Garg A., Krca S., Gupta R.C. (1989b) DNA adducts as biomarkers in genotoxic risk assessment in the aquatic environment. Marine Environ. Res. 28:317-321

Malins D.C., Krahn M.M., Brown D.W., Rhodes L.D., Myers M.S., McCain B.B., Chan S.L. (1985) Toxic chemicals in marine sediment and biota from Mukiteo, Washington: relationships with hepatic neoplasms and other hepatic lesions in English sole (*Parophrys vetulus*). J. Natl. Cancer Inst. 74:487-494

Malins D.C., McCain B.B., Brown D.W., Myers M.S., Krahn M.M., Chan S.L. (1987) Toxic chemicals, including aromatic and chlorinated hydrocarbons and their derivatives, and liver lesions in white croaker (*Genyonemus lineatus*) from the vicinity of Los Angeles. Environ. Sci. Technol. 21:765-770

Malins D.C., McCain B.B., Landahl J.T., Myers M.S., Krahn M.M., Brown D.W., Chan SL, Roubal W.T (1988) Neoplastic and other diseases in fish in relation to toxic chemicals: an overview. Aquat. Toxicol. 11:43-67

Masfaraud J.F. (1992) Activités enzymatiques à cytochrome P450 comme marqueur biologique de pollution: relation entre l'activité 7-éthoxyrésorufine O-dééthylase et la formation d'adduits à l'ADN chez la truite (*Oncorynchus mykiss*) exposée au benzopyrène. Thèse Doctorat Univ Lyon I

McCain B.B., Pierce K.V., Wellings S.R., Miller B.S. (1977) Hepatomas in marine fish from an urban estuary. Bull. Environ. Contam. Toxicol. 18:1-2

McCarthy J.F., Jacobson D.N., Shugart L.R., Jimenez B.D. (1989) Pre-exposure to 3-methylcholanthrene increases benzo(a)pyrene adducts on DNA of bluegill sunfish. Marine Environ. Res. 28:323-328

McMahon G., Huber L.J., Moore M.J., Stegeman J.J., Wogan G.N. (1990) Mutations in c-Ki-*ras* oncogenes in diseased livers of winter flounder from Boston Harbor. Proc Natl. Acad. Sci. 87:841-845

Murchelano R.A., Wolke R.E. (1985) Epizootic carcinoma in the winter flounder *Pseudopleuronectes americanus*. Science 228:587-589

Myers M.S., Landahl J.T., Krahn M.M., McCain B.B. (1991) Relationships between hepatic neoplasms and related lesions and exposure to toxic chemicals in marine fish from the U.S. West Coast. Environ. Health Perspect. 90:7-15

Myers M.S., Rhodes L.D. (1988) Morphologic similarities and parallels in geographic distribution of suspected toxicopathic liver lesions in rock sole (*Lepidopsetta bilineata*), starry flounder (*Platichthys stellatus*), Pacific staghorn sculpin (*Leptocottus armatus*), and Dover sole (*Microstomus pacificus*) as compared to English sole (*Parophrys vetulus*) from urban and non-urban embayments in Puget Sound, Washington. Aquat. Toxicol. II:410-411

Pierce K.V., McCain B.B., Wellings S.R. (1978) Pathology of hepatomas and other liver abnormalities in English sole (*Parophrys vetulus*) from the Duwamish river estuary, Seattle, Washington. J. Natl. Cancer Inst. 60:1445-1453

Sarokin D., Schulkin J. (1992) The role of pollution in large-scale population disturbances Part 1: Aquatic Populations. Environ. Sci. Technol. 26:1476-1483

Shugart L., Bickham J., Jackim G., McMahon G., Ridley W., Stein J., Steinert S. (1992) DNA alterations. In Huggett R.J., Kimerle R.A., Mehrle P.M. Jr, Bergman H.L. (eds.) Biomarkers: Biochemical, Physiological, and Histological Markers of Anthropogenic Stress. pp:125-153

Sloan S.R., Newcomb E.W., Pellicer A. (1990) Neutron radiation can activate K-*ras* via a point mutation in codon 146 and induces a different spectrum of *ras* mutations than does gamma radiation. Mol. Cell Biol. 10:405-408

Slooff W. (1983) A study on the usefulness of feral fish as indicators for the presence of chemical carcinogens in dutch surface waters. Aquat. Toxicol. 3:127-139

Smith C.E., Peck T.H,. Klauda R.J., McLaren J.B. (1979) Hepatomas in Atlantic tomcod, *Microgadus tomcod* (Walbaum), collected in the Hudson River estuary in New York. J. Fish Dis. 2:313-319

Varanasi U., Reichert W.L., Stein J.E. (1989) ^{32}P-Postlabeling analysis of DNA adducts in liver of wild English sole (*Parophrys vetulus*) and winter flounder (*Pseudopleuronectes americanus*) Cancer Res. 49:1171-1177

Vethaak A.D., Bucke D., Lang T., Wester P.W., Jol J., Carr M. (1992) Fish disease monitoring along a pollution transect: a case study using dab *Limanda limanda* in the German bight. Mar. Ecol. Prog. Ser. 91:173-192.

Vethaak A.D. (1993) Fish disease and marine pollution. Thesis Universiteit van Amsterdam

Wrisberg M.N., Bilbo C.M., Spliid H. (1992) Induction of micronuclei in hemocytes of *Mytilus edulis* and statistical analysis. Ecotoxicol. Environ. Saf. 23:191-205

Zimmerman M.R. (1977) An experimental study of mummification pertinent to the antiquity of cancer. Cancer 40:1358-1362

11. ASSESSMENT OF ECOSYSTEM HEALTH: DEVELOPMENT OF TOOLS AND APPROACHES

Peter-Diedrich Hansen
Berlin University of Technology, FB 7- Institute for Ecology
Department for Ecotoxicology
Keplerstrasse 4-6 (1.Stock), D-10589, Berlin
Germany

INTRODUCTION

There are in excess of 100,000 anthropogenic pollutants in waterways, making it impossible to rely solely on chemical detection in environmental monitoring. The monitoring of biological effects is an important tool for assessing the survival of individuals, populations and species. Environmental effects monitoring approaches provide information on environmental response which serves as a sensitive early-warning "signal" for understanding and predicting environmental change. Importantly, effects-based approaches provide a measure of the integrated effect of substances released to the environment.

To more fully understand the complex processes involved in environmental response to pollutants, we have to direct our efforts to rapid and cost-effective biochemical and cytotoxic testing methods as well as on-line monitoring systems (i.e. biosensors). In addition to physico-chemical methods, biological and biochemical methods are of increasing importance for the integrated monitoring of pollutants. As an example, the hepatic mixed function oxygenases (MFO) activity, as indicated by 7-ethoxyresorufin O-de-ethylase (EROD) and other measurements of Cytochrome P4501A1, is a very sensitive indicator of the ability of fish to detoxify certain pollutants. MFO activity can be related quantitatively to the extent of pollutant exposure, and can be used as a bioassay to identify the effects of pollutants. A vast amount of expertise exists in the area of environmental mapping of MFO-activities (Quality Status Report of the North Sea 1993) and genotoxic potential (i.e. DNA-unwinding and umu-C-assay). Measurements of phagocytosis (immunological defence activity of organisms) also has high potential for biological effects monitoring. The use of *in situ* bioassays offers an opportunity to evaluate "signals" and to anticipate changes in the environment. The "signals" from the effects monitoring investigations are important in promoting environmentally sensitive and sustainable use of waterways and coastal zones.

This article outlines the application of the currently available biomonitoring tools to generate information on early warning signals of ecosystem damage due to both man made and natural pollution. Emphasis is put on the use of recently developed "on-line monitoring" approaches as a biotechnology approach for assessing environmental damage and managing ecosystem health.

ECOLOGICAL REQUIREMENTS

In order to establish effective pollution control measures for aquatic ecosystem conservation, both emissions and their levels must be taken into consideration to improve our understanding of ecological responses to environmental impacts. Knowledge of small scale individual physico-chemical processes is essential for developing general models. Principles associated with the different scales of biochemical processes relating to ecosystem function are summarized in Figure 1 which shows the structural and functional hierarchies of biological interactions as they relate to ecosystem complexity. In order to understand the complex interaction across these hierarchical scales, it is necessary to reduce the system to its respective parts for study (McIntosh 1981). However, in order to fully understand overall ecosystem function and ecosystem health, it is ultimately necessary to use both this reductionist approach together with holistic approaches (Odum 1969).

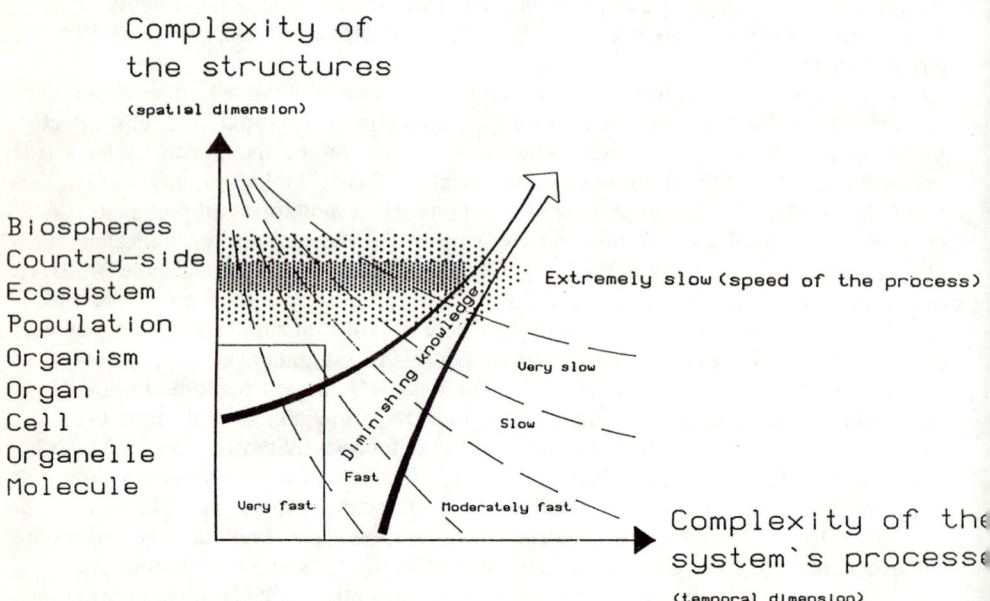

Figure 1. Structural and functional hierarchies in biological systems (after Hansen 1992a).

Figure 2 outlines how biological systems respond to sublethal levels of environmental pollutants at suborganismic, organismic, population and ecosystem levels. Rapid reaction generally occurs at the suborganismic level (1-3 days), where symptoms relating to initial stages of detoxification and regulation processes are manifested. Examples of monitoring tools used to assess biochemical responses at this level include Mixed Function Oxygenase (MFO) activity, acetylcholinesterase inhibition (AChE) and immunosuppression (Phagocytosis). Overlapping with this functional level, effects at the organismic level can be observed when organisms are exposed to environmental pollutants for a longer period. Effects at this level would eventually result in contaminant genotoxic effects at macromolecular levels such as DNA-damage and mutagenesis. Since environmental genotoxicity assessment approaches provide information on such effects, they have a high potential as environmental monitoring tools. Table 1 summarizes the effects of DNA damage at different levels of biological organization. Extended exposure of organisms to environmental genotoxins would result in several physiological disorders such as reproductive impairments and other related abnormalities. Therefore, measurements of reproductive response are essential for assessing the effects of anthropogenic sources of pollutants. For such assessment approaches, incubation periods of 20 to 120 days are required. This could be the beginning of the adaptation process in the organism which would reflect changes (feedback) in the ecosystem. These effects can also be monitored at population levels so that information on the overall functional potential of the ecosystem can be obtained. This assessment would include such processes as reorganization, redevelopment and structure of the system.

Responses in biological systems

[min.]

10^6 [> 2a] **Ecosystem Level:** alteration in ecosystems - redevelopment of the system's elements and structure

[0.5-1a] **Population Level:**
[0.5-1a] population dynamics - self organization - reorganization
10^5 [1-12 mon.] change in growth and adaptation of the system

[20-120 d] **Organismic Level (exposure): "in-situ Bioassays"**
- growth, reproduction, ELST (Early Life Stage Test), accumulation,
10^4 biotransformation: MFO, reaction with macromolecules, DNA-damage and repair, mutagenesis

Organismic / Suborganismic Level:
[1-3d] Symptons in individuals: detoxification and regulation processes
10^3 -biochemical response ("MFO", AChe, Phagocytosis)
- change in behavior

10^2

10^1 [10 min.] Early Warning Systems - On Line Monitoring - "Biosensors"

10^0 **Input of pollutants   (sublethal level)**

Figure 2. Time-scale responses in biological systems to sublethal pollutants.

Table 1. Consequences of DNA damage at different levels of biological organization.

Level of Biological Organization	Effects
DNA	mutations
Cell	cell death disordered proliferation and differentiation neoplastic transformation
Tissue / Organ	functional defects malformations tumours
Organism	reduced viability reduced fertility
Population	reduction of population size extinction
Ecosystem	reduction of species diversity

In order to obtain meaningful information of overall ecosystem response, an assessment period of at least two years is required. Biological effects monitoring over this period is analogous to taking the "pulse" of the ecosystem, providing "signals" of larger events. These "signals" are based on measurable responses of biota to environmental stressors such as detoxification (MFO), immunosuppression, and ability of the organism to reproduce. Use of "biosensors" is important in detecting the initial onset of the environmental deterioration process (Hansen 1992b). Biosensors are defined here as selective biological systems (enzymes, antibodies, organelles, cells) combined with a transducer (thermistor, potentiometric and amperiometric electrode, piezoelectric and optical receivers) which generate continuous information. The immunoassay technique promises to be an important biosensor and progress has already been made in the use of enzyme and bacterial bioluminescence detection systems. Once biosensor approaches are readily available, effective control measures can be easily applied.

BIOLOGICAL RESPONSES AND IMPACTS OF ENVIRONMENTAL STRESSORS

There are several examples of the effects of stressors from anthropogenic sources on biological systems. A conceptual model of the response of an organism to xenobiotic pollutants is shown in Figure 3. The initial exposure of xenobiotic pollutants and their uptake in biota precede their distribution and accumulation in organisms. Several internal systems respond to these distributed and concentrated substances. The site specific responses within the organism are controlled by these substances. For example, the basic mechanism of toxic action of phosphoruspesticides and carbamates is the inhibition of the acetylcholinesterase activity (AChE) in the nervous tissue. This results in the accumulation of the pollutants and eventual disruption of the nervous function.

Detoxification responses to these substances within organisms mainly occur in liver tissues and kidneys. Measurement of hepatic Mixed Function Oxygenase (MFO) activity is a sensitive indicator of the organisms ability to detoxify these pollutants. These enzyme systems are located in the endoplasmic reticulum (ER) of the hepatic cells. There are two phases in the metabolism of these xenobiotic substances. Phase I includes hydroxylation, de-ethylation, de-alkalization and de-amination. Primary oxidation products from phase I are excreted or transformed into water soluble products by the series of conjugating phase II enzymes (Gelboin 1980). Phase II is the real detoxification stage where the polarity of these substances increases. However, the main process is the oxidation reaction of phase I products.

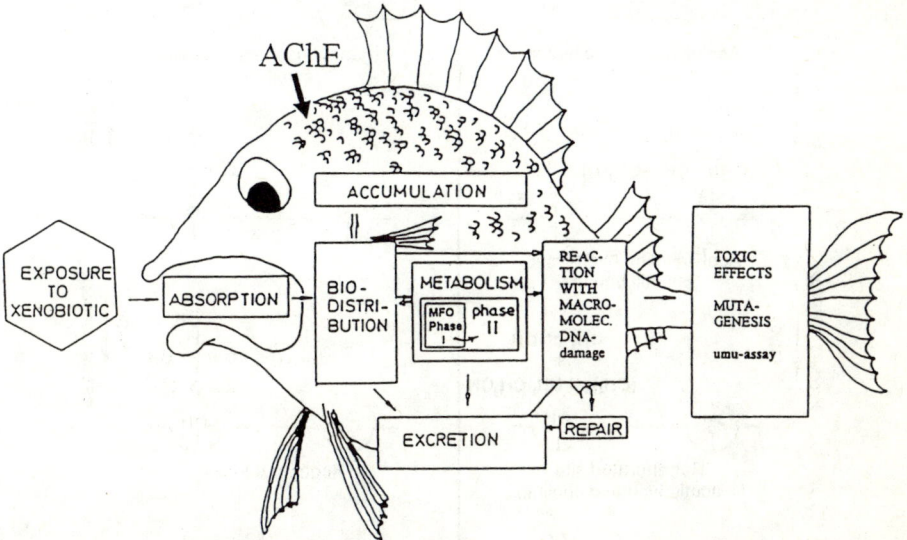

Figure 3. Responses by the biological system. AChE - Acetycholinesterase inhibition; MFO - Mixed Function Oxygenase activity.

In addition to the detoxification systems, there are other systems such as phagocytosis which measure and quantify the immunological defense activities. This procedure allows us to establish the stress of the immunological defence mechanism of organisms exposed to chemical pollutants. Similarly, information on Deoxyribo Nucleic Acid (DNA) damaging activity provides information on genotoxic effects of these pollutants through macromolecular reactions (Figure 3).

Enzymatic Inhibition (Cholinesterase)

The basic mechanism of the toxic action of organophosphates and carbamates includes accumulation of these pollutants at nerve endings with the eventual disruption of nerve functions. This process is outlined in Figure 4. Essentially, recovery of AChE activity in an organism that survives acute effects is dependent on the spontaneous but slow dephosphorylation of the inhibited site and the synthesis of new AChE. This monitoring approach is not only helpful in identifying 'hot spots', and pollution gradients near inputs, but also in characterizing the impact of pollutants. The effects of the substances as detected by cholinesterase inhibition are quantified as paraoxon equivalents and the inhibition constant (Table 2).

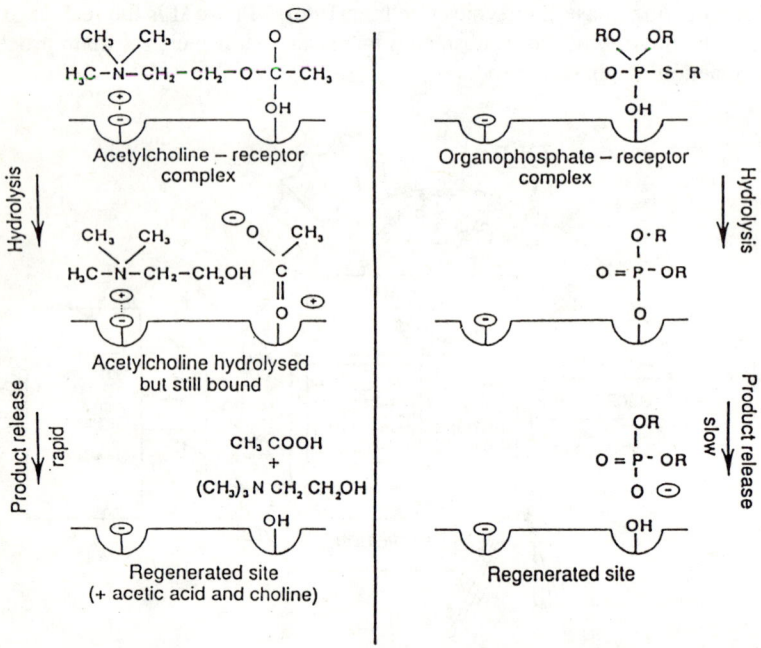

Figure 4. Mode of action of inhibition of Acetylcholinesterase (after Peakall 1992).

The phosphoruspesticides and insecticide carbamates inhibit cholinesterases at different levels. The well known cholinesterases are acetylcholinesterase and butyrylcholinestrerase. The first one was isolated from bovine and human erythrocytes and from electric eel (*Electrophorus electricus*); the second from serum from horses and humans. The strength of the inhibition effect is expressed as the inhibition constant. The inhibition constants of the most important phosphoruspesticides and insecticide carbamates have been published by Herzsprung et al. (1989). A standard protocol is currently available (Beutler 1993) which outlines the details of a procedure applicable to water samples and with slight modifications, for samples from organisms such as fish and worms.

Table 2. Paraoxon Equivalents and Inhibition-constant (Beutler 1993). 1 µg Paraoxon Equivalent/L = 1 µg/L Paraoxon-Ethyl/L).

Substance equivalent	Values expressed in 1 µg/L Paraoxon
Paraoxonmethyl	2.0
Aldicarb (sulfoxid)	0.8
Azinphosethyl	0.5
Azinphosmethyl	0.6
Carbofuran	0.6
Carbaryl	25.0
Chlorfenvinphos	60.0
Demeton-S-methylsulfon	800.0
Diazinon	13.0
Methamidophos	360.0
Omethoat	550.0
Parathionethyl	1.1
Paraoxonethyl	1.0
Parathionmethyl	2.1
Primicarb	90.0
Propoxur	3.0

Enzymatic Induction (Detoxification Activity MFO)

MFO (MFO/EC-No. 1.14.14.1) is widely used as monitoring tool to detect impacts of pollutants on fish (Addison et al. 1991a,b; Hansen et al. 1993a,c) and more recently on mussels (Michel et al. 1993). It is a sensitive indicator of the presence and toxicity of certain pollutants. MFO activity can be used as a quantitative measure of the extent of exposure to pollutants, and as a bioassay to detect sublethal effects. As shown in Figure 5, the most common substrate used for the determination of the MFO induction are 7-Ethoxyresorufin (EROD) and 7-Ethoxycoumarin (ECOD).

Factors affecting MFO-induction in fish include their physiology (sexual maturity or reproductive status and nutrional status) and other environmental factors such as pH, temperature and oxygen. Salinity apparently has no effect on the MFO induction (Pluta 1992). Application of MFO induction system for monitoring environmental samples of the coastal North Sea waters is shown in Figures 6 and 7.

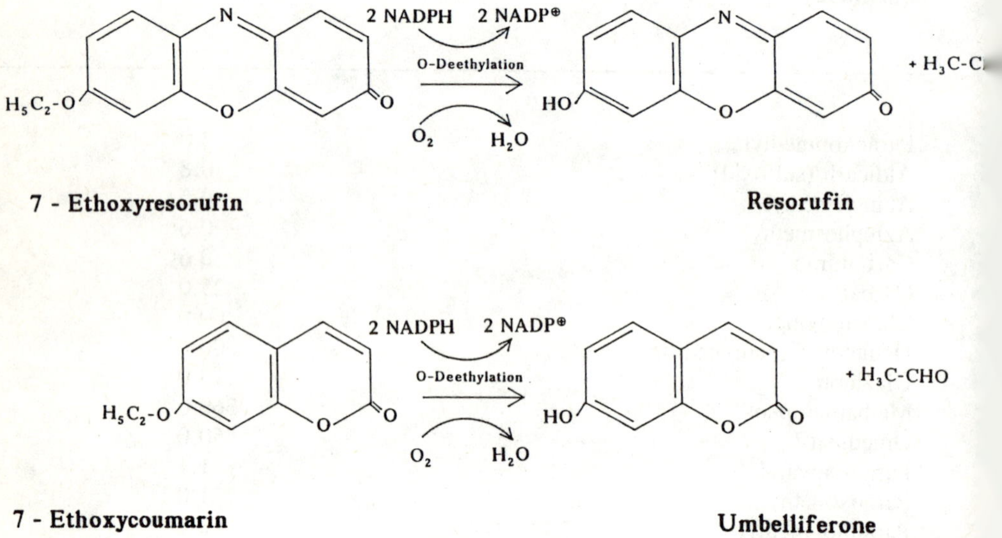

Figure 5. Determination of the Mixed Function Oxygenase activity using 7-Ethoxyresorufin (EROD) and 7-Ethoxycoumarine (ECOD). The measured products of the enzymatic reaction are Resorufin and Umbelliferone respectively.

Figure 6 shows elevated MFO activity near input sources and declining activity with distance. However, at a distant site (station 9) an observable increase in MFO activity was seen due to coastal zone input. Such data support the usefulness of the parameter as biomonitoring tool. Figure 6 also indicates that the sex of the fish is a contributing factor in MFO induction. Adult females (columns in the front rows) were spawning at that time and consequently the MFO levels were extremely low because of the elevated levels of steroid hormones. Male fish are therefore prime candidates for monitoring MFO activity. Figure 7 shows the usefulness of MFO-activity as a screening tool for detecting the areas where fish are induced due to anthropogenic impact. This approach is an ideal strategy for determining impact effects and managing ecosystem health. A standard protocol for MFO-induction is currently available including data on inter-laboratory comparisons (Stagg and Addison 1994). A measurement parameter that would provide information about the effects of multiple environmental components is of extreme importance in environmental monitoring. Such parameters need to be standardized for uniform application.

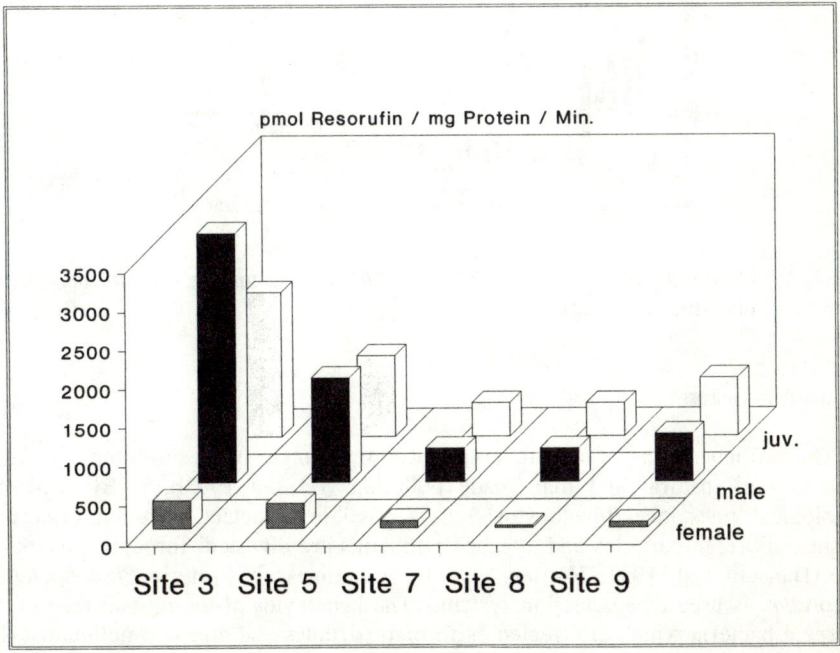

Figure 6. MFO activity (detoxification) in liver microsomes of fish (*Limanda limanda*) exposed to the pollution of the River Elbe estuary and sea as far as the Dogger Bank. The columns in the back are juvenile fish, the dark bars represent male fish, while the front row represents female fish. The product resorufin is calculated as mg microsomal protein (ERODM).

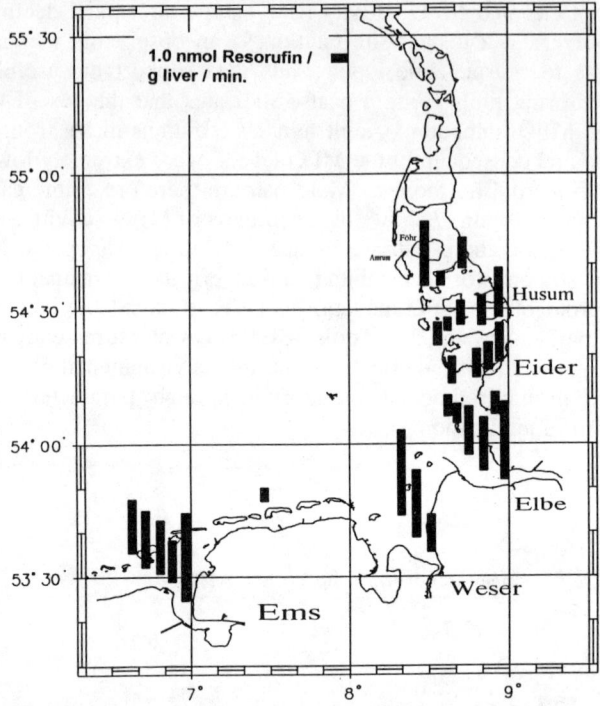

Figure 7. MFO activity in Flounder (*Platichthys flesus*). Sampling sites in the German Wadden Sea and estuaries (August 1989).

Immunosuppression

The resilience of an organism is influenced by changes in the environment which are due to both natural and man made pollution (Hansen 1993b,c). By recording immunological resistance (phagocytosis) it is possible to detect pollution effects on organisms. Foreign particles and attached pollutants are digested through phagocytic activity (Hansen et al. 1991; Hansen 1992c). A luminescent bacteria *Photobacterium phosphoreum* is used as a detection system. The hemocytes of the mussel feed on the luminescent bacteria which are treated as foreign particles and attached pollutants. The phagocytic activity is directly related to a decrease in luminescence (Figure 8). Hemocytes play a major part in the immunological defence system of many invertebrates. Measurement of phagocytic activity offers ample opportunity for detecting unknown biotoxins through their influence on mussel immunology.

A good example for application of the phagocytic bioassay is the detection of algal toxins in marine and freshwater environments. The number of blooms of toxic algae has increased in recent time because of eutrophication of freshwater and marine ecosystems. These blooms pose a threat not only to natural ecosystems and fisheries, but also to human health. Since biotoxins excreted by algae have not been identified, a bioassay system such as phagocytosis could be extremely useful for detecting their effects. Environmental effects monitoring using this assay to record the integrated effects of biotoxins would be highly useful.

The phagocytosis analysis from an experiment in which mussels were fed with toxic algae (*Chrysochromulina polylepis*) and non-toxic algae (*Isochrysis galbana*), is reported by Krumbeck et al. (1994). Here the phagocytosis was performed using a microplate technique. Currently, developments in this direction are underway using FITC-conjugated yeast cells and the final measurements of the phagocytosis index using a fluorescence microplate reader. This procedure has potential for becoming an on-line immunotoxicity biosensor system.

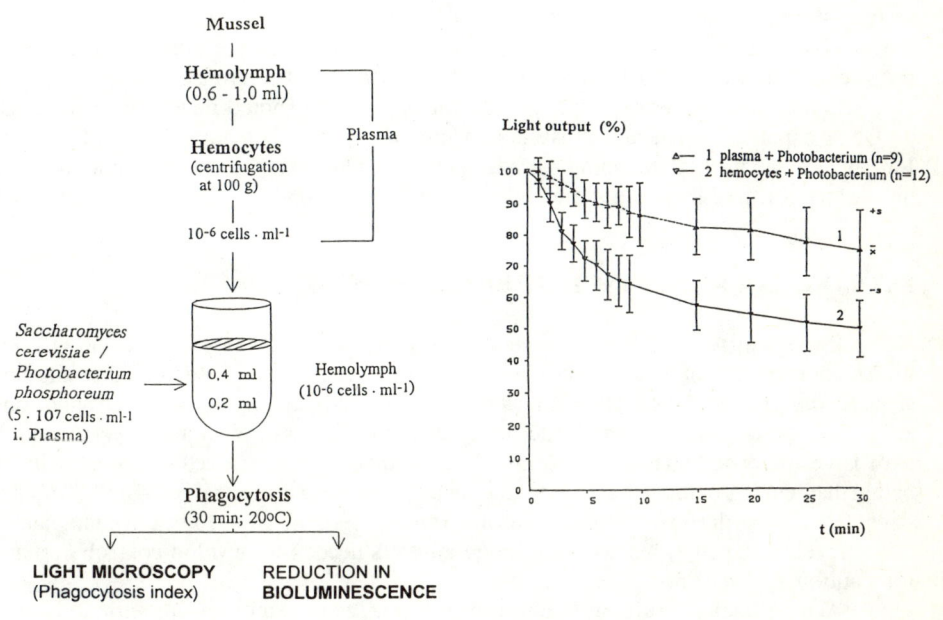

Figure 8. Protocol for studying immunosuppression responses in mussels (*Mytilus edulis*).

Genetic Damage

Monitoring of the effects of environmental toxic and genotoxic substances has gained increased importance in recent times. Assessment of environmental samples for the presence of genotoxins (carcinogens and mutagens) has become a valuable approach for evaluating the genotoxic potential of environmental pollutants. Several biomonitoring approaches currently exist. In Germany there are three protocols for environmental genotoxicity assessment. These are: DNA unwinding (Shugart 1988; Herbert 1990; Herbert and Zahn 1990, Herbert and Hansen 1992), AFE (alkaline filtration elution) technique (Waldmann 1993) and umu-assay (Oda et al. 1985). The German Institute for Standardisation (DIN) has recently adapted the umu-C assay as an official protocol for monitoring environmental genotoxicity.

The DNA unwinding assay is a promising tool for detecting DNA damage due to environmental genotoxins in aquatic animals. Extensive work is currently underway to develop a basis for uniform application of this bioassay to environmental samples. Recent work (Figure 9) on the application of this protocol (Herbert 1990) has clearly indicated its usefulness as a genotoxic monitoring tool. Figure 9 shows DNA damaging activity in mussel hemocytes (*Mytilus edulis*) after 96 hour static exposure to genotoxic chemicals. Figure 10 indicates that the DNA damage in *Mytilus edulis* is increased in the stressed area in the harbour compared to an unpolluted reference station.

In the future, environmental monitoring approaches should include measurements of DNA damage in fish and mussels and umu-assay of other environmental samples. Finally, environmental mapping of hot spots for genotoxic potential should be made mandatory for assessing and maintaining ecosystem health.

ECOLOGICAL RELEVANCE OF BIOLOGICAL RESPONSES

Biological responses in ecosystems to environmental stress could provide us with information that signals potential damage. These responses, if perceived at an early stage, could prevent the eventual deterioration of the ecosystem. On the other hand, once ecosystem damage has occurred, the remedial action process for their recovery could be expensive and pose logistical problems. Prevention of ecosystem deterioration is always better than curing damaged ecosystems. Ideally, "early warning signals" of ecosystem stress, as can be derived using biosensors (Hansen 1992b), would not only indicate the initial level of damage, but would provide answers needed to develop control strategies (precautionary measures).

Acute toxicity results in organism selection, genotoxicity results in mutagenicity and physiological impairments (genetic disease syndromes), and induction of MFO (biochemistry) provides data on the effects of specific chemicals (warning signals). Genotoxic damage endpoints have high ecological significance as they relate directly to reproductive potential. Stress responses at population levels have direct ecological implications even though they exhibit low specificity (Figure 11). There should therefore be a harmonized ecosystem assessment approach where the overall information (high specificity to low specificity) is considered in parallel.

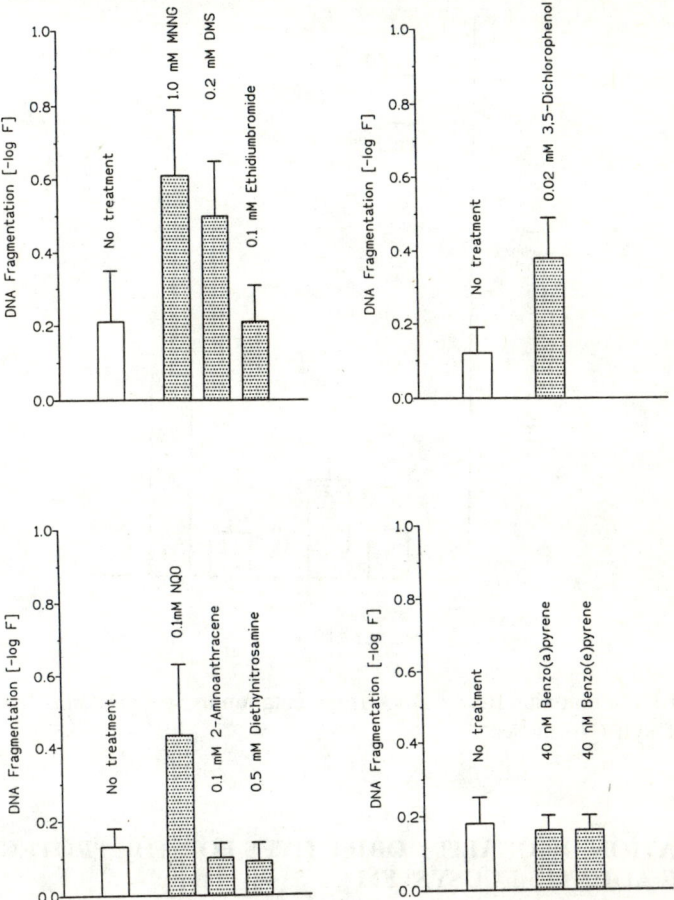

Figure 9. Genotoxic effects: DNA fragmentation in *Mytilus edulis* from laboratory studies. MNNG = N-Methyl-N'-nitro-N-Nitrosoguanidine; DMS = Dimethylsulfate, NQO = 4-Nitroquinoline-1-Oxide.

In considering the impact of either natural stress or man made stress we always encounter detoxification, disease defence, regulation and adaptation processes. This situation makes the assessment approach rather complicated. On the other hand, symptoms analysis, including functional (behaviour, activity and metabolism) and structural changes in organisms (cellular, tissue and organs), have high ecological assessment potential.

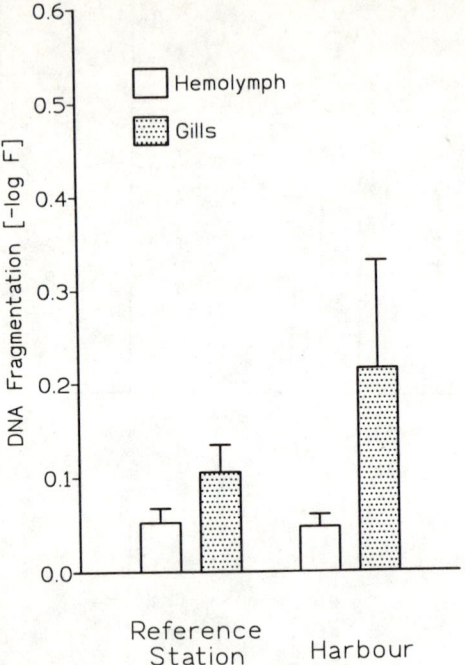

Figure 10. Genotoxic effects: DNA fragmentation in *Mytilus edulis* from two sites at the Island of Sylt (North Sea).

DERIVATION OF QUALITY OBJECTIVES FOR THE PROTECTION OF THE AQUATIC ECOSYSTEM

Substances dangerous to ecosystem and human health, as well as substances suspected of being such, must, in so far as possible, be kept out of the environment and away from waterways. The concept of establishing water quality objectives can be applied when the goal is to manage the water in such a manner that it is available for a wide variety of uses in addition to being the source of potable water. Among these uses are commercial and sport fishing and general outdoor recreation.

Water management practices and objectives play a crucial role in determining the extent to which individual surface waters, or parts thereof, are to be protected. This report thus also directs its attention to the importance of establishing water quality objectives which can serve as management tools for protecting the waterways as a valuable natural resource supporting both commercial and sport fishing and outdoor recreation as well as general ecosystem health.

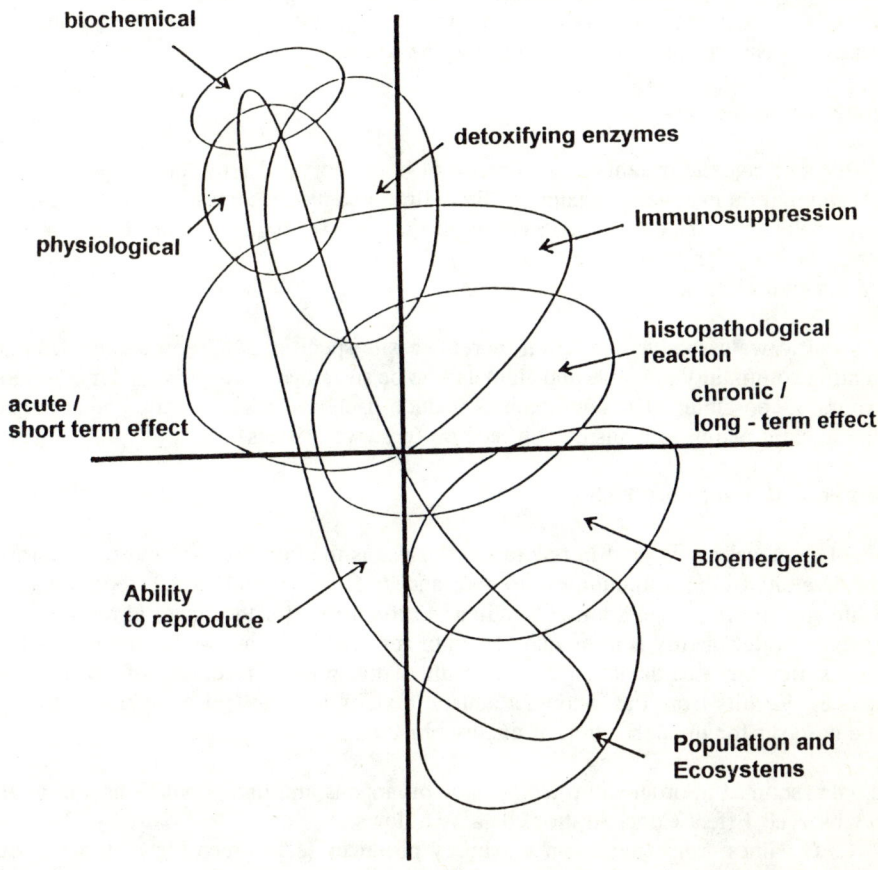

Figure 11. Ecological response levels of biological markers.

Water quality objectives are defined as being specific values that serve to specify the water quality to be attained in order to sustain a specified beneficial use. The water quality objectives based on this concept are also important as a planning instrument for setting use priorities for a given waterway and in establishing water quality standards to be attained and thereafter enforced. They place in the hands of water management officials, objective means by which regulations and ordinances pertaining to water quality can be established and enforced.

Criteria are defined as concentrations of substances in water or organisms which, when reached or exceeded, can be scientifically proven to cause adverse effects. Criteria serve as the basis for the stipulation of quality objectives. In relation to dangerous substances, criteria take into consideration the following:

Aquatic Communities

- Effects on aquatic organisms as a result of both short and long term exposure.
- Accumulation in aquatic organisms. By definition a bioconcentration in the organisms are relevant, if the bioconcentration factor is higher than BCF = 100 (bioconcentration factor = BCF values).
- Mutagenic effects.

In a specific water body, or section thereof, a site-specific, self reproducing, and self regulating community of plants and animals is to be sustained or restored under conditions as natural as possible. The communities include bacteria, lower aquatic plants (algae), higher aquatic plants, organisms fish feed on (e.g. water fleas).

Commercial and Sport Fishing

The management goal in relation to this use is the maintenance and/or restoration of the diversity of the communities of importance to fishery, such as fish-food organisms and site-specific fish populations, including the conditions for their natural reproduction. Moreover, water quality is to be maintained or restored in such a way that human beings do not suffer any health damage as a result of the pollutant content of the fish they consume. Results from the following aquatic toxicity tests should be available to serve as the base set for the derivation of quality objectives.

The data required in order to protect aquatic organisms and fish populations are NOEC (No Observed Effect Concentration) data as follows:
- NOEC values from studies on a primary producer (e.g. green algae in a test over several cell generations - 72 hours), using a recognised test method.
- NOEC values from studies on a primary consumer (e.g. *Daphnia magna* in a 21 day reproduction test), using a recognized test method.
- NOEC values from studies on a secondary consumer (e. g. test on one fish species of at least 28 days duration, including reproduction, alternatively an early life stage test),

using a recognised test method.
- NOEC values from studies on a reducer (e.g. the bacterium *Photobacterium phosphoreum* or the bacterium *Pseudomonas putida* in a test over several cell generations -16 hours), using a recognised test method.

Recognised test methods include methods developed by international agencies such as DIN, ISO, OECD or EU. As a general principle, the lowest test result for the most sensitive species is to be used as the starting point for the risk assessment and for the derivation of the water quality objectives in relation to manganese. There are some reports in the literature on the effects of manganese on metabolism and thus such data should also be taken in account. The toxicity data used for the risk assessment have to be examined critically with respect to validity and relevance. A compensation factor $F^\wedge$ has been introduced to take into account the uncertainty associated with extrapolating results to the real environment. $F1=0.1$, if NOEC values are available for each of the trophic stages (primary producer, primary consumer, secondary consumer, reducer). Furthermore, a compensation factor $F2$ ($F2 = 0.1$) can be used, if additional risk factors exist such as bioaccumulation and genotoxic potential or mutagenicity.

In order to achieve the goal of having a water supply which is safe for human consumption, it is necessary to establish quality objectives for waterways as laid down in public health guidelines and ordinances (e.g. Germany's Drinking Water Ordinance). The values of the quality objectives for waterways are determined as follows:

Using the limits value for drinking water (TG) and the reduction factor F:

$$QO = F \cdot TG$$

For xenobiotic substances the reduction factor F may not exceed a value of 0.5 because the concentrations of substances in water bodies vary, depending on the discharge and other factors such as substance-specific seasonal fluctuations. The water quality objectives must be the more stringent in cases where substances cannot be eliminated through conventional water treatment processes.

Water quality objectives for deleterious anthropogenic substances are listed in Tables 3, 4 and 5.

To establish quality objectives aimed at the safeguarding of commercial and sport fishing, a tiered procedure is used. Where legally binding guide or limit values (maximum permissible quantities = w_F, see Table 4) for the pollutant content in fish have been established to protect human health, these are employed together with the bioaccumulation factors (BCF) to derive quality objectives (QO). The BCF values are to be obtained from the literature or determined in laboratory experiments, using OECD Guideline 305E. BCF values below 100 are not taken into account as such substances are not assumed to accumulate in fish to any dangerous degree.

Table 3. Water Quality Objectives for the protection of inland waters against dangerous substances (n.r.=not relevant; r.=relevant, but no data available; dw=dry weight).

	Aquatic communities [µg/l]	Fishery [µg/l]	Drinking Water Supply [µg/l]	Sediments [µg/kg dw]	Water Quality Objectives [µg/l]
2-Chloroaniline	0.1	n.r.	1.0	n.r.	0.1
4-Chloroaniline	0.01	n.r.	0.1	n.r.	0.01
3,4-Dichloroaniline	0.1	r.	0.1	n.r.	0.1
1,4-Dichlorobenzene	10	r.	1	n.r.	1
Hexachlorobenzene	0.001	0.001	0.1	40	0.001
Hexachlorobutadien	0.01	r.	1	n.r.	0.01

Table 6 shows clearly that there is a safety buffer between laboratory and field data. The determination of the NOEC data or threshold values, serves to protect the fundamental functions of ecosystem health such as metabolism, growth, reproduction and regulatory performance. The water quality objectives cannot be applied universally but are to be evaluated for each individual body of water. The water quality objectives help determine whether sufficient measures have been taken to protect the individual body of water and its uses from "dangerous substances". In order to protect the functioning and stability of an aquatic ecosystem, or to actually re-establish these qualities, ecological knowledge is necessary concerning the interaction of the function and structure at different levels of biological organization as well as nutrient cycles and energy fluxes. This is the only way of recognize the multifactorial burden on a ecosystem caused by hazardous substances, and of establishing restoration measures.

Where water quality objectives can be implemented by way of water management measures, e.g. water management plans, this is a national task. If implementation can only be achieved through substance-related regulations, e.g. prohibition of substances, this is primarily a task to be tackled internationally. Further intensification of the research and international cooperation with respect to the derivation of water quality objectives for the various uses is urgently required. However, in view of the many thousands of substances reaching water bodies from both anthropogenic and natural sources, the application of quality objectives has to remain confined to a manageable, finite number of individual substances. The monitoring and evaluation of water quality therefore has to be supplemented by appropriate parameters and especially by suitable biological tests covering not only acute, but also longer-term effects on aquatic organisms. For the derivation of quality objectives using NOEC-data from the laboratory and the field some longer-term effects data are listed in Table 6.

Table 4. Maximum permissible pesticide residues (Pflanzenschutzmittel-Höchstmengen-Verordnung-PHmV) and the maximum permissible pollutant residues in food stuffs (Schadstoff-Höchstmengenverordnung-SHmV), bioconcentration factor (BCF), and water quality objectives for fisheries.

	Maximum permissible quantities PHmV / SHmV [mg / kg]	Bioconcentration Factor (BCF)	Water Quality Objectives - Fishery [mg / l]
Lindane	2.0	350	0.6
α, β, δ - HCH	0.5	100,000	0.0005
Endrine	0.01	3,000	0.003
Heptachloroepoxyd	0.01	2,000	0.005
PCB IUPA - Nr.			
28	0.2	100,000	0.002
52	0.2	100,000	0.002
138	0.3	100,000	0.003
180	0.08	100,000	0.0008

RECOMMENDATIONS

From the practical standpoint there are several concepts that can be used as powerful tools for management of ecosystem health. There are grounds for determining the water quality requirements to protect inland surface waters against dangerous substances. These water quality requirements have to be derived from scientific approaches, for example NOEC data. There is currently a battery of test approaches available for assessing the trophic levels of food chain using bacteria, algae, crustacean and fish. In addition, we need to apply other assessment concepts which include effects parameters such as AChE-Inhibition, MFO-induction, and the Phagocytosis Index. Among the existing genotoxicity assessment approaches, the umu-C-assay has become a DIN standard (DIN 384/\5, part 3) and will become part of the Federal Regulatory Programme after the Federal Water Act (§ 7a WHG).

Table 5. Derivation of Water Quality Objectives[1,2] for the Protection of Inland Surface Waters Against Heavy Metals (n.r. = not relevant).

	Aquatic communities [mg/kg dw]	Aquatic communities [µg/l]	Fishery [µg/l]	Drinking Water Supply [µg/l]	Sediments (susp. solids) [mg/kg dw]
Pb	100	3.4	5	40	100
Cd	1.2	0.07	1	5	1.5
Cr	320	10	n.r.	50	100
Cu	80	4	n.r.	50	60
Ni	120	4.4	n.r.	50	50
Hg	0.8	0.04	0.1	1	1
Zn	400	14	n.r	3,000	200

1 - Derivation of Water quality Objectives (1989) Concept for the Derivation of Quality Objectives for the Protection of Inland Surface Waters Against Dangerous Substances, Developed by the Federal Government/Federal States' Working Group "Quality Objectives" (BLAK QZ).
2 - Drinking Water Ordinance of 22 May 1986, BGB1. I, p. 760

Table 6. Water Quality Objectives for the Protection of Aquatic Communities in comparison with NOEC-Data (NOEC = No Observed Effect Concentration) from laboratory and field investigations.

	NOEC Laboratory [µg/l]	NOEC Field [µg/l]	Water Quality Objectives FRG [µg/l]	Netherlands Maximum Tolerable Risk Level — Limit value (guideline) [µg/l]	Netherlands Maximum Tolerable Risk Level — Intervention value [µg/l]	U.S. EPA Extropolation [µg/l]
Azinphosmethyl	0.1	0.25	**0.01**	0.0007	0.02	0.01
Parathionethyl	0.002	0.1	**0.0002**	0.00005	0.005	0.002

REFERENCES

Adams S.M., Shepard K.L., Greeley M.S. Junior, Jimenez B.D., Ryon M.G., Shugart L.R., McCarthy J.S., Hinton D.E. (1989) The use of biomarkers for assessing the effects of pollutant stresses on fish. Mar. Env. Res. 28: 459-464

Addison R.F., Hansen P.-D., Wright E.C. (1991a) Hepatic mono-oxygenase activities in American Plaice (*Hippoglossoides platessoides*) from the Miramichi Estuary, N.B. Canadian Technical Report of Fisheries and Aquatic Sciences No. 1800

Addison R.F., Pluta H.J., Hansen P.-D. (1991b) Absence of MFO-Induction in several species of North Sea flatfish (*Platichthys flesus*, *Limanda limanda* and *Solea solea*) following treatment with PCB substitute with Ugilec 141, Mar. Env. Res. Vol. 31:137-144

Beutler H.O. (1993) Cholinesterase-Hemmtest. In Biochemische Methoden zur Schadstofferfassung im Wasser - Möglichkeiten und Grenzen. Hrsg. Fachgruppe Wasserchemie in der GDCh, VCH Verlagsgesellschaft, Weinheim

Gelboin H.V. (1980) Benzo (a) pyrene metabolism, activation, and carcinogenesis: Role and regulation of mixed-function oxidases and related enzymes. Physiol. Rev. 60: 1107-1165

Hansen P.-D., Bock R., Brauer F. (1991) Investigations of phagocytosis concerning the immunological defence mechanism of *Mytilus edulis* using a sublethal luminescent bacterial assay (*Photobacterium phosphoreum*). Comp. Biochem. Physiol. Vol. 100C, No 1/2: 129-132

Hansen P.-D. (1992a) Suborganismische Testverfahren - Anwendungsbereich und Bewertungsfrage. In Steinhäuser K.G., Hansen P.-D. (eds.) Schriftenreihe Wasser -Boden- und Lufthygiene, Bd. 89, Gustav. Fischer Verlag Stuttgart, Jena, New York (1992)

Hansen P.-D. (1992b) On-Line Monitoring mit Biosensoren am Gewässer zur ereignisgesteuerten Probenahme. Acta hydrochim. Hydrobiol. 20, 2: 92-95

Hansen P.-D. (1992c) Phagocytosis in *Mytilus edulis*, a system for understanding the sublethal effects of anthropogenic pollutants and the use of AOX as an integrating parameter for the study of equilibria between chlorinated organics in *Dreissena polymorpha* following long term exposures. In: Neumann and Jenner (eds.) The Zebra Mussel *Dreissena polymorpha*. Limnologie aktuell, Vol 4. Gustav Fischer Verlag, Stuttgart, Jena, New York (1992)

Hansen P.-D. (1993a) Enzymatische Verfahren zur Erfassung der Biotransformation (Entgiftungsaktivität) in der Fischleber - ein Beitrag zum biologischen Effektsmonitoring. In Biochemische Methoden zur Schadstofferfassung im Wasser - Möglichkeiten und Grenzen. Hrsg. Fachgruppe Wasserchemie in der GDCh, VCH Verlagsgesellschaft, Weinheim (1993)

Hansen P.-D. (1993b). Schadstoffwirkungen auf das Immunsystem. In: Biochemische Methoden zur Schadstofferfassung im Wasser - Möglichkeiten und Grenzen. Hrsg. Fachgruppe Wasserchemie in der GDCh, VCH Verlagsgesellschaft, Weinheim (1993)

Hansen P.-D. (1993c) Regulatory significance of toxicological monitoring by summarizing effect parameters. In: Mervyn Richardson (ed.) Ecotoxicology Monitoring. VCH Publishers, New York (1993)

Herbert A., Zahn R.K. (1990) Monitoring DNA damage in *Mytilus galloprovincialis* and other aquatic animals. II. pollution effects on DNA denaturation characteristics. Angewandte Zoologie 77, 1:13-33

Herbert A. (1990) Monitoring DNA damage in *Mytilus galloprovincialis* and other aquatic animals. III. A case study: DNA damage in fish from a Florida Marsh. Angewandte Zoologie 77, 2:143-150

Herbert A. & Hansen, P.-D. (1992) Erfassung des erbgutverändernden Potentials von Gewässern durch Messung von DNS-Schäden mittels alkalischer Denaturierungsverfahren. In Steinhäuser K.G., Hansen P.-D. (ed.) Schriftenreihe Wasser-, Boden- und Lufthygiene, Bd. 89, Gustav Fischer Verlag; Stuttgart, Jena, New York (1992)

Herzsprung P., Weil L., Quentin K.E. (1989) Bestimmung von Phosphorpestiziden und insektiziden Carbamaten mittels Cholinesterasehemmung, Mitt 1: Hemmwirkung von Phosphorpestiziden und insektiziden Carbamaten auf immobilisierte Cholinesterase. Z. Wasser-Abwasser-Forsch. 22,:67-72

Krumbeck H., Elbrächter M., Herbert A. and Hansen P.-D. (1994) Effects of algae toxins on the phagocytic activity of mussel hemocytes. In: Lassus (ed.) 6th International Conference on Toxic Marine Phytoplankton, Nantes 1993

McIntosh R.P. (1981) Succession and ecological theory. In: Wset D.C., Sugart H.H., Botkin D.B. (eds.) Forest Succession: Concepts and Application. Springer, New York pp:10-23

Michel X.R., Cassand P.M., Narbonne J.F. (1993) Activation of benzo[a]pyrene and 2-aminoanthracene to bacteria mutagens by mussel digestive gland postmitochondrial fraction. Mutation Research 301:113-119

Oda Y., Nakamura S.I., Oki I., Kato T., Shinagawa H. (1985) Evaluation of the new system (umu-test) for the detection of environmental mutagens and carcinogens. Mutation Research 147:219-229

Odum E.P. (1969) The strategy of ecosystem development. Science 164: 262-270

Peakall D. (1992) Biomarkers of the nervous systems. In Animal Biomarkers as Pollution Indicators. Ecotoxicology Series I, Chapman & Hall pp: 20-45

Pluta H.-J. (1992) Joint Monitoring in der Nordsee (North Sea Task Force, NSTF). MFO-Untersuchungen als Teil des Deutschen Beitrages zum Quality Status Report (QSR). In Steinhäuser K.G., Hansen P.-D. (eds.) Schriftenreihe Wasser-, Boden- und Lufthygiene, Bd. 89, Gustav Fischer Verlag; Stuttgart, Jena, New York (1992)

Quality Status Report of the North Sea (1993) Report on Sub-Region 7a, Ed. R. Salchow, Bundesamt für Seeschiffahrt und Hydrographie (BSH), 124

Shugart L.R. (1988) Quantitation of chemically induced damage to DNA of aquatic organisms by alkaline unwinding assay. Aquatic Toxicology 13: 43-52

Stagg R.M. and Addison R.F. (1994) An inter-laboratory comparison of measurements of ethoxyresorufin O-de-ethylase activity in dab (*Limanda limanda*) liver. Mar. Env. Res. (in press)

Waldmann P. (1993) Die Alkalische Filterelution als Methode zur Erfassung von gentoxischen Potentialen in Gewässern. - Dissertation Universität Mainz, Fachbereich Biologie

RAPPORTEUR'S REPORT

Valery Forbes
Institute for Biology and Chemistry
Roskilde University, P.O. Box DK-4000 Roskilde
Denmark

SUMMARY OF PRESENTATIONS

The objective of this session is to identify scientifically robust indices that can be used to detect change in the state of ecosystems (or their component parts) relative to reference points in space or time. The authors present a variety of criteria for assessing ecosystem structure and function and discuss the strengths and limitations of each. The concept of scale is a recurring theme, particularly variability across time and space and across levels of biological organization (i.e. biosphere - region - ecosystem - community - population - individual - physiological - cellular -molecular).

Gray portrays the 1990's as a period of increasing public awareness of environmental issues. Likewise, he notes that anthropogenic effects on ecosystems are being detected at very large scales, and new techniques suggest that even the open oceans may show evidence of human impact. Gray also presents case studies to demonstrate that indices of impact applied to the same system can differ in non-obvious, yet significant ways. For example, estimated impacts of the Ekofisk oil platform on benthic community structure ranged from an area of 3 km^2 to 27 km^2, depending on the type of statistical analysis applied to the data. An important theme of his presentation was the need to place greater emphasis on designing statistically sound sampling programs and on incorporating statistical power analyses into assessment schemes.

An array of techniques is rapidly being developed to detect responses of molecular, cellular and physiological systems to human impact. These so-called biomarkers are specifically suited for detecting change in response to chemical pollutants. Vasseur presents examples of the types of molecular biomarkers (e.g. changes in the structure and function of DNA) that can be used to predict carcinogenic risk and discussed the limitations inherent in such measures.

Although impacts of human activities may be detected at the ecosystem level of organization, it is generally difficult to identify cause-effect relationships at this level. Dethlefsen outlines a suite of changes that have occurred in the North Sea on regional and local scales and at many different levels of the biological hierarchy. These include decreases in the abundance of planktonic diatoms and benthic bivalves, increases in the abundance of planktonic flagellates and benthic polychaetes, increases in the frequency of toxic blooms and fish diseases, and malformations of pelagic fish embryos. He stressed the difficulties in estimating the extent to which such changes have arisen in response to anthropogenic versus natural factors and suggested that indices of ecosystem change may be useful as qualitative, but not quantitative, measures of human impact.

For the purpose of understanding the mechanistic bases of ecosystem change, linking responses of ecosystem components occurring at different levels of spatio-temporal and biological organization is critical. For the purposes of assessing or predicting system change, the elements and processes upon which studies are focused should be determined by the objectives of the study. Finding mechanistic links to higher or lower levels of organization may not be as important as ensuring that a suite of indicators is applied which are able to detect change at several levels. However, because we need not only to estimate the potential for change but the seriousness of the change as well, linkages across scales of measurement are ultimately necessary.

While it is recognized that changes in ecosystem components vary in sensitivity and information content, such differences remain to be explicitly quantified. For example, an important assumption of the application of various molecular, cellular and physiological biomarkers is that these parameters are more sensitive (i.e. respond after a shorter duration of impact or to a lower degree of impact) than changes at the community or ecosystem level. However, this assumption has not been rigorously tested.

Whereas ecologists recognize that critical ecosystem elements and processes operate at different scales, we know very little about how linkages across scales vary among ecosystems. Costanza discusses a new research program that he and coworkers designed to quantify linkages among ecosystem components across spatial and temporal scales. The Mutltiscale Experimental Research Center consists of artificial ecosystems built in a range of sizes and incorporating different degrees of complexity. Models predicting ecosystem performance are constructed for each system; model predictions are tested within and across systems as well as in natural ecosystems. This program aims to provide an explicit understanding of the role of spatial and temporal constraints on ecosystem performance and to determine the extent to which linkages among scales are ecosystem dependent.

It is widely agreed that indices of ecosystem change have been far too heavily concerned with chemical measures to the detriment of important biological criteria. Karr provided convincing evidence that the narrow focus on chemical water quality in attempts to fulfil mandates of environmental legislation has not succeeded in protecting ecosystem elements and processes. He attributed this to a historically narrow focus of environmental legislation on human health and to the fact that chemical parameters can often be precisely measured. Karr emphasizes that because ecosystems are multivariate, a multivariate approach for ecosystem assessment is required. He also argues that the standards against which ecosystem impacts are assessed should be regional, rather that universal, because systems vary geographically. Karr has applied these principles in developing an index of biotic integrity. The index, which was originally developed to evaluate human impacts on American midwestern streams, is calculated as a sum of 12 biological attributes (i.e. species types and numbers, trophic position of resident species, and condition) of a resident fish assemblage. This approach is being expanded to other assemblage and ecosystem types and provides a substantial improvement over chemically-based water quality measures.

DISCUSSION

It is a fact that although science plays an important part in the assessment of ecosystem change, values and acceptability are critical issues in ecosystem health assessment. Thus enhancing communication between scientists and other concerned groups was a central theme in many of our discussions. A possible scenario for integrating the important players would be to involve scientists, policy makers, and the public (including representatives from special interest groups) in defining acceptability criteria a priori (i.e. before a management decision is taken). The entire group would then decide on a feedback loop that includes a clear strategy if the acceptability criteria are exceeded - again a priori. This type of approach has two benefits. It clarifies the role of science in environmental management, and it forces us as a society to decide in what type of world we want to live.

CONCLUSIONS

Major points emerging from the discussion session were:

1. Biological indices of change must incorporate a range of spatio-temporal and biological hierarchical scales.
2. More attention should be focus on proper statistical design of sampling schemes and on quantification of the statistical power of the detection program.
3. Investigating links from higher to lower levels of biological organization can provide explanations of the mechanistic basis of observed changes. Thus, a traditional reductionist approach can be effective for determining cause-effect relationships of ecosystem change.
4. Developing quantitative links from lower to higher levels of biological organization, particularly for predictive purposes, is a much more difficult (some would say fundamentally unattainable) goal. However, the idea that one can use changes at lower levels of biological organization (e.g. molecular or physiological) as predictive indicators of changes occurring at the ecosystem level is an important assumption of the biomarker approach.
5. Evidence from various case studies indicates that changes in ecosystem structure can occur without corresponding changes in ecosystem function, but that changes in function rarely occur without corresponding changes in structure. The implication is that protecting ecosystem structure should also protect ecosystem function.
6. We need to develop tighter feedback loops among scientists, policy makers, the public and various non-governmental representatives during all stages of management actions. An important part of this is to achieve a priori agreement on acceptability criteria as well as responses to be taken if the criteria are not met.

The themes emphasized by this session on quantitative indices for ecosystem health assessment highlight several key areas for future focus. They can be summarized by the following questions:

1. To what extent can we quantify linkages across spatial and temporal scales and across biological hierarchical levels?
2. Under what circumstances can ecosystem function alter while ecosystem structure is maintained?
3. How well do the available approaches and criteria for detecting ecosystem change transfer among different types and sizes of ecosystems?
4. Do responses occurring at very low levels of biological organization provide a more sensitive measure of ecosystem change than responses occurring at high levels of organization?

Session III

DIAGNOSTIC APPROACHES

INTRODUCTION

Chair: Kenneth Sherman

The session will deal with reviewing and discussing strategies for measuring changes in large biome areas of the globe and smaller more manageable units at the regional ecosystem scale. Examples of biomes under stress include rivers, semi-enclosed seas, forests, deserts, and large marine ecosystems.

At present, efforts for reducing ecosystem stress and promoting the sustainability of resources are underway by single nations. Unfortunately, on a global scale, no single international regime has been empowered to assess, monitor, and manage natural resources in a strategic context that reconciles the needs of individual nations with those of the community of nations in taking appropriate mitigating actions in response to disturbances. The need for a regional approach to implement ecological research, monitoring, and stress mitigation at less than the global level has been recognized. The ecological concept that critical processes controlling the structure and function of biological communities can best be addressed on a regional basis has been argued cogently by a number of ecologists.

Although large ecosystems are not amenable to the usual controlled experimental approach, they are perfectly amenable to the comparative method of science as described by Mayr. It is within this context that the case study approach can be most instructive as a comparison among terrestrial, marine, and freshwater ecosystems in different ecological states. Findings from comparative studies may be used to forecast or predict future conditions among similar systems. For example, the recent observations that the biodiversity of the fish community of the Black Sea ecosystem has been reduced from the stress of massive mortalities resulting from a long-series of eutrophication-driven events like harmful algal blooms and associated anoxic and biotoxic conditions may be pertinent to other ecosystems. It has been reported that these conditions were accompanied by significant increases in the biomass of jellyfish. Similar observations have recently been reported for the Yellow Sea ecosystem where depleted fish stocks may have been undergoing replacement by large biomass of jellyfish during the past several years.

The session will explore the extent to which similar replacement events have been observed in changing the productivity and community structure of forest, river, and desert ecosystems.

12. DEGRADATION AND REHABILITATION OF RIVERS: A NOTE ON THE ECOSYSTEM APPROACH

Wim Admiraal
Department of Aquatic Ecotoxicology
University of Amsterdam
Kruislaan 320, 1098 SM Amsterdam
The Netherlands

INTRODUCTION

River systems, with their many ramifications and their complex hydrology, provide a special habitat for aquatic and semi-terrestrial communities of the riparian zone. The ecology of large river systems has been reviewed (Davies and Walker 1986; Whitton 1984) and it was found that our understanding of these waters, so vitally important for all the continents, lags much behind that of stagnant water bodies. Much of our insight in the functioning of lotic communities stems from research on smaller streams, where the interactions of the hydrology, the chemistry and the biology of such small streams and their catchments have been thoroughly analyzed (Likens 1992). Research efforts on large rivers may have been limited partly by methodological problems and partly by tradition. It is difficult to sample a large river and cause and effect relationships may extend over a wide geographic scale.

It may seem impractical to use river systems in a search for techniques to assess the "well-being" of ecosystems. Basic data on the composition and functioning of the biota may be missing and the dynamic nature of rivers may not allow for reflections on natural ecosystem stability and man-made perturbation. Yet, it was recognized over the last few years that large rivers and their catchments are necessary units for environmental management because of the intense use and abuse of these systems. Action plans have been designed for the rivers Rhine (Europe), St. Lawrence (North America) and Ganges (Asia), and many others. The ongoing construction of dams which neglect the ecological and cultural basis of local economies, such as on the R. Narmada (India), has given rise to political conflicts at the highest level. In Europe, the chemical and toxicological quality of river water is now seen as a touchstone for environmental management (Newman et al. 1992).

Actions in river management require well-structured input by ecologists. In this paper some observations on the ecological degradation of the rivers Rhine and Don will be described, and the potential for recovery discussed. The orientation of ecological and ecotoxicological studies needed to identify the state of riverine ecosystems and as a basis for comprehensive river basin management will be discussed.

BASIC CHARACTERISTICS OF RIVER ECOSYSTEMS

River systems show a startling degree of variation in relation to stream size, flow-rate, climate and vegetation (Burt 1992). An extreme case of variability is provided by rivers that do not contain water at all during the dry season. It is therefore difficult to indicate universal characteristics of rivers, nor is it easy to delineate riverine ecosystems.

Much of the earlier literature on the ecology of rivers was concentrated on the within stream biological communities (Hynes 1970; Whitton 1975). The production biology of these communities over longitudinal gradients (Admiraal and Van Zanten 1988; Cummins 1979) as well as detailed analysis of food relations among riverine organisms (Allan 1983), also deserve further elaboration. On the other hand, the intricate physical, chemical and biological relationships between small streams and their catchments (Likens 1992) provides convincing arguments for treating land and water as one ecosystem unit. While the material flow in the uplands is mostly directed to the main stream, this is not the case in the lower reaches of larger rivers. Here the water distributes its silt load over the flood plains and sometimes shapes wide delta regions, such as that in the River Rhine which created The Netherlands. The integrity of the aquatic/terrestrial habitats on the lower river reaches is not as well analyzed as it is in the upper courses. In particular, the seasonal floods in large tropical rivers produce a riparian zone of extreme ecological complexity (Junk 1986). Finally, the interaction of river flow and sea water produces characteristic estuarine habitats, which have been studied for decades. The importance of riverine material input for the biogeochemistry of coastal seas has also been recognized (Mantoura et al. 1991).

What entity related to rivers should be considered as 'ecosystem'; and for what purpose? For reasons of environmental management, an entire river system including the catchment and receiving coastal waters is an essential unit. It addresses the hierarchy of concerns, ranging from global and continental down to regional and local questions. However, scientific assessments of ecological quality in entire river systems are likely to be based on assessments for the very different aquatic and terrestrial sub-systems. Furthermore, the causes of ecosystem change are likely to be analyzed in connection with local perturbations, before the results are extrapolated to larger units. The question of delineating ecosystems and the scale of the analysis arises here and must be dealt with.

Observations in smaller streams have revealed a strong resilience of lotic communities towards local perturbation. Yount and Niemi (1990) have reviewed numerous case studies describing the recovery of these communities after disturbance, ranging from point sources of organic pollution or toxicant input, to clear cutting of forests. An inherent capacity to survive natural and man-made catastrophes, such as extreme floods, is probably caused by rapid recolonization from upstream reaches or tributaries. Furthermore, the constant flushing of the habitat may carry away dissolved and particulate substances preventing the long-term alteration through accumulation of nutrients and toxicants as occurs in lakes. A large-scale spill of toxicants in the R. Rhine in 1986, caused by a fire at the Sandoz chemical factory in Basel (Switzerland), caused mass mortality of macro-invertebrates and fish. However, the biota had recovered within the following year (Tittizer et al. 1990).

Long-term changes and diffuse inputs are likely to have a persistent impact on the riverine biota. Numerous organisms, indigenous to rivers, have special requirements for a stable (low) temperature, a constantly high oxygen content and show morphological adaptations to the constant shear of the water. Examples are given for the anadromous fish that depend very much on intact river systems and the rich aquatic insect fauna in streams (Ward 1992). Therefore, benthic invertebrates have been extensively used to classify the quality of rivers (De Pauw et al. 1992; Schiller 1990; Sweeting et al. 1992). Similarly, periphytic algae especially the diatoms, have been used (Whitton et al. 1992). The rationale for these assessments of ecological quality is that they indicate in a sensitive way the local constraints on the benthic community (e.g. organic enrichment or acidification), and allow monitoring of changes over years. Further evaluation is needed on how these assessments can be used to quantify ecosystem quality in a more general sense.

TRENDS IN THE DEGRADATION AND REHABILITATION OF LARGE RIVERS

The ecological degradation of the world's rivers is generally caused by a number of interacting environmental perturbations. Dam construction has changed many small streams through the creation of multiple impoundments, and many large rivers have been turned into shipping canals through embankment and straightening of their beds (Petts 1989). Altered hydrological regimes, due to deforestation or irrigation, has intensified the downstream transport of soils (Meybeck et al. 1989; Milliman and Meade 1983) and salts (Walker, this volume; Williams 1987) from the catchments. Salts are also released through mining and industrial processes. Intensified agriculture has accelerated the cycling of nutrients over vast areas and introduced a variety of agricultural chemicals to river systems. This has resulted in a greatly enhanced leaching of potentially deleterious substances, especially nitrates and pesticides, to groundwater and rivers (Billen et al. 1991; Meybeck et al. 1989). Eutrophication of riverine backwaters and the coastal seas (Dethlefsen, this volume; Sherman, this volume) is rapidly proceeding and is changing local ecosystems. The development of large cities and industries on river banks has led to massive inputs of untreated sewage and toxic effluent into rivers. This caused virtual extinction of the higher fauna in reaches of rivers turned into open sewers, such as occurred in the Thames River in the sixties (Andrews 1984).

Meybeck and Helmer (1989) indicated that the wax and wane of these problems occurs in some sort of systematic way. Organic pollution of rivers due to the discharge of untreated sewage from growing cities creates a health risk which is then managed through purification of the water. Similarly, metals have been discharged by factories until the toxic effects became too evident, and better technologies were used to manage pollution. However, many of the environmental problems are not cured as simply as indicated above and ecological principles to guide us to 'redesign' new 'sustainable' environments are urgently required as part of existing and new action plans. Comparative studies on river ecosystems under different degrees of impact by man are potentially helpful to identify these ecological principles.

Examples from the Rivers Rhine and Don

A tentative comparison of two rivers, one in western Europe: the River Rhine (Switzerland, France, Germany, The Netherlands), and one in eastern Europe: the River Don (Russia, Ukraine), is attempted here to clarify the relation between perturbation of similarly sized river systems and ecological degradation. The aim is to develop a perspective for ecological recovery, as it is initiated on the R. Rhine. Both cases have been described elsewhere, the R. Rhine in Friedrich and Müller (1984) and in Admiraal et al. (1993) and the R. Don/Azov Sea in Volovik (in press) and Volovik et al. (1993), and in Kuzin and Sukhorukov (1993). Both systems show similar dimensions and population size (Table 1); the relatively low discharge of the R. Don reflects the semi-arid climate as opposed to the atlantic climate governing the R. Rhine.

Table 1. Basic features of the catchment of the R. Rhine and R. Don/Azvow and the number of inhabitants.

	R. Rhine	R. Don/Azov Sea
Average water discharge km^3y^{-1}	70	40
Catchment area 1000 km^2	180	420
Numbers of inhabitants 10^6	50	30

The hydrological regimes of the rivers Don and Rhine are strongly modified by man. On the main stream of the R. Don, a 2700 km^2 large reservoir has been built to generate electricity. This Tsymlyanski reservoir provides also water for irrigation (Figure 1). In the delta of the R. Rhine, roughly 2000 km^2 of freshwater basin has been created. These lakes (IJsselmeer and Haringvliet) are close to the natural mouth of the Rhine and provide protection against flooding by the sea and safeguard a freshwater buffer in The Netherlands (Figure 1). In addition to these major waterworks, the flow of water in the Don and Rhine river systems is interrupted by hundreds of smaller dams and weirs. Extensive damming in the R. Don and tributaries reduced the river flow and this consequently shifted the salt gradient in the Azov Sea. These conditions greatly reduced spawning areas of anadromous fish species such as sturgeon (*Acipencer* spp.).

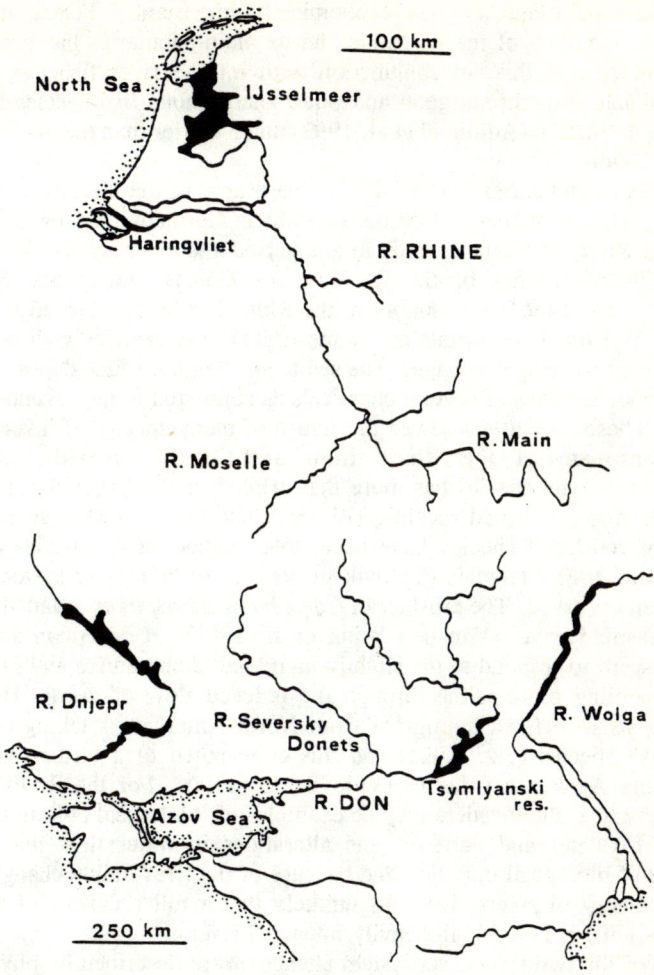

Figure 1. Map of lower parts of the rivers Rhine and Don showing the location of large reservoirs (in black).

As a consequence, the economically important fisheries are declining. The extensive damming of the R. Rhine and its main tributaries has been completed in the 19th century and this, in conjunction with intense river fisheries and the upcoming pollution had brought sturgeon and other anadromous fish species like salmon (*Salmo salar*) to extinction (Admiraal et al. 1993) much earlier than the decline of migratory fish in the R. Don.

Heavy industries occupy the Donbass area in the Ukraine and Russia on the R. Seversky Donets as they did on the R. Ruhr in Germany until the seventies. The water of the R. Rhine was acutely toxic in this period and this may also be the case now in the very polluted reaches of the R. Seversky Donets (Kuzin and Sukhorukov 1993). Improvement of the water quality in the Rhine has been achieved by drastic reductions of industrial inputs of metals and some organic toxicants as well as through extensive treatment of municipal sewage. The sediments that the Rhine deposits nowadays contain lower concentrations of several chemicals as compared to the seventies (Beurskens et al. 1993). These measures allowed the return of many species of invertebrates and fish in the mainstream of the Rhine from about 1975 onwards. The numbers of macroinvertebrate species has more than tripled in 1987 (Schiller 1990). The mayfly, *Ephoron virgo*, returned recently (Bij de Vaate et al. 1992). In the Don, substantial losses of red-listed species have been noted including 41 species of insects, 28 bird species and four mammals (Volovik in press). In both river systems invading species have been observed. The crustacean *Corophium curvispinum* invaded the Rhine from the Ponto-Caspian area (Van den Brink et al. 1991). *Corophium* and invading shrimp species seem to respond to the slightly increased temperatures and salinity. In the Azov Sea, becoming more saline through the reduced flow of the R. Don, the ctenophore *Mnemiopsis leydyi*, an immigrant from North America, is taking over the position of small fish species (e.g. Kilka) and this contributed to a virtual collapse of the once flourishing Azov Sea fisheries (Volovik, in press). For the R. Rhine and for the R. Don/Azov Sea, the invaders may be examples of 'biological pollution', but the changing load of toxicants and nutrients and alterations in water flow may also be involved. Because of biological pollution and because of the irreversible changes in the hydrology and chemistry of rivers, it seems unlikely that a full recovery of the original species ('restoration') is possible in heavily impacted rivers.

For the two rivers, ecosystem changes were described by physical, chemical and biological parameters. Difficulties arise when these widely divergent parameters are to be integrated into a criterium of "ecosystem quality". An approximation has been reached for the river Rhine by selecting about 25 key biological parameters and organizing these into a radial diagram that enables a comparison of current abundance of key species with a reference condition (i.e. conditions at the beginning of this century) as shown in Figure 2. This procedure, referred to as the AMOEBA-technique (Ten Brink et al. 1991), allows a quantitative expression of the difference between two conditions of an ecosystem. The technique also enables a comparison of current status with the biotic conditions or goals desired for that ecosystem in the future, and of the progress towards these goals (i.e. ecological objective). The approach is pragmatic and, through the choice of the parameters, flexible. Insight into complex ecological interactions of the key species with their abiotic and biotic environment is, of course, essential in the selection of the parameters. A fundamental understanding of the ecology of these species should be available or should be developed when implementing an AMOEBA approach. .

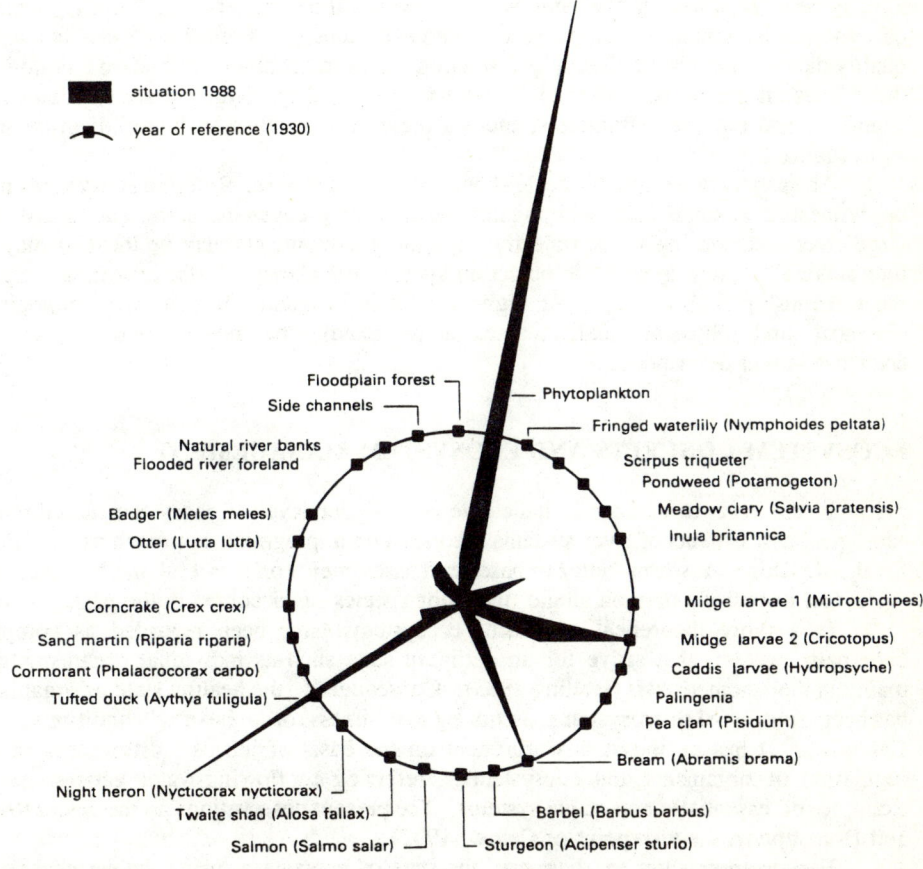

Figure 2. Radial diagram, comparing ecological parameters measured in the Dutch part of the Rhine in 1988 with that of ca. 1930. Parameter values similar in 1988 and 1930 are plotted on the circle (100%). Zero values (e.g. absence of salmon in 1988) are in the centre. Increased abundance of organisms (>100%) is plotted as a "spine". The irregular black figure indicates a disproportional development in the river community (after Admiraal et al. 1993).

The technique described above aims at assessments of biological quality that are assumed to reflect the physical and chemical changes of the habitat. However, in the case of rivers, where the change of the morphology and hydrology is so prominent and the input of toxicants and nutrients is substantial, a quantitative description of the abiotic environment should complete an ecological assessment. In Figure 2, morphological parameters such as the surface area of floodplain and of undisturbed river banks has already been included. In the same way, the chemical parameters, e.g. salinity or metal concentrations, could be compared to reference values. Frequently, Russian water quality data are expressed already as a fraction of the environmental standard. Following these lines, it seems necessary to document ecological quality by a set of biological, chemical, and physical parameters, each of these three synthesized into a diagram such as in Figure 2.

The aspect of geographic scale should be reiterated here. Riverine ecosystems may be delineated as small units within catchments or may cover the entire catchment of a large river. Accordingly, the objective of quality assessments may be local or may be international. There is no basic objection against any choice of dimension, as long as the relationship with the lower or higher scales is included. In summary, biological, chemical and physical quality assessments should be nested into an overall river/catchment description.

ECOSYSTEM CONCEPTS AND ECOSYSTEM MANAGEMENT

In the absence of criteria indicative of a higher level of ecosystem functioning, ecological assessments of river systems are based on a pragmatic approach as was done for the R. Rhine It seems better to base such assessments on a careful and broad choice of several organisms than on single 'indicator species' as occurred in the past.

In a more theoretical approach, ecosystems have been regarded as complex cybernetic systems that strive for an optimum state such as individual organisms that maintain their homeostasis (Holling 1983). Consequently, the healthy state of organisms has been transferred to ecosystems, assuming that 'illness' or 'unhealthy' conditions exist. Calow (1992) has criticized this argument on the basis of intrinsic differences in the regulation of organisms and ecosystems, referring to flowing-water ecosystems as examples of essentially non-stable systems. The present observations on the rivers Rhine and Don support the viewpoint of Calow (1992).

Earlier approaches to 'diagnose' the state of ecosystems using the six criteria of Rapport et al. (1985), reflected the different manifestations of ecosystems much better than a strictly kinetic analysis. These criteria included nutrient cycling, primary production, species diversity, succession/retrogression, organism size and disease frequency. These criteria have been applied to aquatic ecosystems in the flood plain of the R. Rhine (Admiraal et al. 1989) and although the perception of the ecosystems' state may be clarified, a factual, let alone a quantitative analysis, could not have been achieved. The absence of straightforward theory-based ecological assessments underlies the current need to use pragmatic multi-species evaluations, such as the AMOEBA-technique.

Since the early eighties a certain notion of ecosystems and their functioning has pervaded the management of Dutch surface waters, as it did in North America (Regier 1992). This notion, perhaps not very well defined, also developed in relation to international bodies. In the case of the Rhine Action Plan (IRC 1987) the original orientation on water quality has been adapted to ecology-driven objectives such as the return of species and restoration of habitats. An UN-ECE guideline (ECE 1990; Reynolds 1993) on ecosystem approaches to water management expresses a comprehensive practical guideline on the integration of societal needs, action plans, environmental monitoring, information supply, and research. It is remarkable to see how humans and human activities are incorporated in a conception of ecosystems, that has been put into practice in several cases of river rehabilitation.

At the same time, theoretical approaches to ecosystems seem to exclude an anthropogenic component, while disturbances through man do not necessarily equate with destabilization (Chapman 1992). Qualifications like 'healthy', such as used in the present NATO Workshop, could be applied to ecosystems in its broad sense i.e. including exploitation by man, and including physical and chemical aspects.

NOTE ON RESEARCH PRIORITIES

It is now evident that river ecosystems constitute a vital unit in environmental management, and it is essential to indicate priorities for future research. This research should provide the multi-disciplinary basis for river management (Statzner and Sperling 1993). The Dutch program 'Ecological Rehabilitation of the River Rhine' provides support to the Rhine Action Plan through its research on many ecological and ecotoxicological subjects (Van Dijk and Marteijn 1993). In line with these actions, it is imperative to fill the gaps in our knowledge of large rivers. This research could be organised under the following themes:

1. Analysis of the natural geochemical and biogeochemical structure of catchments, rivers and coastal waters and comparison of the "natural" and man-made fluxes of materials. This should facilitate evaluation of the systems' capacity to absorb man-made input.

2. Description of the riverine habitats and their quantitative relation to the hydrology and morphology of the catchment. The connectivity of in-stream habitats and the riparian zone is a vital aspect, which needs attention in view of the progressive reconstruction of river systems. Autecological studies on biological species from different trophic levels, such as migrating fish, invertebrates (e.g. sensitive insects), periphytic or planktonic algae (responding to eutrophication), and macrophytes (reflecting the water regime) are needed. These studies should reveal cause and effect relationships with

the abiotic environment and allow a certain degree of predictive capacity as to ecological rehabilitation. Ultimately, an ecological approach to river engineering should emerge.

3. Analysis of relations between diffuse or point sources of toxicants and biological effects. The uncertainties in assessing the availability of toxicants in water and sediments to organisms (Persoone 1992; Van de Guchte 1992) are to be resolved for individual rivers rather than for laboratory cultures and the sequence of events ranging from immediate intoxication and physiological adaptation to long-term selection of species needs to be elucidated. Toxicant impact is to be related to other 'ecological' constraints in order to design sound guidelines.

A classical scientific approach to these three themes is perfectly able to cope with the new dimensions of the ecosystem approach to river management. However, a closer cooperation of the disciplines, as well as a more intense cooperation of scientists, technicians, policy makers and the public remains to be established. In this respect, the case studies on the R. Rhine and the R. Don/Azov Sea reveal a major difference. For the R. Rhine, the difficult cooperative action has recently been initiated through the Rhine Action Plan and has led already to a partial recovery of the R. Rhine. Ukrainian and Russian scientists in cooperation with Dutch colleagues, have made proposals in 1993 for a Don/Azov Action Plan, following the approach developed for the R. Rhine. It is expected that the action on the Don/Azov will gain support by the national and international community so that some of the acute environmental problems can be cured effectively.

ACKNOWLEDGEMENTS

I would like to thank Dr. M.H.S. Kraak and Dr. C. Davids and two unknown reviewers for their critical remarks on the manuscript.

REFERENCES

Admiraal W., Van Zanten B. (1988) Impact of biological activity on detritus transported in the lower river Rhine: an exercise in ecosystem analysis. Freshwater Biology 20: 215-225

Admiraal W., De Ruyter van Steveninck E.D., De Kruijf A.M. (1989) Environmental stress in five aquatic ecosystems in the floodplain of the River Rhine. The Science of the Total Environment 78:59-75

Admiraal W., Van der Velde G., Smit H., Cazemier W.G. (1993) The rivers Rhine and Meuse in The Netherlands: present state and signs of ecological recovery. Hydrobiologia 265:97-128

Allan J.D. (1983) Predator-prey relationships in streams. In Barnes J.R., Wayne G.W. (eds.) Stream-ecology. Plenum Press, New York. pp:191-229

Andrews M.J. (1984) Thames estuary: pollution and recovery. In Sheehan P.J., Miller D.R., Butler G.C., Bourdeau P.H. (eds.) Effects of Pollutants at the Ecosystem Level, SCOPE, John Wiley and Sons Ltd. pp:195-227

Beurskens J.E.M., Mol G.A.J., Barreveld H.L., Munster B. van, Winkels H.J. (1993) Geochronology of priority pollutants in a sedimentation area of the Rhine River. Environmental Toxicology and Chemistry 12:1549-1566

Bij de Vaate A., Klink A., Oosterbroek F. (1992) The mayfly, *Ephoron virgo* (Olivier), back in the Dutch parts of the rivers Rhine and Meuse. Hydrobiol. Bull. 25: 237-240

Billen G., Lancelot C., Meybeck M. (1991) N, P, and Si retention along the aquatic continuum from land to ocean. In Mantoura R.F.C., Martin J.M., Wallast R. (eds.) Ocean Margin Processes in Global Change pp:19-44

Burt T.P. (1992) The hydrology of headwater catchments. In Calow P., Petts G.E. (eds.) The Rivers Handbook (vol. 1) pp:3-38

Calow P. (1992) Can ecosystems be healthy? Critical consideration of concepts. Journal of Aquatic Ecosystem Health 1:1-5

Chapman D. (1992) Ecosystem health synthesis: can we get there from here? Journal of Aquatic Ecosystem Health 1:69-79

Costanza R. (1992) Towards an operational definition of ecosystem health. In Costanza R., Norton B.G., Haskell B.D. (eds.) Ecosystem Health - New Goals for Environmental Management. Island Press Washington, D.C. pp:239-256

Cummins K.W. (1979) The natural stream ecosystem. In Ward J.V., Stanford J.A. (eds.) The ecology of regulated streams. Plenum Press, New York. pp:7-24

Davies B.R., Walker K.F. (eds.) (1986) The Ecology of River Systems. Dr. W. Junk Publishers, Kluwer Acad. Publ., The Netherlands. 793 pp.

De Pauw N., Ghetti P.T., Manzini P., Spaggiari R. (1992) Biological assessments methods for running water. In Newman P.J., Piavaux M.A., Sweeting R.A. (eds.) River Water Quality, Ecological Assessment and Control, Official Publications of the Commission of the European Community, Brussels pp:217-248

ECE (1990) Ecosystems Approach to Water Management. Report of the senior advisors to ECE governments on environmental and water problems. Economic Commission for Europe (United Nations) Geneva. 25 pp.

Friedrich G., Müller D. (1984) Rhine. In Whitton B.A. (ed.) Ecology of European Rivers, Blackwell, Oxford pp:265-315

Holling C.S. (1983) Resilience and stability of ecological systems. Annual Review of Ecology and Systematics 4:1-23

Hynes H.B.N. (ed.) (1970) The Ecology of Running Waters. Liverpool University Press. 555 pp.

IRC (1987) Rhine Action Programme. Report of the 'Technisch-wissenschaftliches Sekretariat der Internationale Kommission zum Schutze des Rheins', Koblenz

Junk W.J. (1986) Aquatic plants of the Amazon systems. In Davies B.R., Walker K.F.

(eds.) The Ecology of River Systems. Dr. W. Junk Publishers, Kluwer Acad. Publ., The Netherlands pp:319-337
Kuzin A.K., Sukhorukov G.A. (1993) Ecological problems in the Severski Donets River Basis. Report of the Ukrainian Scientific Centre for Protection of Water, Kharkov. 7 pp.
Likens G.E. (1992) The ecosystem approach: its use and abuse. In Kinne O. (ed.) Excellence in Ecology. Ecology Inst., Luhe, Germany. 166 pp.
Mantoura R.F.C., Martin J.M., Wollast R. (eds.) (1991) Ocean Margin Processes in Global Change. John Wiley and Sons Ltd.
National Research Council (1992) Restoration of Aquatic Ecosystems. National Academy Press, Washington, D.C. 552 pp.
Meybeck M., Chapman D.V., Helmer R. (eds.) (1989) Global environment monitoring systems, Global Freshwater Quality, Great Britain Alden Press, Blackwell, Oxford pp:139-157
Meybeck M., Helmer R. (1989) The quality of rivers: from pristine stage to global pollution. Palaeogeography, Palaeoclimatology, Palae-oecology (Global and Planetary Change Section) 75:283-309
Meybeck M., Friedrich G., Thomas R., Chapman D. (1992) Rivers. In Chapman D. (ed.) Water quality assessments. A guide to the use of biota, sediments and water in environmental monitoring. Chapman and Hall, University Press, Cambridge
Milliman J.D., Meade R.H. (1983) World-wide delivery of river sediment to the oceans. The Journal of Geology 91:1-21
Newman P.J., Piavaux M.A., Sweeting R.A. (eds.) (1992) River water quality, ecological assessment and control. Official Publications of the Commission of the European Community, Brussels 751 pp.
Persoone G. (1992) Ecotoxicology and water quality standards. In Newman P.J., Piavaux M.A., Sweeting R.A. (eds.) River water quality ecological assessment and control, Official Publications of the Commission of the European Communities, Brussels pp:461-482
Petts G.E. (ed.) (1989) Historical Change of Large Alluvial Rivers: Western Europe. John Wiley pp:167-182
Rapport D.J., Regier H.A., Hutchinson T.C. (1985) Ecosystem behavior under stress. Amer. Natur. 125: 617-640
Regier H.A. (1992) Ecosystem integrity in the Great Lakes Basin: an historical sketch of ideas and actions. Journal of Aquatic Ecosystem Health 1: 25-37
Reynolds C.S. (1993) The ecosystems approach to water management. The main features of the ecosystems concept. Journal of Aquatic Ecosystem Health 2:3-8
Schiller W. (1990) Die Entwicklung der Makrozoobenthonbesiedlung des Rheins in Nordrhein-Westfalen im Zeitraum 1969-1987. In Kinzelbach H.R., Friedrich G. (eds.) Biologie des Rheins. Gustav Fischer Verlag, Stuttgart-New York pp:259-276
Statzner B., Sperling F. (1993) Potential contribution of system-specific knowledge (SSK) to stream-management decisions: ecological and economic aspects. Freshwater Biology 29:313-342

Sweeting R.A., Hale P., Lowson D., Wright J.F. (1992) Biological assessment of rivers in the UK. In Newman P.J., Piavaux M.A., Sweeting R.A. (eds.) River water quality ecological assessment and control, Official Publications of the Commission of the European Communities, Brussels pp:319-326

Ten Brink B.J.E., Hosper S.H., Colijn F. (1991) A quantitative method for description and assessment of ecosystems: the AMOEBA-approach. Mar. Poll. Bull. 23:265-270

Tittizer T., Schöll F., Schleuter M. (1990) Beitrag zur Struktur und Entwichlungsdynamik der Benthalfauna des Rheins von Basel bis Düsseldorf in den Jahren 1986 und 1987. In Kinzelbach H.R., Friedrich G. (eds.) Biologie des Rheins, Gustav Fischer Verlag, Stuttgart-New York pp:293-323

Van den Brink F.W.B., Van der Velde G., Bij de Vaate A. (1992) Amphipod invasion on the Rhine. Nature 352:576

Van Dijk G.M., Marteijn E.C.L. (eds.) (1993) Ecological Rehabilitation of the River Rhine 1988-1992. Report of the project 'Ecological Rehabilitation of the rivers Rhine and Meuse', Report no. 50, Bilthoven. 62 pp.

Van de Guchte C. (1992) The sediment quality triad: an integrated approach to assess contaminated sediments. In Newman P.J., Piavaux M.A., Sweeting R.A. (eds.) River water quality ecological assessment and control, Official Publications of the Commission of the European Communities, Brussels pp:417-424

Volovik S.P. (in press) The effect of environmental changes caused by human activities on the biological communities of the river Don (Azov Sea basin). Water Science and Technology

Volovik S.P., Dubinina V.G., Semenov A.D. (1993) Water medium, dynamics of populations and fisheries in the Azov Sea. Report Azov Research Institute of Fishery Problems, Rostov, USSR. 90 pp.

Ward J.V. (1992) Aquatic Insect Ecology 1. Biology and Habitat. Wiley, New York. 437 pp.

Whitton B.A . (1975) River Ecology. Blackwell Sci. Publ., Oxford. 725 pp.

Whitton B.A. (ed.) (1984) Ecology of European Rivers. Blackwell Sci. Publ., Oxford. 530 pp.

Whitton B.A., Rott E., Friedrich G. (eds.) (1991) Use of algae for monitoring rivers. Proceedings of International Symposium held at the Landesamt für Wasser und Abfall Nordrhein-Westfalen. Studia Studentenförderungs-Ges.m.b.H., Austria. 192 pp.

Williams W.D. (1987) Salinization of rivers and streams: an important environmental hazard. Ambio 16:180-185

Yount J.D., Niemi G.J. (1990) Recovery of lotic communities and ecosystems from disturbance – A narrative review of case studies. Environmental Management 14: 547-569

13. AN INFLUENCE DIAGRAM APPROACH TO THE DIAGNOSIS AND MANAGEMENT OF THE BALTIC SEA

Mikael Hildén
National Board of Waters and the Environment
Unit of Environmental Impact Assessment
P.O.BOX 250, FIN-00101 Helsinki
Finland

INTRODUCTION

The complexities of the diagnosis and management of a large sea area such as the Baltic Sea call for a systematic framework. Within such a framework it should be possible to use knowledge of the functioning of the system and to examine both quantitative and qualitative information on its state, including risks to the integrity of the ecosystem. This diagnosis can be used for analyzing possible management actions aimed at sustaining, and whenever possible, rehabilitating the system. In this study Bayesian influence diagrams were used for this purpose. The analysis suggests that influence diagrams could become a useful tool in providing the scientific background needed for negotiations on the future of the Baltic Sea.

The Baltic Sea

The world's largest brackish water body, the Baltic Sea, is 415266 km^2 in area including the Kattegat and the Belt Sea, and has a total volume of 21 721 km^3. Its catchment area covers 1.67 million km^2 which is inhabited by about 80 million people. The area is highly industrialized and its agriculture is intensive.

The environmental problems of the Baltic Sea (Brügmann 1993) have prompted the countries of the catchment area to agree on a Convention for the Protection of The Baltic Sea (1974, 1992). There is also a Convention on Fishing and Conservation of the Living Resources in the Baltic Sea and Belts (1973). Specific plans for the reduction of emissions have also been formulated (SNV 1990, HELCOM 1992a).

In the Baltic Sea area there is a consensus that the health of the Baltic ecosystem could be better and that sustainability should be a goal for resource management. Key decision makers have thus had to find practical solutions to theoretical problems related to an assessment of the health of an ecosystem documented by Costanza et al. (1992). The Baltic Sea Joint Comprehensive Environmental Action Programme (HELCOM 1993) is an ambitious plan for the rehabilitation of the Baltic Sea and is based on a large body of environmental studies.

Although the goal to improve the state of the Baltic Sea is generally accepted, many potentially controversial decisions will have to be made (Wulff and Niemi 1992). Several uses compete for scarce resources, especially in countries undergoing transition of their economic systems. The necessary financial commitments may be difficult to achieve, despite backing by high level conferences (HELCOM 1993). Decisions on resource allocation and on the types of measures contributing to the restoration of the state of the Baltic Sea are difficult, because among other things development in the catchment creates incentives for new activities. The Baltic Sea and its catchment are not in a stable state. Information and management strategies thus have to be revised from time to time. Negotiations concerning the exploitation of Baltic resources have to be repeated annually.

It is possible to reach agreements on Baltic environmental issues in an *ad hoc* fashion, but since all decisions are intimately linked through their effects on the Baltic ecosystem, it is worthwhile to look for a common framework. In this study I analyze general requirements for a diagnostic approach which can be connected to policy level decision making. Influence diagrams are used as the methodological framework; they have been developed within decision theory and have been used in many applications involving complex decision making problems (Oliver and Smith 1990). Influence diagrams have also successfully been applied to environmental and resource management problems (Varis et al. 1990; Kuikka and Varis 1992).

GENERAL REQUIREMENTS FOR A POLICY-ORIENTED DIAGNOSTIC APPROACH

The decisions on environmental policy that have to be made call for a systematic approach to the wealth of information that is available. The Baltic Marine Environment Bibliography 1986-1990 contains altogether 2,316 references (HELCOM 1992b), suggesting that more than 500 documents related to the Baltic Sea are produced annually. More than a 100 environmental variables are regularly monitored in the Nordic countries, in addition to monitoring of point source polluters, populations of fish, seals and birds (Nordisk havovervågningsprogram 1993). It is obvious that this information can be used in policy level decision making only after considerable aggregation and synthesis.

The connection to decision making processes demands an adaptive and pragmatic approach to diagnosis which should be applicable on different scales and to different types of problems. Some decisions, for example those related to physical restructuring of shorelines or the seabed, concern local changes which may have wider impacts. A prime example is the plan to construct a fixed link across the Öresund between Sweden and Denmark (Öresundkonsortiet 1992). The fixed link could have an impact on water exchange between the Baltic proper and the Kattegat and therefore a decision on authorization of the project must take into account local, regional and basinwide effects. Other decisions, for example the 1988 Ministerial Declaration to reduce land based and airborne inputs of nutrients, heavy metals and toxic or persistent organic substances by 50% by 1995 from the level of 1987 (HELCOM 1990) or decisions to ban certain substances, define a general policy without reference to concrete projects.

A difficulty with diagnostic work is that the degradation of a large ecosystem may show up as many different types of changes. Rapport et al. (1985) identified several characteristics of stressed ecosystems which appear to be general, but many of the features of the ecosystem distress syndrome appear only when the degradation is all too evident. A diagnosis should also be able to identify more subtle changes (Rapport 1992).

Ideally the diagnostic approaches should be able to continuously incorporate new information on the state of the ecosystem. The updated diagnosis could thus provide feedback to environmental management and allow an assessment of the relative merits of specific measures to improve the state of the system.

Few, if any, environmental decisions will be free from uncertainty or ignorance despite the mass of new information produced. Diagnostic approaches should therefore be able to deal explicitly with concepts of risk and probability, including subjective judgement. In this way the precautionary principle can be made operational. The Helsinki Convention (1992) defines it as the principle "to take preventive measures when there is reason to assume that substances or energy introduced...may create hazards to human health, harm living resources and marine ecosystem, damage amenities or interfere with other legitimate uses of the sea even when there is no conclusive evidence of a causal relationship between inputs and their alleged effects".

The foregoing suggests that for policy decisions the present state and its potential changes is a basic starting point. For policy level decisions a diagnostic approach that is able to identify changes is often sufficient and it can avoid some of the pitfalls of the 'absolute' health definitions recognized by Costanza (1992). A second starting point is that the relevant information for policy decisions is very diverse. There are quantitative data on primary production, fish yields and currents, results of laboratory and field experiments and perceptions of the state of the sea by its many users. All this information cannot be treated in the same way or given the same weight in a diagnosis, but it is not rational to disregard any information that cannot be converted to $C\ gm^2y^{-1}$, as required for example by the analysis of ascendancy developed by Ulanowitz (1992). The possibility of quantification cannot be the only criterion for the value of information when decisions are made on environmental policy.

INFLUENCE DIAGRAMS AND BELIEF NETS AS DIAGNOSTIC TOOLS

Belief nets and influence diagrams are graphical presentations of probabilistic relationships which are based on Bayes' formula (Pearl 1988). They are formulated as networks consisting of nodes and directed arcs with no cycles. The nodes can be decision nodes, chance nodes, deterministic nodes or value nodes (Figure 1). Belief nets differ from influence diagrams in that they lack explicit representation of decision variables or utilities. Decision nodes represent points at which choices have to be made between well-defined alternatives and arcs directed to a decision node depict information available at the time of making the decision. Chance nodes represent uncertain events. When they precede decision nodes the decision maker has no control over their outcome. Chance nodes following decision nodes can partially be affected by the decision maker's action. A deterministic node is completely determined by the nodes with arcs pointing into it. Value nodes represent decision-making criteria that can be expressed on a common scale such as money or utility. The arcs leading from decision nodes to chance nodes or deterministic nodes, including value nodes, are called conditioning arcs because they express the impact of an event on subsequent events. When influence diagrams are used the problem structure is specified in the form of nodes and arcs between the nodes. The relationships between different nodes can then be specified in the form of functions or matrices giving conditional probabilities for events in a node, given specific events in the predecessors.

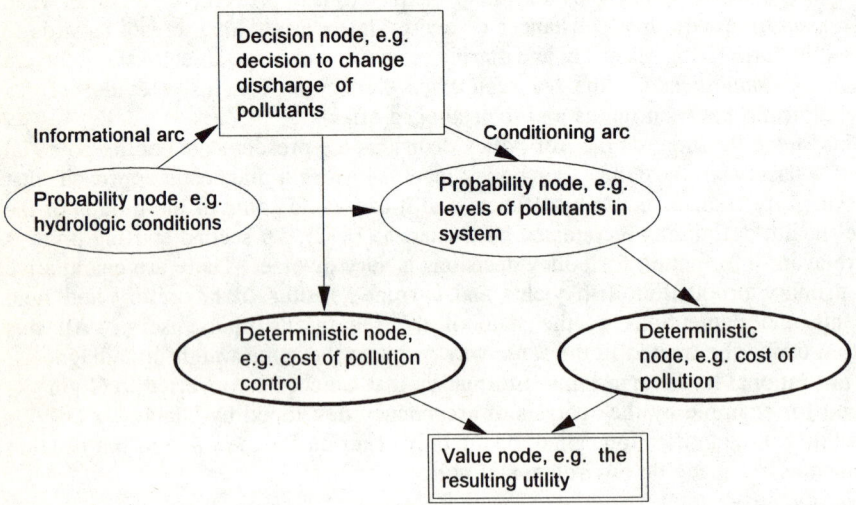

Figure 1. A simple influence diagram illustrating the different types of nodes in an influence diagram and the graphical codes used in this paper. Boxes represent decision nodes, ovals - chance nodes, ovals with heavy borders - deterministic nodes, and boxes with double borders - decision making criteria.

Influence diagrams and belief nets have been used in several medical applications (Shachter et al. 1990). They have been used as a basis for diagnosis using large databases and complex sets of findings indicating the presence or absence of a particular disease (Henrion 1990). A similar approach could be used for a large ecosystem, provided that a compendium or database of ecosystem diseases were available; for the time being, however, such information does not exist (Haskell et al. 1992). As noted by Haskell et al. (1992) the nature of 'ecosystem diseases' differs considerably from that of diseases of individuals, making the medical approach difficult to implement, especially on large-scale systems. One can identify a 'sewage plant disease', a 'dammed river disease' or a 'chemical installation disease' and collect information for prognosis and treatment. While such an approach can be useful for local or regional environmental management or single court cases, it is not directly applicable to general policy decisions, which demand a synthesis of general systemic changes.

Problems related to identification of the state of a whole system are encountered in the management of automatic machinery on the basis of readings from numerous sensors which give information on the state of the machinery and the process (Agogino et al. 1990). The information from the sensors is diverse and rapidly accumulating, and may also include professional judgement. The state of the system can, however, be characterized by relatively few classes. When material is assembled for decisions on environmental policy for large-scale ecosystems a similar simplification of the information occurs.

For an ecosystem, classes of its state can be identified by emphasizing different aspects of the ecosystem distress syndrome (Rapport et al. 1985). The state can be described as healthy, eutrophied, poisoned, physically disrupted or overexploited. Most large systems probably represent various combinations of states. The classes of health and their combinations can form a basis for a diagnosis in an influence diagram.

An influence diagram can also explicitly illustrate the coupling between diagnosis and management. The problem of the diagnostician is to correctly identify the state of a system. The problem of the manager is to choose an acceptable course of action on the basis of an imperfectly known state and uncertain predictions of future changes that follow from a particular action. A diagnosis does not have to be evaluated in terms of utilities or monetary gains, whereas the results of management measures will be judged against chosen objectives or set in a policy framework (Sagoff 1992). In an influence diagram this can be modelled explicitly by taking into account costs and gains, or by defining the utilities associated with particular combinations of management measures and the true state of the system.

AN INFLUENCE DIAGRAM FOR THE BALTIC SEA

A diagram illustrating the diagnosis and management of the Baltic Sea can become very complex. There are many different possible representations of the system depending on which parts and which types of decisions one wishes to analyze. For this study I developed a diagram which illustrates policy level decision problems (Figure 2). There are, however, several possible formulations and therefore the primary aim is to present the approach and the questions it raises. This particular diagram focuses on intermediate term policy decisions and the corresponding changes in the system. It was implemented using the InDia software (Decision Focus Inc. 1991).

The first step was to determine the overall structure of the diagram; thereafter, the states and corresponding probabilities of the nodes were specified. For nearly all nodes three possible states were identified (i.e. increase, status quo or decrease), because the focus of the intermediate term policy level decisions was considered to be the change relative to the present state.

The probabilities of each state in nodes without predecessors (i.e. in nodes with no arcs pointing into them), represent imperfectly known background information; for example fish stock assessments have strongly suggested that the cod productivity is declining (Thurow 1993) and monitoring results indicate that the nutrient pool in the Baltic Sea is increasing.

Conditional probabilities were determined for those nodes which had conditioning arcs pointing into them. In an influence diagram the number of conditional probabilities in a particular node depends on the number of predecessors and their states. Thus a node with three predecessors with three possible states each requires nine 3 by 3 matrices.

To achieve consistent judgements of the conditional probabilities of nodes with several predecessors I compared the probabilities pairwise using the analytic hierarchy process (Saaty 1988); for example the change of zooplankton biomass was assumed to be determined by the change in abundance of planktivores and the change in primary production. The probability of a decrease, status quo and a decline of the zooplankton are assessed conditional upon all possible combinations of the predecessors. By making pairwise comparisons in a systematic way it is possible to avoid inconsistencies. Such inconsistencies could take the following form: p(increase of zooplankton|increase of primary production, decline of planktivores) < p(increase of zooplankton|increase of primary production, status quo of planktivores) > p(increase of zooplankton|increase of primary production, increase of planktivores). In pairwise comparisons between all alternatives these inconsistencies in the conditional probabilities can be avoided.

Management Decisions

Several different types of management decisions are made concerning the Baltic Sea environment. To cover a range of different decisions I focus on decisions to reduce pollution load and decisions on the exploitation of fish resources. In the influence diagram I assume that the decisions to reduce the pollution load are made relative to decisions already made. A status quo decision thus means that present reduction goals are kept but not increased further. The other extreme represents a full implementation of the Baltic Sea Joint Comprehensive Environmental Action Programme developed by HELCOM (1993). An intermediate option is a partial implementation of the programme. The decisions on exploitation also represent changes relative to the present. Contrary to decisions on pollution it is conceivable that exploitation can be increased by deliberate policy decisions. Environmental decisions affect changes in nutrients and hazardous substances, but managers do not have complete control over the changes. Imperfectly known present and future states and ecological processes will determine the actual change. The costs associated with a particular management decision can, however, be fairly well estimated, in comparison with the other items of the influence diagram.

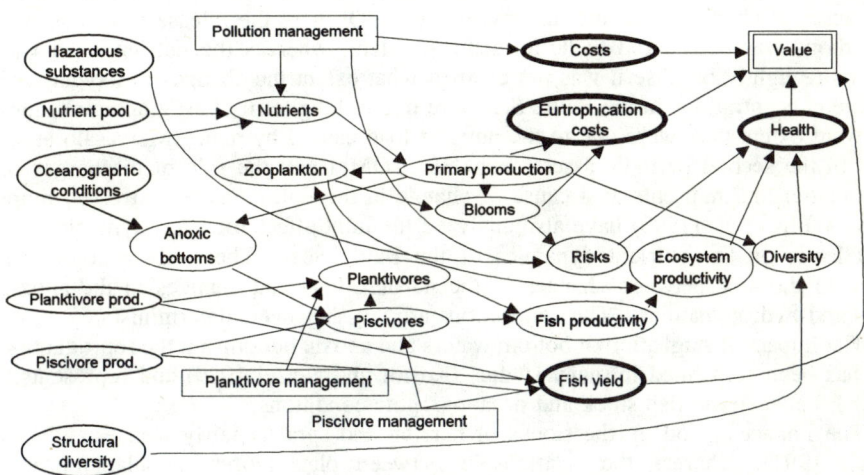

Figure 2. An example of an influence diagram for Baltic Sea diagnosis and management. Probabilistic nodes without predecessors are given in the left hand column. These reflect the boundary conditions for present decisions, but cannot be influenced by the decisions. The graphical codes for the different nodes are given in Figure 1. Note that the nodes influenced by the initial conditions and management actions focus on changes relative to the present; e.g. the node "planktivores" illustrates changes in planktivore stocks, not absolute stock levels.

The fishery management decisions will result in specific yields for which net values can be calculated, assuming that quotas will be taken. The decisions will also induce changes in fish resources, but the changes depend on the present state of the stocks, their interactions and the environmental conditions; for example anoxic bottoms have a strong impact on the spawning success of cod and the future state of the stock.

Connections Within the System

The connections between different chance nodes in the influence diagrams are based on a simplification of the current understanding of the Baltic ecosystem. Recent overviews were presented at a Dialogue meeting by ICES (1993a).

The first assessment of pollution on the natural resources of the Baltic Sea by the Helsinki Commission (HELCOM) and the International Council for the Exploration of the Sea (Melvasalo et al. 1981) considered the release of harmful or toxic synthetic organic chemicals, organic wastes, nutrients, trace elements and hydrocarbons to be major causes of change in the marine environment. Of these the release of chlorinated hydrocarbons was considered to be an acute problem, whereas the nutrient load was treated more lightly because it was not clear to what extent the changes in oxygen and phosphate concentrations in the Baltic Sea were due to long-term hydrographic changes and to what extent they were due to the nutrient load caused by man (Melvasalo et al. 1981). In the second periodic assessment (HELCOM 1990) the role of discharges of organic matter and nutrients as a cause of change in the Baltic Sea was stressed more strongly. Wulff et al. (1990) have also analyzed the joint effects of hydrodynamics and nutrient loads on the nutrient dynamics of the Baltic Sea. These observations are reflected in the arcs between changes in the nutrient load and changes in the anoxic bottoms and hydrodynamics. The connections are not, however, deterministic.

The impact of stagnation of bottom waters and anoxic bottoms on the reproduction of cod has been confirmed in many studies (Kosior and Netzel 1989) and represents a direct link between the fish stock and oceanographic conditions.

The impact of cod on the stocks of herring and sprat is fairly well established (Sparholt 1991), whereas the relationship between planktivores, zooplankton and phytoplankton is less clear. The traditional 'bottom-up' assumption is that an increase of primary production is channelled via an increased zooplankton production to increased fish production (Nehring et al. 1989). Alternatively 'top-down' effects may also be important. Predation on zooplankton is intense in the Baltic Sea, and suppression of zooplankton by planktivorous fish may thus have an impact on primary production (Rudstam et al. 1994).

Different assumptions on the structure and functioning of the system influence the diagnosis. If only 'bottom-up' relationships are considered important, increased phytoplankton production and blooms point directly to excessive nutrient loads and 'simple' eutrophication. With 'top-down' control the same changes may also indicate a lack of piscivores due to overexploitation or destruction of spawning areas. Although both assumptions lead to the conclusion that the health of the system has degraded, they may also lead to different predictions of the recovery and the effectiveness of management measures. In a 'top-down' controlled system changes in symptoms such as algal blooms may at least in an intermediate term perspective depend on the development of the planktivorous and piscivorous fish stocks (Rudstam et al. 1994).

Health Changes

Costanza (1992) suggested that at least three dimensions are needed to describe ecosystem health: vigor, organization and resilience. All are aggregates of certain characteristics of the ecosystem. In the influence diagram I have operationalized the health dimensions by considering productivity, diversity and risks. Total system productivity could be calculated, although it is not clear whether it is more or less sensitive to changes than the productivity of individual components in the system. Here I assume that increased fish productivity and zooplankton productivity is beneficial to the system. Total system diversity is in theory measurable, but in practice the task is nearly impossible and definitely not applicable for routine purposes. Structural diversity is, however, one important element and its changes relative to the present are reasonably easy to determine; for example conversion of shoreline to harbour areas certainly reduces structural diversity as does a loss of macroalgae due to increased turbidity (Brügmann 1993).

The limited view on diversity and productivity I use is a rather incomplete measure of system health and unlikely to respond quickly to changes. However, even a more extensive use of data on the Baltic Sea leaves scientists in doubt regarding future changes in and the list of gaps in the knowledge is long (Brügmann 1993). To overcome this problem, without solving it for the time being, I have included a general risk node. The justification is that although the exact impact on the health of the system, or even on an individual species or population is not known accurately, it is possible to state with some confidence whether the risk of future adverse changes increases or decreases. The risk node is thus an operationalization of ecosystem risk factors proposed by Rapport (1992). The influence diagram approach provides a method of combining different types of risks and is not limited to the case of a single dominant stress.

The Value Node

Gains in fish yield and costs of environmental protection can often be estimated with reasonable accuracy. The dockside value of the total annual yield from the Baltic Sea is around ECU 0.5 billion (Thurow 1993) and the investment costs of the Baltic Sea Joint Comprehensive Environmental Action Programme are at least ECU 18 billion over a twenty-year period (HELCOM 1993). With a yearly investment rate of ECU 1 billion and a discount rate of five percent the present value of the investments are about ECU 13 billion. The present value of the fish yield over 20 years is ECU 6.5 billion when dockside values are used. The present net value of the commercial fish catch is obviously lower. The value of the catch of recreational fishers is, however, often considered more valuable although it is not the catch but the entire fishing experience that carries the value. Different estimation methods yield very different values. In this study these types of values are treated as part of the value of the health of the system (see below).

Some of the costs of pollution can be determined fairly accurately; for example point source polluters pay indemnities to fishermen and owners of waters according to the

Finnish Water Act. The indemnities ordered paid by a major industrial polluter have been more than ECU 1 million in a recent court case (Decision 43,44/1993/3 by the Water Court of Western Finland). The 132 'hot spots' identified by HELCOM (1993) are generally worse polluters, suggesting higher cost of pollution. Costs of non-point source pollution have, however, not been estimated although it is evident that many recreational fisheries have found that non-point sources of pollution have negative impacts even in the least polluted areas of the Baltic Sea, the Bothnian Bay (Lappalainen and Hildén 1993).

The conversion of the health of the ecosystem to utilities or monetary units is controversial. If one would aim only at a diagnosis of the state of the system, the arcs and nodes related to the value of health could be neglected in the influence diagram. The influence diagram is, however, also concerned with the problem of the manager and thus some form of a comparison between utilities associated with a particular state of the ecosystem and the costs and gains of particular management regimes seems inevitable. By including these nodes I do not, however, propose that the value of a particular state of the Baltic Sea ecosystem can be determined objectively. The value of a health change depends largely, as Serchuk (1993) observed, on the answer to the question "what kind of a world do we want to live in?" The value node for health represents one way of specifying this answer. Because of this it is appropriate to examine a range of possible values for health and to analyze at which point the management implications change. To illustrate this I used three different ranges of health values, ECU -10 to 10 billion, ECU -50 to 50 billion and ECU -100 to 100 billion. In each range the extreme values corresponded respectively to a maximal decline of the health and a maximal improvement of the health of the system relative to the present.

Evaluation of the Influence Diagram

With three alternative courses of action for each decision node the influence diagram of this study offers altogether 27 decision alternatives. An evaluation of the influence diagram can proceed in many ways. The most straightforward evaluation gives the expected value of each management decision, which in theory gives an optimal solution to the management problem. A single evaluation is not, however, particularly useful because complex management decisions are usually arrived at through negotiation; in these subjective preferences and judgements play an important part and therefore it is important to present alternatives.

An influence diagram can aid judgements and negotiations by offering a systematic way of examining to what extent different assumptions imply a change of the preference order of alternative management decisions. This is illustrated in Figure 3 which gives the expected values of all decision alternatives under different assumptions for the value of a health change. The optimal environmental strategy changes from status quo environmental policy to a full implementation of the joint comprehensive environmental programme with increasing absolute values of health changes. The shift of the optimum has an impact on all related decision alternatives. The set of decision alternatives that involves a status quo environmental programme is most sensitive to a shift in the value of a health change, whereas the set involving a full implementation of the environmental programme is least sensitive (Figure 3).

The relative difference between different fisheries management alternatives also changes as a function of the value of health changes. The high initial probability of a decline of the piscivorous productivity combined with the relationships modelled in this influence diagram suggests that a reduction of the exploitation of piscivorous resources is a robust decision alternative. In general increased exploitation of planktivores also appears beneficial, but with high values of health changes a status quo policy for the exploitation of planktivores gives slightly higher utilities.

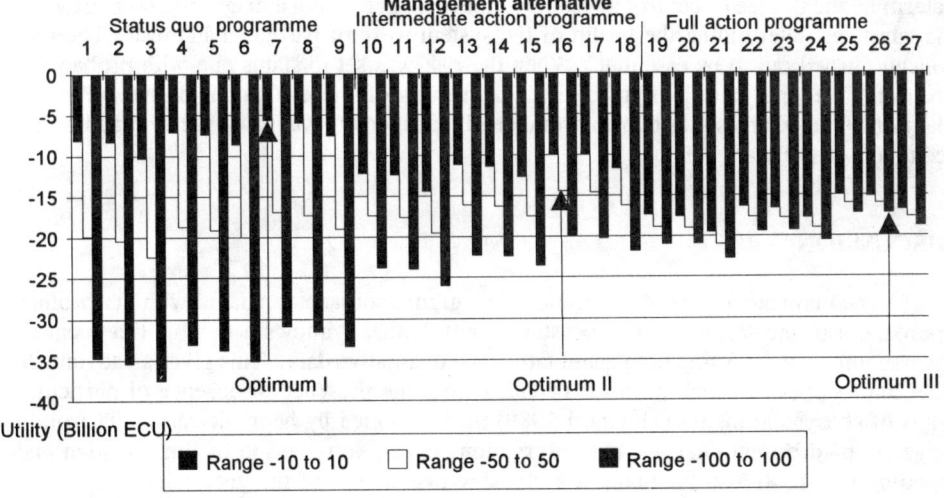

Figure 3. The 27 decision alternatives for the Baltic Sea management influence diagram under three different assumptions of value of ecosystem health changes. Decision alternatives 1-9 are all based on a status quo environmental programme. Alternatives 10-18 assume an intermediate, and alternatives 19-27 a full implementation of the joint comprehensive environmental programme. Within each set the first three alternatives assume increased exploitation of piscivores, and within each such triad the planktivore fisheries policy goes from increasing exploitation over the status quo to decreasing exploitation. The following three alternatives assume status quo piscivore exploitation, and the last three alternatives in each set of nine assume decreasing piscivore exploitation.

A further analysis of the diagram reveals the significance of other implicit assumptions. I have assumed that negative net values of fish yields arise when exploitation increases on weak stocks. If all catches are given positive values, optimal solutions suggest increasing exploitation. An increase of the exploitation of piscivorous fish (in practice cod) is contrary to current scientific advice (Thurow 1993), but corresponds to actual management decisions made by the Warsaw Commission (Hildén 1994).

The sensitivity analysis was also extended to the chance nodes without predecessors, i.e. nodes completely outside the control of any management decision. For example the role of oceanographic conditions was examined. The conclusions regarding the management action were robust against alterations between assumed stagnation or assumed inflow of oxygenated ocean water, although the relative differences between decision alternatives changed.

In general, structural changes and changes in the implicit assumptions of intermediate nodes may be more important than changes in the initial nodes, because they determine the degree of control available to the manager. From a diagnostic point of view the variables determining the health of the system were of particular interest. The risk concept turned out to be essential. When the risk was set to status quo with probability one for all events contributing to risk, the results of the evaluation of the influence diagram changed and status quo environmental management was favoured over additional reduction of pollutants.

DISCUSSION

The main advantage of the influence diagram approach compared with many other approaches to the diagnosis of ecosystem health is that it allows a flexible treatment of diverse information, using both quantitative and qualitative data. This gives an advantage over approaches in which health is deduced from the presence or absence of particular signs of ecosystem distress (Rapport 1989) or determined by heuristic reasoning from a large set of different changes and observations over a long period of time (Hildén and Rapport 1994). It also facilitates a systematic evaluation of the uncertainties involved in both diagnosis and management.

The influence diagram approach can easily accommodate new information on interrelationships or new insights that change the perception of the functions of the system. New data on toxic algal blooms, given a particular nutrient level and primary production, allow an adjustment of the probabilities in the node 'blooms'. If a definite proof emerges that the reproductive failure of salmon due to M74 (ICES 1993b) is caused by pollutants, it can be taken into account by adding an arc from the node 'hazardous compounds' to the node 'piscivores'. The probability matrices can also be replaced by functions or models which directly specify the relationship between nodes.

Flexibility is important for an approach to ecosystem diagnosis because of the characteristics of ecosystems (Norton 1992). At the same time a clear structure is advantageous in aiding predictions of future change because empirical data are difficult to use without a general model structure of how the system operates (Hannon 1992). Thus the relationship between environmental and fisheries management has been extensively discussed, but unless the relationship can be explicitly formulated it is difficult to arrive at any well-founded management conclusions.

Recently systems for providing feed-back from monitoring results to environmental management action has been developed in environmental impact assessments (Gray 1994). Influence diagrams can be used develop these feed-back systems by focusing on the informational arcs leading to the decision nodes. By using influence diagrams both probabilistic and deterministic rules for action can be developed.

The complexity of management problems creates difficulties which also affect the use and construction of influence diagrams. There is no unique correct structure, but instead different conceivable structures. The choice between different possible formulations will depend on the problem at hand and subjective judgement which stresses the connection between health assessments and policy formulation (Sagoff 1992). Explicit modelling of the choice between point source pollution management and non-point source management would lead to a different influence diagram. The time horizon is also a matter of judgement, which can affect the influence diagram formulation. In this study the emphasis is on intermediate time frames of several years to a decade, but it is evident that some of the environmental management decisions have implications beyond this time horizon. The conversion to net present value and the choice of a suitable discount rate can accommodate some of these concerns, but they are clearly not uniquely solvable. Therefore influence diagrams should not be regarded as a method by which environmental experts provide decision makers with definite recommendations for action. Instead influence diagrams should be used interactively as a method that facilitates systematic communication between experts, decision makers and other interested parties.

A practical difficulty with using influence diagrams is the need to formulate connections between system components in probabilistic terms. The uncertainty regarding many of these relationships is considerable (HELCOM 1990, Brügmann 1993), and scientists may be reluctant to specify anything. In this case the influence diagram can be used as an exploratory tool to determine to what extent it matters whether we have better data or not. An uncertain influence diagram may thus turn out to be useful by highlighting those cause and effect relationships having the greatest significance for diagnosis and future management of the Baltic Sea.

ACKNOWLEDGEMENTS

This paper has benefitted from comments by Wim Admiraal, Kai Kaatra, Sakari Kuikka, Eeva-Liisa Poutanen, David Rapport, Kenneth Sherman, Liisa Tuominen-Roto and Robert Williams, and from discussions at the NATO Advanced Research Workshop in October 1993. James Thompson kindly checked the English language.

REFERENCES

Agogino A.M, Ramamurthi K. (1990) Real time influence diagrams for monitoring and controlling mechanical systems. In Oliver R.M., Smith J.Q. (eds.) Influence Diagrams, Belief Nets and Decision Analysis J Wiley & Sons, Chichester pp:199-228

Brügmann L. (1993) Environmental review for the Baltic Sea. ICES Coop. Res. Rep. 186:3-18

Costanza R. (1992) Towards an operational definition of ecosystem health. In Costanza R., Norton B.G., Haskell B.D. (eds.) Ecosystem Health - New Goals for Environmental Management. Island Press Washington DC pp:3-20

Costanza R., Norton B.G., Haskell B.D. (eds.) (1992) Ecosystem Health - New Goals for Environmental Management. Island Press Washington DC pp:239-256

Decision Focus Inc. (1991) InDia Version 2.0, El Camino Real, Los Altos CA.

Gray J.S. (1994) Using environmental impact statements to better protect the environment: the role of feedback monitoring. Integrated Environmental Management 27:8-10

Hannon B. (1992) Measures of economic and ecological health. In Costanza R., Norton B.G., Haskell B.D. (eds.) Ecosystem Health - New Goals for Environmental Management. Island Press Washington DC pp:207-222

Haskell B.D., Norton B.G., Costanza R. (1992) What is ecosystem health and why should we worry about it? In Costanza R., Norton B.G., Haskell B.D. (eds.) Ecosystem Health - New Goals for Environmental Management. Island Press Washington DC pp:3-20

HELCOM (1990) Second periodic assessment of the state of the marine environment of the Baltic Sea, 1984-1988; general conclusions. Balt. Sea Environ. Proc. 35A

HELCOM (1992a) Activities of the Commission 1991 including the 13th Meeting of the Commission held in Helsinki 3 - 7 February 1992. Balt. Sea Environ. Proc. 42

HELCOM (1992b) Baltic marine environment bibliography 1986-1990. Balt. Sea Environ. Proc. 43

HELCOM (1993) The Baltic Sea Joint Comprehensive Environmental Action Programme. Balt. Sea Environ. Proc. 48

Henrion M. (1990) Toward efficient probabilistic diagnosis in multiply connected belief networks. In Oliver R.M., Smith J.Q. (eds.) Influence Diagrams, Belief Nets and Decision Analysis J Wiley & Sons, Chichester pp:385-410

Hildén M. (1994) Boundary conditions for the sustainable use of major fish stocks in the Baltic Sea. Ecological Economics (in press)

Hildén M., Rapport D.J. (1994) Four centuries of cumulative impacts on a Finnish river and its estuary: an ecosystem health approach. J. Aquat. Ecosystem Health (in press)

ICES (1991) Reports of the ICES Advisory Committee on Fishery Management 1990, part 1. ICES Coop. Res. Rep. 173

ICES (1993a) Eighth ICES dialogue meeting: How to use the sea: management interactions with special reference to the Baltic and its fisheries. ICES Coop. Res. Rep. 186

ICES (1993b) Report of the Baltic salmon and trout working group ICES C.M. 1993/Assess:14

Kosior M., Netzel J. (1989) Eastern Baltic cod stocks and environmental conditions. Rap. pp-v. Réun. Cons. int. Explor. Mer 190:159-162

Kuikka S., Varis O. (1992) Use of Bayesian influence diagram in fisheries management - the Baltic salmon case. ICES C.M. 1992/D:5

Lappalainen A., Hildén M. (1993) Recreational fishing and environmental impacts in the Archipelago Sea and the Finnish part of the Gulf of Bothnia. Aqua Fennica 23:29-37

Melvasalo T., Pawlak J., Grasshoff K., Thorell L., Tsiban A. (eds.) (1981) Assessment of the Effects of Pollution on the Natural Resources of the Baltic Sea, 1980. Balt. Sea Environ. Proc. 5 A

Nehring D., Schulz S., Rechlin O. (1989) Eutrophication and fishery resources in the Baltic. Rap. pp-v. Réun. Cons. Int. Explor. Mer 190:198-205

Nordisk havovervågningsprogram (1993) Nordic Council of Ministers, Nord 1993:14

Norton B.G. (1992) A new paradigm for environmental management. In Costanza R., Norton B.G., Haskell B.D. (eds.) Ecosystem health - New Goals for Environmental Management. Island Press Washington DC pp:23-41

Oliver R.M., Smith J.Q. (eds.) (1990) Influence Diagrams, Belief Nets and Decision Analysis. J Wiley & Sons, Chichester

Öresundkonsortiet (1992) Environmental impact assessment for the fixed link across the Öresund. SVEDAB June 1992

Pearl J. (1988) Probabilistic Reasoning in Intelligent Systems: Networks of Plausible Inference. Morgan Kaufmann, San Mateo CA

Rapport D.J. (1989) Symptoms of pathology in the Gulf of Bothnia (Baltic Sea): ecosystem response to stress from human activity. Biol. J. Lin. Soc. 37:33-49

Rapport D.J., Regier H.A., Hutchinson T.C. (1985) Ecosystem behaviour under stress. Am. Nat. 125:617-640

Rapport D.J. (1992) Evaluating ecosystem health. J. Aquat. Ecosyst. Health 1:15-24

Rudstam L., Aneer G., Hildén M. (1994) Top-down control in the Baltic Sea. Dana (in press)

Saaty T.L. (1988) Multicriteria decision making - the analytic hierarchy process: planning, priority setting, resource allocation. ISBN 0-07-054371-2, USA

Sagoff M. (1992) Has nature a good of its own? In Costanza R., Norton B.G., Haskell B.D. (eds.) Ecosystem Health - New Goals for Environmental Management. Island Press Washington DC pp:57-71

Serchuk F.M. (1993) Summary and comments to the eighth ICES dialogue meeting. ICES Coop. Res. Rep. 186:86-89

Shachter R.D., Eddy D.M., Hasselblad V. (1990) An influence diagram to medical technology assessment. In Oliver R.M., Smith J.Q. (eds.) Influence Diagrams, Belief Nets and Decision Analysis J Wiley & Sons, Chichester pp: 321-350

SNV (1990) Marine pollution '90 - Action programme. Swedish Environmental Protection Agency Informs. Sweden
Sparholt H. (1991) Multispecies assessment of Baltic fish stock. ICES Mar. Sci. Symp. 193:64-79
Thurow F. (1993) Fish and fisheries in the Baltic Sea. ICES Coop. Res. Rep. 186:20-36
Varis O., Kettunen J., Sirviö H. (1990) Bayesian influence diagram approach to complex environmental management including observational design. Comp. Stat. Data Anal. 9:77-91
Wulff F., Niemi A. (1992) Priorities for the restoration of the Baltic Sea - a scientific perspective. Ambio 21:193-195
Wulff F., Stigebrandt A., Rahm L. (1990) Nutrient dynamics of the Baltic Sea. Ambio 19:126-133

14. EVALUATION OF NEW TECHNIQUES FOR MONITORING AND ASSESSING THE HEALTH OF LARGE MARINE ECOSYSTEMS

Robert Williams
Plymouth Marine Laboratory, Prospect Place, West Hoe Plymouth
Devon PL1 3DH
England

MONITORING CHANGE WITHIN THE MARINE ECOSYSTEM

Until recently to effectively monitor the marine biota and associated environmental variables both temporally and spatially throughout ocean basins or Large Marine Ecosystems (LMEs) (Sherman 1993) was a prohibitively costly and almost an impossible logistical task. Biological processes within a pelagic ecosystem range from seconds to years with a spatial domain of centimetres to tens or even hundreds of kilometres. Developments in the last 15 years in marine instrumentation and remotely sensed measurements of water colour from space together with increased data processing and storage capabilities have allowed us for the first time to approach this task with any realism. The observation made by Aebischer et al. (1990) that changes in one section of the marine community, over several decades, could be representative of the 'whole ecosystem' was significant. Their data on long term trends (33 years, 1955-1987) across four marine trophic levels in the north west North Sea of phytoplankton, zooplankton, herring, kittiwake breeding and one of climatic data, westerly weather, all showed a similar parallel pattern. The data series were examined by means of spectral and cross spectral analysis and showed remarkable parallelism. The authors were cautious about the causal mechanisms but they stated that "the results posed a major challenge to modellers to reproduce such patterns which could be used in interpreting global and climate change." It was probably the first supportive evidence that it was not necessary to monitor 'every' variable within the system and that detailed information of a selection of 'key' variables, over time, could indicate the general patterns of the change within a Large Marine Ecosystem. This still implies that we require a holistic approach to sampling the marine ecosystems because ecosystem health is assessed in the overall performance of the biological systems and not just in production of selected species (Costanza 1992; Sherman and Solow 1992). Slow change or fluctuations within the pelagic community implies 'natural' change while degradation is brought about by anthropogenic impacts on the system which accomplish destruction of species and complexity of ecological systems. This is primarily true in that anthropogenic impacts usually alter the system relatively quickly and these changes can be monitored against the background of natural fluctuations within slow-changing ecosystems.

It can be said that there is no well developed, generally accepted body of theory as to why biomass and productivity of pelagic organisms vary greatly in space and time. According to McGowan (1989), this can be attributed to a lack of adequate data on the descriptions of the frequency spectra of changes in biological, physical and chemical properties of the system. If we are to understand the consequences of large scale climatic variations or anthropogenic impacts then we must understand the response of the multispecies pelagic biota to change. Significant time scales of long-term variability in plankton and environment have been demonstrated from Continuous Plankton Recorder (CPR) data (Colebrook and Taylor 1984; Radach 1984; Colebrook 1986; Dickson et al. 1988; Aebischer et al. 1990). The main body of information on the effects of climatic change on the marine ecosystem comes from these long-term studies of plankton (Holligan and Reiners 1992) and the fact that these 'signals' are now an anticipated feature of marine pelagic biota is due, in part, to the results obtained over the last forty years by the CPR. The interpretation of these time series observations remains difficult although the detection of consistent patterns of relationships are beginning to allow the development of empirical predictive models (McGowan 1990).

If we follow the definition of Costanza (1992) of 'health' of an ecosystem as a characteristic of the complex natural systems then we have to monitor the change within the biota over time to be able to develop an understanding of these systems (Sherman and Solow 1992). There are essentially three methods in which data are collected from the pelagic ecosystem, ie. from surveys using research ships, remotely by aircraft or satellite and in future by unmanned autonomous vehicles and finally, using ships-of-opportunity. These methods are not mutually exclusive and are complimentary in providing time series and data necessary for monitoring and predicting the changing states of large marine ecosystems. A full review of all methodologies cannot be accomplished but specific technologies required to effectively sample a LME will be assessed.

MEASUREMENTS FROM RESEARCH SHIPS

Large international multidisciplinary research cruises and programmes are now directed at many problems in the ocean. Although of limited duration these cooperative ventures by the international scientific community are invaluable in addressing biological, hydrological, geological and pollution problems. In reality, with the limited resources available, it is now the only means whereby a full range of modern state-of-the-art techniques can be brought to bear on these complex oceanographic problems. Certain of these techniques are relevant to sampling LMEs.

Acoustic and Optical Techniques

The state-of-the-art of acoustical and optical sampling methods were reviewed in the GLOBEC Workshop held in Woods Hole, USA in April 1991 (GLOBEC 1991). It was recognised that different system parameters such as frequencies, beamwidths, signal and processing algorithms were required to examine the different size spectra of biological material from zooplankton to fish. The main conclusion from the Workshop was that there is no all encompassing technique available to measure the full size spectral range of marine organisms and that further technological development as well as experimental studies are required.

Acoustic Techniques

A review of the application of acoustic technology to LMEs is given by Holliday (1993). The authour's conclusions were that the main sensor technology had already been completed and that a new system could be built, to address the questions of fisheries biologist and biological oceanographers, if resources were made available.

There are two main acoustic techniques for quantifying distribution and abundance of marine organisms. These systems are either based one beam per frequency or multiple beams per frequency. The Multi Frequency Acoustic Profiling System (MAPS) (Greenlaw 1979; Holliday 1980; Holliday and Pieper 1980; Pieper and Holliday 1984 Holliday et al. 1989) relies on frequency-dependent backscattering models to partition volume backscattering data into size classes (Herman 1992; Herman et al. 1993). The multiple beams per frequency allows analysis of single echoes and target strength estimation. There are two methods: the dual-beam technique (Traynor and Ehrenberg 1979; Richter 1985; Greene et al. 1989); and the split-beam technique (four beams per channel) (Ehrenberg 1979), both developed for detection and quantification of fish schools although the former has been used in studies of zooplankton (Greene et al. 1989; Wiebe et al. 1990). The main problems with acoustic techniques is the variability of target strength because of the different orientation of the animals and the necessary characterization of the targets through net hauls and taxonomic identification of species. Software to extract structure from multidimensional data, which allows classification and identification of fish schools, is constantly being improved (Scalabrin et al. 1992). Acoustics remains an invaluable technique for fast quantification of abundance and distribution of pelagic species (especially fish populations) and is a pre-requisite for evaluation of resources within an LME.

Photo-optical Systems

The use of non-video optical instruments for studying distributions of mesozooplankton abundance and distributions have centred around the optical plankton counter (OPC) of Herman (1988). The unit, which can be towed attached to a Batfish (Herman and Dauphinee 1980; Herman 1985; Herman et al. 1993), detects scattered or transmitted light and can measure abundance and size of organisms ranging from 250 μm to approximately 20 mm equivalent spherical diameter at rates up to 200 s^{-1}, towed at speeds from 0.5 to 4.0 m s^{-1} up to depths of 1000 m (Sprules et al. 1992). The value of this technology is already proven (Herman et al. 1991; Herman et al. 1993) but there are limitations. These are problems associated with coincident counts at high particle densities, errors in size measurements due to translucency, orientation of the organisms in the sampling window and the inability of the system to distinguish living from non living particulate (Sprules et al. 1992). All these difficulties are solvable with further development or linking with other instrumentation (video and acoustics) but it must be emphasised that the OPC is a major tool in gathering essential information on particle size and abundance on basin wide scales.

Bio-optical properties of the water, chlorophyll florescence, upwelling and downwelling irradiance, photosynthetically available radiation, beam transmission etc., can be measured from towed or moored systems using a variety of sensor (Dickey 1988, 1991). These data are essential at various locations to validate satellite imagery and to provide additional vertical information which cannot be measured by satellite colour scanners.

One of the most important sensors to be developed in recent years is the pump and probe fluorometer (Falkowski and Kobler 1990) which enables the *in situ* mapping, both spatially and temporally of primary productivity in the ocean. Another flash stimulated fluorescence technique based on the same fluorescence/photosynthesis relationship as the pump and probe is the fast repetition rate fluorometer (Kobler and Falkowski 1992) which measures the photosynthetic rates *in situ*. This method also measures several other photosynthetic parameters, effective absorption cross section, photoconversion efficiency, and turnover time of photosynthesis and relates them to primary productivity (Kobler and Falkowski 1992). Both systems can be used on moored or profiling systems and development is proceeding to build a unit for the Undulating Oceanographic Recorder for use in monitoring LMEs.

Towed Net Systems

Many of these advanced plankton net system (MOCNESS - Wiebe et al. 1976; BIONESS - Sameoto et al. 1980) are used as carriers for a variety of sensor packages and to calibrate acoustic and optical systems. The Video Plankton Recorder (Davis et al. 1992b) is a case in point, it can be towed at speeds $0.5-3.0m\ s^{-1}$ in shelf and oceanic waters providing data on plankton on scales from micrometres to kilometres with the nets catching plankton to calibrate the data. These systems can be used on transects of 100km taken in 14 h at towing speed of $2\ m\ s^{-1}$. The VPR is currently rated to 300m and a pressure case rated to 2000m is under construction (Davis and Gallager 1993)

Molecular Biological Techniques

The cellular and molecular techniques now being applied to marine ecology and biological oceanography are set to revolutionise our approach to biological oceanography, especially when coupled with the information gained from satellite remote sensing (Powers 1993). It is postulated that the use of these techniques will help our understanding of the mechanisms responsible for biological variability in our oceans. The value and application of molecular techniques using isozymes, nucleic acid sequencing techniques, deoxyribonucleic acid probes and immunological methods are aptly summarised by Powers (1993). He postulates that it will be possible to use these techniques to "trace water masses, to examine energy and carbon flow in the coastal and ocean food webs, to unravel the complexities of trophic dynamics and to delineate the intricacies of marine community structures". These prophetic concepts already have their followers and it is an area of marine science which will show great expansion in the coming years especially with the continued development of automated molecular analyses techniques.

Underway Profiling Systems

The Continuous Plankton Recorder and associated instrumentation (Williams and Aiken 1990) was recommended by the 1991 Cornell workshop (Sherman and Laughlin 1992) as one of the instruments to provide the data for monitoring the changing states of Large Marine Ecosystems. Similarly the Undulating Oceanographic Recorder/ Aquashuttle (Aiken and Bellan 1990) and the versatility of its sensor payload has a major role to play in this capacity, especially when configured for bio-optical measurements. Both of these towed vehicles have the added advantage of being small, robust and deployable from ships-of-opportunity but are capable of carrying the 'state-of-the-art' in sensor technology. The SeaSoar (Pollard 1986) and BATFISH are both towed undulating data acquisition vehicles with large payload capacities with the capability of being towed at speeds of 10 knots. These instruments require the back-up of a research ship but are invaluable for ocean basin scale surveys and can provide real-time high resolution data acquisition from the surface to 500m and are being used in World Ocean Circulation Experiment (WOCE) and the Joint Geochemical Ocean Flux Studies (JGOFS) and in the UK's Biogeochemical Ocean Flux Studies (BOFS). These vehicles use faired cable to obtain maximum amplitude and minimum wave length of undulation and carry a multisensor package (CTD, fluorometer, nephelometer and an OPC (Herman and Dauphinee 1980; Herman et al. 1991) together with other sensors). One of the most important sensors to be added to the package is the pump and probe fluorometer or the Fast Repetition Rate (FRR) fluorometer.

MOORED AND DRIFTING BUOYS

Many, if not all, of the developed sensors and technologies for towed vehicles can be modified for use on moored or drifting buoys. Examples are seen in the present OPC which is capable of measuring zooplankton population abundance and size, as well as the new development of a laser-based OPC which is smaller and more compact and will allow multiple arrays on a mooring (Herman 1993) for long term monitoring. A new *in situ* underwater particle analyzer based on the principle of diffraction pattern analysis of a laser beam, measuring particles in the range of 1 to 500 um, is already in use in French coastal waters (Gentien and Lunven 1993). The Video Plankton Recorder system can also be either moored or towed. Davis et al. (1992ab) give the description and design of this equipment.

The multi-variable moored system (MVMS) such as one deployed in the Biowatt (Bio-optical Variability on Seasonal Time Scales in the Sargasso Sea) Program (Dickey 1991; Dickey et al. 1991) are necessary to provide data on the temporal and spatial variability of the optical properties of the upper ocean. The data collected in this programme in 1987, from eight depths between 14 and 160 m, constituted the longest time series of high resolution measurements of bio-optical and physical variables in the open ocean (Dickey et al. 1991). A similar multi-sensor buoy system which was free drifting, the R/P FLIP (Dickey et al. 1986), was used during the Optical Dynamics Experiment (ODEX) in the North Pacific subtropical gyre in 1982 to measure *in situ* oceanographic measurements of physical and bio-optical variables.

There are many moored and drifter type of buoys deployed in the ocean equipped with meterological and environmental sensors with their data transmitted via satellite. Free-drifting satellite telemetered buoys are essential for providing long-term circulation data of ocean basins and should included in the sampling equipment necessary for monitoring long term changes within LMEs.

REMOTE SENSING

The value and importance of satellite remote sensing of ocean colour with regard to the understanding and monitoring of LMEs cannot be emphasised enough. Remote sensing provides the only source of 'truly' synoptic data over large areas. The application of this technology began with the launch of NASA's NIMBUS-7 satellite equipped with the multispectral Coastal Zone Color Scanner (CZCS) which was operational, following its launch in 1978, from 1979 to 1986. Details of its six spectral bands (three were available for pigments), limitations of spectral resolution and sensitivity of the imagery and general problems with the sensor (it was only operational for approximately 10% of the time) are given in Yoder et al. (1988), and Yoder and Garcia-Moliner (1993). These authors concluded that the work completed on these data from the CZCS sensor will provide the necessary groundwork for the full exploitation of the improved color scanner, the Sea-Viewing Wide-Field-of-View Sensor (SeaWiFS), due to be launched by NASA in 1994 (SeaWiFS Working Group 1987; General Sciences Corp. 1991). Observations on ocean colour have been used to measure chlorophyll concentrations, define frontal zones and upwelling regions as well as estuarine suspended particulate concentrations and scattering by coccolithophores and as a tracer for upper ocean dynamics (Robinson 1990). The biological activity of whole ocean basins could be monitored in temporal sequence and research cruise activities placed in a larger context. Examples of the use of CZCS data are seen in Oceanography from Space (Gower 1981), the South African Ocean Colour and Upwelling Experiment (Shannon 1985), CZCS Atlas of Water Optical Properties in the Alboran Sea (Arnone and Oriol 1985) and the North Sea Satellite Colour Atlas (Holligan et al. 1989). The use of remote sensing with respect to biological oceanography is reviewed by Aiken et al. (1992) who concluded that the CZCS sensor, despite limitations, "provided the basis for new approaches for the estimation of global marine primary productivity" and that new sensors such as SeaWiFS will provide the necessary data for new studies on oceanographic and climatic models. A detailed programme of ocean optical measurements for validation and calibration of the SeaWiFS sensors is already in place (Aiken in press) but will require a combination of deployed sensors including those of the UOR to achieve the goals of SeaWiFS of 5% accuracy of water-leaving radiance and 35% accuracy in chlorophyll **a** measurements. The properties of the surface ocean waters and of the marine atmosphere can also be modified by the optical and biochemical properties of marine organisms (Holligan and Reiners 1992). The authors concluded that models should be extended to include relevant optical parameterization of both the surface ocean and marine atmosphere.

Much of the current work in the UK is concentrated on relationships between the optical properties of marine waters and their biogenic constituents (phytoplankton biomass, chlorophyll accessory pigments and dissolved organic compounds). This involves the analysis of airborne and satellite imagery of sea surface temperature and ocean colour including the development of processing methods and atmospheric correction algorithms, especially the development of algorithms of sea colour in terms of its biogenic constituents. The role of towed vehicles, equipped with sensors for acquiring data on the optical properties of the surface layers, deployed from research ships and ship-of-opportunity vessels is central to the effective survey of large ocean regions.

AUTONOMOUS VEHICLES

Possibly in excess of 100 Autonomous Underwater Vehicles (AUVs) are being developed around the world primarily aimed at acquiring high quality data, both temporal and spatial, across ocean basins. To date only a handful have actually been deployed. Many of these projects involve development of sensor technology and instrument systems, and are directed towards addressing major global environment questions. The Natural Environment Research Council's (U.K.) programme on AUTOSUB is such a project. One of its sensor payload configurations DOLPHIN (Deep Ocean Long Path Hydrographic Instrument) will allow it to undulate across an ocean basin, carrying out physical, chemical and biological measurements throughout the water column. The system will surface at intervals to position itself and allow data to be telemetered via satellite (NERC 1992). This system could be configured to sample biological, physical and chemical variables within the euphotic zone of a Large Marine Ecosystem and has the added advantage that its track could be configured to provide detailed data on large and small scale features within the LME. Another UK project funded under the Department of Trade and Industry's 'Wealth from Oceans' programme to a British industrial consortium have constructed and 'flown' an Autonomous Underwater Vehicle developed for hydrographic and environmental survey under the Arctic ice. This low cost system has a depth capability of 300 m and an approximate 2 day, 300 km, duration with

ARGOS tracking and GPS position fixing with a payload volume for its sensors of 0.45 m^3. In the future, such systems with extended duration, would be ideal tools for monitoring Large Marine Ecosystems or Exclusive Economic Zones in conjunction with a series of moorings.

SHIPS-OF-OPPORTUNITY

Underway Profiling Systems

The future of ocean science may be suites of moorings with every 'state of the art' sensor attached with AUVs roaming the ocean basins monitored by satellite and research ships. What we must address now is the implementation of an affordable monitoring strategy designed to answer the problems facing the management of ocean and continental shelf resources.

Currently, the cheapest and most effective way to monitor environmental variables, biological diversity and changing community structure, both spatially and temporally, of the marine plankton in large sea areas requires the sampling strategy embodied in the ship-of-opportunity deployments of the Continuous Plankton Recorder (CPR) and the Undulating Oceanographic Recorder(UOR) or Aquashuttle. Description of the temporal variability in marine populations obtained from such surveys is one way to gain information on the whole ecological processes acting in the sea (Cushing 1988). CPR surveys are now the only measure of the general state of oceanic plankton, over large space and time scales, available to the marine scientific community. The peak coverage of the U.K. CPR survey was from the mid 1960s to the 1980s when it covered the North Sea, English Channel and the northern North Atlantic Ocean between 35° and 65°N (Colebrook 1982). As part of the routine analysis, a total of 391 plankton entities (sub-species, species, developmental stages and higher taxa) are analyzed and counted. A partner Survey has been operated from the National Marine Fisheries Service Laboratory at Narragansett, RI, on the east coast of the U.S. from late 1960's to the present.

Continuous Plankton Recorder

The Continuous Plankton Recorder (CPR) was designed as a plankton sampling instrument to be towed behind merchant ships at a constant depth along their regular trading routes. The CPR Survey was started in 1931. For many years no physical measurements were taken with the CPR which imposed limitations on the interpretation of the plankton data collected. Recording thermographs measuring temperature were employed on selected routes during the 1960s and early 1970s and these were superseded by electronic systems logging data onto tape cassettes (Aiken 1980). The development of a variety of sensors and the availability of solid state data loggers have made it possible to measure and record a whole range of environmental variables simultaneously with the plankton sampling in the CPR (Williams and Aiken 1990).

The present CPR has been modified to accept a self-contained battery powered solid state logging package. The instrument can measure depth, temperature, conductivity, chlorophyll fluorescence, turbidity and upwelling and downwelling light at two wavelengths (450 and 550 nm, ie. blue and green). A total of 16 channels are available, so additional sensors can be added. The data logger has an eight Mbyte capacity and can accommodate 200,000 measurements of all 16 channels. The scanning rate is user selectable from 60 s to 1 per day. At a ship's speed of 15 kts the distance the vehicle can be towed is 450 nm before battery replacement, which is equivalent to the duration of the plankton sampling system. If longer intervals are selected then the system can be set to power down between measurements thus giving a longer battery life and extended tow duration.

The sensors and data logger are automatically switched on and off by contact with sea water which allows the system to be deployed unmanned in a true ship-of-opportunity mode. The system is currently being deployed on a number of routes in the north east Atlantic. This instrument pack has now been re-designed by Chelsea Instruments Ltd and is sold as the **Aquapack**. As well as having the standard sensors for conductivity, temperature and depth, it has options for turbidity, pH, Redox and dissolved oxygen.

The present CPR Survey is run from Plymouth by the Sir Alister Hardy Foundation for Ocean Science (SAFHOS) and routes are towed monthly throughout the year. A summary of the Survey is given by Gamble and Hunt (1992).

Undulating Oceanographic Recorder (UOR)/Aquashuttle

The UOR, like the CPR, is a self contained instrument which can be deployed from merchant ships. The concept of an undulating towed vehicle dates back to the 1930s but it was the development of electronics to monitor and control the undulation pattern which allowed a practical and reliable system to be developed.

In the current Aquashuttle system the data are logged at pre-set intervals on 16 channels in a solid state data logger with up to 2 Mbyte CMOS memory. The sensor package measures depth, temperature, conductivity, chlorophyll, turbidity, upwelling and downwelling light at 3 wavelengths (412nm, violet; 443nm, blue; 554nm, green, with options for additional wavelengths at 490, 520 and 632nm), photosynthetically available radiation (PAR) and two aspects of performance of the towed vehicle, alternator output and the angle of the crank; which determines the position of the diving plane. The system has been configured during the JGOFS North Atlantic Bloom Experiment in 1989 with sensors to measure in-water reflectances (R_{410} R_{445}, R_{490}, R_{520}, R_{560}) to compare with NASA aircraft measurements of upwelling radiance in 32 wavebands between 380 and 740 nm (Hoge et al. 1986ab; Aiken et al. 1992). The ship and aircraft measurements were comparable over a track spanning 8° latitude and provided validation for the aircraft laser induced chlorophyll fluorescence measurement of chlorophyll. The towed vehicle can also carry as options a plankton sampling system to collect phytoplankton, zooplankton and samples for fish eggs and larvae, a sensor package to measure dissolved nitrate and nitrite, a bioluminescence sensor, a hydrocarbon sensor, blue and green transmissometers, pH electrode, dissolved oxygen electrode and, it is hoped, a pump and probe fluorometer to derive *in situ* photosynthetic rates. The technical specifications of the UOR are given by Aiken et al. (in press). The Plymouth Marine Laboratory instrument is now marketed by Chelsea Instruments Ltd. as the Aquashuttle.

The Aquashuttle is an ideal vehicle for deployments from high-speed merchant ships (ships-of-opportunity) to collect data underway of environmental variables (Aiken et al. in press). The vehicle can be programmed to tow horizontally or undulate from the surface to a maximum depth of 100m, provided a short length of faired cable is used, with a frequency of 800 to 4000m. Since 1989, the UOR has been deployed regularly on ship-of-opportunity tows between Grimsby and Aberdeen; along the north east coast of U.K. The use of the UOR, equipped with a suite of light sensors for measurement of the optical properties of the oceans for interpretation of satellite remotely sensed images of ocean colour has been discussed by Aiken and Bellan (1990), Aiken et al., (1992, in press), Trees et al. (1992), and for Airborne Oceanographic Lidar (AOL) by Yoder et al. (1993). Aiken and Bellan (1990) discussed the difficulties of co-incident measurements especially inaccuracies brought about by geo-referencing the data sets but illustrate an excellent coincident set of data from Advanced Very High Resolution Radiometry (AVHRR) and UOR reflectances for 3 UOR tows in the northeast Atlantic in June 1987. The UOR, as a carrying vehicle for a variety of sensor packages, is a relatively low cost system which can be used for monitoring LMEs and will also provide the necessary high-resolution, quasi-synoptic measurements required to complement satellite observations.

SAMPLING A LARGE MARINE ECOSYSTEM USING SHIPS-OF-OPPORTUNITY

It is necessary to set up a network of routes to initiate a sampling programme of a LME. These routes are selected after full consultation with the shipping companies involved and outline permission obtained from their management and Captains. The specific intention of the deployments is to provide essential information on the geographical distribution and seasonal changes in abundance and population structure of the major larger phytoplankton and zooplankton species (excluding fragile gelatinous plankton) within the LME in relation to the environmental data measured. In addition, the information from the towed sensors will indicate the timing and locality of phytoplankton blooms (as indicated by changes in chlorophyll levels and turbidity). The chlorophyll data can be used with the irradiance data to improve the algorithms for the interpretation of satellite images and to look at the distribution of particles and pigments in relation to estimated potential primary production. There is also the potential to use ship-board sensors to measure nutrients, hydrocarbons and to take particulate samples for heavy metal analysis, pesticide and radioactivity. The alpha particle radioactivity can be monitored from selected plankton samples collected throughout the year; detection limits for alphas is about 0.05 Bq Kg (Hamilton et al. 1991). Detailed information will be provided on the marine food web and trophodynamics by evaluation of species composition, community structure (Williams et al. 1993), geographical distribution, biomass and seasonal succession of the dominant species in relation to the physical, chemical and biological variables (Lindley and Williams 1994). These data are essential for understanding food web responses and variations in relation to temporal and spatial resolution for validation of key ecosystem processes.

CONCLUSIONS

A core marine ecosystem monitoring programme has been suggested by Sherman and Laughlin (1992) which is based on transects sampled by the CPR and UOR to be supplemented by satellite remote sensing and systematic trawl and acoustic surveys from research ships. The candidate parameters for such a core sampling programme can be readily obtained from fully instrumented towed vehicles accompanied by on-board flow-through monitoring equipment. The importance of this core monitoring proposal is that it is affordable, cost effective and utilises available and newly developed sensors, towed vehicles and ships-of-opportunity for providing the data series of environmental, chemical and biological variables. These time series data will improve our understanding of marine coastal and ocean processes and are essential to assist in the development of a variety of indices, such as resilience, vigour and organisations, necessary to obtain an operational definition of ecosystem health (Costanza et al. 1992; Sherman and Solow 1992). If these basic monitoring programmes can be instigated now, at selected sites around the world, they will provide the necessary time series data and background to incorporate these newer technologies and sensors as they become available in the final stages of this century.

ACKNOWLEDGEMENTS

This work forms part of the Laboratory Research Project 3 of Plymouth Marine Laboratory a component body of the Natural Environment Research Council.

REFERENCES.

Aebischer N.J., Coulson J.C., Colebrook J.M. (1990) Parallel long term trends across four marine trophic levels and weather. Nature, Lond. 347:753-755
Aiken J. (1980) A marine environmental recorder. Mar. Biol. 57:238- 240
Aiken J. (in press) Special requirements for the validation of ocean colour information. Tech. Conf. on Space-based Ocean Observation. Bergen, Norway, 5-10 September 1993

Aiken J., Bellan I. (1990) Optical oceanography: An assessment of a towed method. In Herring P.J., Campbell A.K., Whitfield M., Maddock L. (eds.) Light and Life in the Sea. Cambridge Univ. Press pp:39-57

Aiken J., Moore G.F., Holligan P.M. (1992) Remote sensing of oceanic biology in relation to global climate change. J. Phycol. 28:579-590

Aiken J., Pollard R., Griffiths G., Bellan I. (in press) Measurements of upper ocean structure with towed profiling systems. Oceanology

Arnone R., Oriol R. (1985) CZCS Atlas of water optical properties in the Alboran Sea. Naval Ocean Research and Development Activity Report 117 NSTL, Mississippi

Colebrook J.M. (1982) Continuous plankton records: seasonal variations in the distribution and abundance of plankton in the North Atlantic Ocean and North Sea. J. Plank. Res. 4:435-462

Colebrook J.M., Taylor A.H. (1984) Significant time scales of long-term variability in the plankton and the environment. Rap. P.-v. Reun. Cons. Perm. Int. Explor. Mer 183:20-26

Colebrook J.M. (1986) Environmental influences on long-term variability in marine plankton. Hydrobiology 142:309-325

Costanza R. (1992) Towards an operational definition of ecosystem health. In Costanza R., Norton B.G., Haskell D.D. (eds.) Ecosystem Health - New Goals for Environmental Management. Island Press, Washington DC pp:239-256

Cushing D.H. (1988) Temporal variability in production systems. In Longhurst A.R. (ed.) Analysis of Marine Ecosystems. Academic Press pp:443-471

Davis C.S., Gallager S. (1993) The Video Plankton Recorder. U.S. Globec News 3:6-9

Davis C.S., Gallager S.M., Berman M.S., Haury L.R., Strickler J.R. (1992a) The Video Plankton Recorder (VPR): Design and initial results. Arch. Hydrobiol. 36:67-81

Davis C.S., Gallager S., Solow A.R. (1992b) Microaggregations of oceanic plankton observed by towed video microscopy. Science 257:230-232

Dickson R.R., Kelley P.M., Colebrook J.M., Wooster W.S., Cushing D.H. (1988) North winds and production in the eastern North Atlantic. J. Plankton Res. 10(1):151-169

Dickey T.D. (1988) Recent advances and future directions in multi-disciplinary *in situ* oceanographic measurement systems. In Rothschild B.J. (ed.) Towards a Theory on Biological-Physical Interactions in the World Ocean. Kluwer Academic, Dordrecht, Netherlands pp:555-598

Dickey T.D. (1991) The emergence of concurrent high resolution physical and bio-optical measurements in the upper ocean and their applications. Revs. Geophys. 29(3):383-413

Dickey T.D., Siegel D.A., Bratkovich A., Washburn L. (1986) Optical features associated with thermohaline structures. Proc. SPIE-Int. Soc. Opt. Eng. 637:308-313

Dickey T.D., Marra J., Granata T., Langdon C., Hamilton M., Wiggert J., Siegel D., Bratovich A. (1991) Concurrent high resolution bio-optical and physical time-series observations in the Sargasso Sea during the spring of 1987. J. Geophys. Res.

96:8643-8663
Ehrenberg J.E. (1979) A comparative analysis of *in situ* methods for directly measuring the acoustic target strength of individual fish. IEEE J. Oceanog. Eng. OE-4:141-152
Falkowski P.G., Kobler Z. (1990) Phytoplankton photosynthesis in the Atlantic Ocean as measured from a submersible pump and probe fluorometer in situ. In Baltscheffsky M. (ed.) Current Research in Photosynthesis IV. Kluwer, London pp:923-926
Gamble J.C., Hunt H.G. (1992) The Continuous Plankton Recorder Survey: a long-term, basin-scale oceanic time series. In Proceedings of the Ocean Climate Data Workshop, 18-21 February 1992, Goddard Space Flight Center, Maryland, USA pp:277-293
General Sciences Corporation (1991) SeaWiFS Science Data and Information System Architecture Report. General Sciences Corporation, Maryland
Gentien P., Lunven M. (1993) New perspectives in coastal marine environment management due to development in instrumentation. ICES Statutory Meeting, Dublin, 22-23 Sept. 1993
GLOBEC (1991) Report No. 4. Workshop on Acoustical Technology and Integration of Acoustical and Optical Sampling Methods. Joint Oceanogr. Instit. Inc. Washington DC
Greene C.F., Wiebe P.H., Burczynski J. (1989) Analyzing zooplankton size distributions using high-frequency sound. Limnol. Oceanogr. 34:129-139
Greenlaw C.F. (1979) Acoustic estimation of zooplankton populations. Limnol. Oceanogr. 24:226-242
Gower J.F.R. (ed.) (1981) Oceanography from Space. Plenum Press, New York
Hamilton E.I., Williams R., Kershaw P.J. (1991) Total alpha particle radioactivity for some components of marine ecosystems. In Kershaw P.J., Woodhead D.S. (eds.) Radionuclides in the Study of Marine Ecosystems. Elsevier, London pp:234-244
Herman A. (1985) Biological profiling in the upper oceanic layers with a Batfish vehicle: A review of applications. In Zirino A. (ed.) Mapping Strategies in Chemical Oceanography. Amer. Chem. Soc. Washington DC pp:293-314
Herman A.W. (1988) Simultaneous measurement of zooplankton and light attenuance with a new optical plankton counter. Cont. Shelf Res. 8:205-221
Herman A.W. (1992) Design and calibration of a new optical counter capable of sizing small zooplankton. Deep Sea Res. 39:395-415
Herman A.W. (1993) Moored Optical Plankton Counter: Long-term monitoring of zooplankton and temperature in Scotian Shelf Waters. US Globec News 3:7-9
Herman A.W., Cochrane N.A., Sameoto D.D. (1993) Detection and abundance estimation of euphausiids using an optical plankton counter. Mar. Ecol. Prog. Ser. 94:165-173
Herman A.W., Dauphinee T.M. (1980) Continuous and rapid profiling of zooplankton with an electronic counter mounted on a Batfish vehicle. Deep Sea Res. 27:79-96
Herman A.W., Sameoto D.D., Shunnian C., Mitchell M.R., Petrie D., Cochrane N.

(1991) Sources of zooplankton on the Nova Scotia Shelf and their aggregations within deep-shelf basins. Cont. Shelf Res. 11:211-238

Hoge F.E., Berry R.E., Swift R.E. (1986a) Active-passive airborne ocean color measurement.1. Instrumentation. Appl. Optics 25:39-47

Hoge F.E., Swift R.E., Yungel J.K. (1986b) Active-passive airborne ocean color measurements 2 Applications. Appl. Optics 25:48-57

Holliday D.V. (1980) Use of acoustic frequency diversity for marine biological measurements. In Diemer F.P., Vernberg F.J., Merkes D.Z. (eds.) Advanced Concepts in Ocean Measurements for Marine Biology. Univ. South Carolina Press, Columbia pp:423-460

Holliday D.V. (1993) Applications of Advanced Acoustic Technology in Large Marine Ecosystems. In Sherman K, Alexander L., Gold B. (eds.), Large Marine Ecosystems: Stress, Mitigation and Sustainability. Amer. Assoc. Adv. Sci. Publ. Washington DC pp:301-219

Holliday D.V., Pieper R.E. (1980) Volume scattering strengths and zooplankton distributions at acoustic frequencies between 0.5 and 3 MHz. J. Acoust. Soc. Am. 67:135-146

Holliday D.V., Pieper R.E., Kleppel G.S. (1989) Determination of zooplankton size and distribution with multi-frequency acoustic technology. J. Cons. Int. Explor. Mer 41:226-238

Holligan P.M., Aarup T., Groom S.B. (1989) The North Sea satellite colour atlas. Cont. Shelf Res. 9:665-765

Holligan P.M., Reiners W.A. (1992) Predicting the responses of the coastal zone to global change. In Woodward F.I. (ed.) The Ecological Consequences of Global Climate Change Adv. Ecol. Res. 22:212-246

Kobler Z., Falkowski P.G. (1992) Fast Repetition Rate (FRR) fluorometer for making *in situ* measurements of primary productivity. Proc. Ocean 92 Conf. IEEE Newport, Rhode Island pp:637-641

Lindley J.A., Williams R. (1994) Relating plankton assemblages to environmental variables using instruments towed by ships-of-opportunity. Mar. Ecol. Prog. Ser.107:245-262

McGowan J.A. (1989) Pelagic Ecology and Pacific Climate. In Peterson D.H. (ed.) Aspects of Climate Variability in the Pacific and Western Americas. American Geophysical Union Geophysical Monograph 55:141-150

McGowan J.A. (1990) Climate and Change in Ocean Systems: the value of time series data. Trends Ecol. Evol. 5:293-299

NERC (1992) The Natural Environment Research Council, Report 1991-1992 NERC Publ. Serv.

Pieper R.R., Holliday D.V. (1984) Acoustic measurements of zooplankton distributions in the sea. Rapp P.-v. Reun. Cons. Perm. Int. Explor. Mer 41:226-238

Pingree R.D., Griffith D.K. (1978) Tidal fronts on the shelf seas around the British Isles. J. Geophys. Res. (Oceans and Atmospheres) 83: 4615-4622

Pollard R.T. (1986) Frontal surveys with a towed profiling conductivity/ temperature/

depth measurement package (SeaSoar). Nature 323:433-435

Powers D.A. (1993) Application of molecular techniques to Large Marine Ecosystems, In Sherman K., Alexander L., Gold B. (eds.) Large Marine Ecosystems: Stress, Mitigation and Sustainability. Amer. Assoc. Adv. Sci. Publ. Washington DC pp:321-352

Radach G. (1984) Variations in plankton in relation to climate. Rapp. P.-v Reun. Cons. Int. Explor. Mer 185:234-254

Richter K.E. (1985) Acoustic scattering at 1.2 MHz from individual zooplankters and copepod populations. Deep Sea Res. 32:149-161

Robinson I.S. (1990) Remote sensing information from the colour of the seas. In Herring P.J., Campbell A.K., Whitfield M., Maddock L. (eds.) Light and Life in the Sea. Cambridge Univ. Press pp:19-38

Sameoto D.D., Jarosynski L.O., Fraser W.B. (1980) The BIONESS - a new design in multiple net zooplankton samplers. J. Fish. Res. Bd. Can. 37:722-724

Scalabrin C., Weill A., Diner N. (1992) The structure of multidimensional data from acoustic detection of fish schools. In Weydert M. (ed.) Underwater Acoustics. Elsevier Applied Science, London pp:141-146

SeaWiFS Working Group (1987) System Concept for Wide-field-of-view Observations of Ocean Phenomenon from Space. Report of the Joint EOSAT-NAS SeaWiFS Working Group

Shannon L.V. (ed.) (1985) South African Ocean Colour and Upwelling Experiment. Sea Fisheries Research Institute Cape Town, South Africa

Sherman K. (1993) Large Marine Ecosystems as global units for mariner resource management - an ecological perspective. In Sherman K., Alexander L., Gold B. (eds.) Large Marine Ecosystems V: Stress, Mitigation and Sustainability, Amer. Assoc. Adv. Sci. Publ. Washington DC pp:3-14

Sherman K., Laughlin T. (eds.) (1992) Large Marine Monitoring Workshop Report (Cornell Univ. July 1991). NOAA Technical Memorandum NMFS-F/NEC-93

Sherman K., Solow A.R. (1992) The changing states and health of a Large Marine Ecosystem. ICES L:38:1-31

Sprules G.W., Bergstrom B.O., Cyr H., Hargreaves B.R., Kilham S.S., MacIsaac H.J., Matsushita K., Stemberger R., Williams R. (1992) Non-video optical instruments for studying zooplankton distribution and abundance. Arch. Hydrobiol. 36:45-58

Traynor J.J., Ehrenberg J.E. (1979) Evaluation of the dual beam acoustic fish target strength measurement method. J. Fish. Res. Brd. Can. 36:1065-1071

Trees C.T., Aiken J., Hirche H-J., Groom S.B. (1992) Bio-optical variability across the Arctic Front. Polar Biol. 12:455-461

Wiebe P.H., Burt K.H., Boyd S.H., Morton A.W. (1976) A mutiple opening/closing net and environmental system for sampling zooplankton. J. Mar. Res. 34:313-326

Wiebe P.H., Greene C.H., Stanton T.K.(1990) Sound scattering by live zooplankton and micronekton: empirical studies with a dual-beam acoustical system. J. Acoust. Soc. Am. 88:2346-2360

Williams R., Aiken J. (1990) Optical measurements from underwater towed vehicles deployed from ships-of-opportunity in the North Sea. In Nielsen H.O. (ed.) Environment and Pollution Measurement, Sensor and Systems. Proc. SPIE 1269 pp:186-194

Williams R., Lindley J.A., Hunt H.G., Collins N.R. (1993) Plankton community structure and geographical distribution in the North Sea. J. Exp. Mar. Biol. Ecol. 173:143-156

Yoder J.A., Esaias W.E., Feldman G.C., McClain C.R. (1988) Satellite ocean color-status report. Oceanogr. Mag. 1:18-35

Yoder J.A., Garcia-Moliner G. (1993) Application of satellite remote sensing and optical bouys/moorings to LME studies. In Sherman K., Alexander L., Gold B. (eds.) Large Marine Ecosystems V: Stress, Mitigation and Sustainability of Large Marine Ecosystems, Amer. Assoc. Adv. Sci. Publ. Washington DC pp:353-358

Yoder J.A., Aiken J., Swift R.N., Hoge F.E., Stegman P.M. (1993) Spatial variablity in near surface chlorophylla fluorescence measured by the Airborne Oceanographic Lidar (AOL). Deep Sea Res. II 40:37-53

15. DESERTIFICATION: IMPLICATIONS AND LIMITATIONS OF THE ECOSYSTEM HEALTH METAPHOR

Walter G. Whitford
Senior Research Ecologist, US-EPA
Environmental Monitoring Systems Laboratory
P O Box 93478, Las Vegas, NV 89193
USA

INTRODUCTION

The semi-arid and arid regions of the world may provide some valuable insights into the applicability of the ecosystem "health" and ecosystem "medicine" metaphors. These regions provide an example of the applications and limitations of the metaphor because they have experienced and continue to experience degradation and change. Attempts at rehabilitation of degraded areas has proven to be expensive or impossible. Those parameters that are frequently examined to determine the "health" of ecosystems may be more robust in ecosystems that are generally considered to be degraded than in those perceived as healthy. Most such areas have suffered loss of productivity as measured by the capacity to support livestock and human inhabitants. Those losses in livestock productivity resulted from changes in the biological and physical characteristics of the ecosystems within these regions.

Such ecosystems are frequently perceived as "fragile" because low intensity single and/or multiple stressors produce large changes in ecosystem structure and properties. It can be hypothesized that fluctuation in the climates of arid and semi-arid regions frequently approach the thresholds for survival of component species and that the addition of other stressors even at low level push key species over those thresholds (Whitford and Steinberger 1989). Thus, in many regions of the world, the original mosaic of ecosystems that constituted the landscapes in recent historical time have been replaced by a mosaic of ecosystems that occupy different space in the landscape and that differ structurally and functionally from those that have been supplanted. The current mosaic of ecosystems that constitute the arid and semi-arid landscapes of the world thus provide a challenge in addressing questions of what constitutes ecological health: are the replacement or "new" ecosystems healthy? and if the desertified ecosystems are not healthy, can those ecosystems be restored to a healthy state?

Herein, I will attempt to address these questions by reviewing the history of desertification in North America, reviewing the available data on the on the characteristics of desertified ecosystems and by presenting some data from experimental studies on the resistance and resilience of one desertified ecosystem.

DEFINITIONS

Desertification can be described as "the diminution or destruction of the biological potential of the land, and can lead ultimately to desert-like conditions. It is an aspect of the widespread deterioration of ecosystems under the combined pressure of adverse and fluctuating climate and excessive exploitation. Such pressure has diminished or destroyed the biological potential, i.e., plant and animal production, for multiple purposes at a time when increased productivity is needed to support growing populations in quest of development." (Verstraete 1986). This definition is the result of a series of United Nations conferences on desertification. The United Nations description of desertification emphasizes the economic implications of desertification and the reduction in management options resulting from degradation of the land resources. Note that the definition of desertification does not mention reduction in net primary productivity. Indeed, in many desertified ecosystems, the grams of carbon fixed per unit time may be higher than that in the undegraded ecosystems. The problem of identifying the changes in net primary productivity associated with desertification is that most of the measurements of primary production on arid rangelands have included only forage species (grasses and herbaceous plants) and the productivity of woody perennials has not been included in the measurements (Le Houerou et al. 1988).

The recognition that desertification has begun occurs when desirable forage species are replaced by species that are largely inedible by livestock (Milton et al. 1994). Desertification progresses along a continuum of erosion and changes in plant cover and species composition to the point where trees, shrubs, and most perennials are gone and the only vegetative production is that of annual plants during wet periods. At this end point of the desertification process, lack of vegetative cover often results in geomorphological changes where dunes, stony pavement slopes and gravel plains replace the original vegetated landforms.

The United Nations definition of desertification focuses on the "loss of options" for use by the human populations inhabiting desertified regions or wishing to use desertified regions in any number of ways. Desertification by this definition is a process in which the biological resources of the ecosystems are diminished thereby reducing current and future use options. Desertification can also be viewed as ill health of arid or rangeland ecosystems. The National Research Council (1994) defines rangeland health as the degree to which the integrity of the soil and ecological processes of rangeland ecosystems are sustained. This definition emphasizes integrity both biological and abiotic. Biological integrity has been defined by Cairns (1977) as "the maintenance of community structure and function characteristic of a particular locale or deemed satisfactory to society." Biological integrity is by this definition is therefore similar but not necessarily equivalent to the concept of ecosystem health (Rapport 1992). Rapport (1992) states that a variety of alternative states might be considered healthy as long as certain basic features, both structural and functional are embodied in those alternative state ecosystems that manifest ecosystem integrity. When ecosystem change results in the loss of options with respect to the use of the system by society, that is generally deemed unsatisfactory or unhealthy. If the integrity of desertified ecosystems has been compromised, what are the structural and functional features of the pre-disturbance ecosystems that have been lost and what can be done to reconstruct those features in the desertified systems? In the context of Rapport's definition of ecological health, some ecosystems that are generally considered to be desertified or degraded could be defined as ecologically healthy, that is supporting a higher diversity of organisms, having higher net primary productivity, resisting drought, etc. The problems of attempting to deal with desertification and desertified landscapes within the construct of an ecological health analog is the central theme of this paper.

The perception of desertified ecosystems as unhealthy systems has largely been based on a commodity yield basis. In North America the focus on arid and semi-arid ecosystems as areas primarily for livestock production, has shifted to now include ecosystem services such as maintenance of ground and surface water supplies, balance of atmospheric gasses and aesthetic characteristics and recreational potentials. How desertification affects these processes and values needs to be evaluated for the whole spectrum of ecosystems that make up the arid and semi-arid lands of the planet.

HISTORICAL PERSPECTIVES

Desertification is a problem with ancient roots. Cutting of forests, overgrazing, and salt accumulation in irrigated lands led to desertification in Mesopotamia, and the lands bordering the Mediterranean more than 2000 years ago. The spread of western technology and culture and rapid growth of human population has spread the desertification process to every continent except Antarctica and in some areas such as the Sahel with devastating effects on the human inhabitants. Desertification is a continuing process in many areas of the world. The productivity of semi-arid lands around the world continues to be reduced by the increase in unpalatable woody plants and reduction in vegetative cover leading to further deterioration of soil structure and soil hydraulic properties (Karrar and Stiles 1984). Thirty five percent of the earth's land area has a high desertification potential (approximately 45 million km^2) and 75% of that area has already been affected by desertification processes (Karrar and Stiles 1984). In this paper I will focus on desertification in North America for most of my examples and data. However, the conclusions drawn should be applicable to most if not all semi-arid to arid regions of the world.

Since the middle of the 19th century, there have been marked changes in vegetation, soils and drainages of the landscapes in the arid and semi-arid regions of North America and in many areas these changes are still occurring (Grover and Musick 1990; Hastings and Turner 1965). The degradation of western rangelands was well recognized by the turn of the 20th century by Griffiths (1901) who described decreased rangeland productivity because of the destruction of grasslands by overstocking. For example: "There were fully 50,000 head of stock at the head of Sulphur Spring Valley and the valley of the Aravipa in 1890. In 1900 there was no more than 1/2 that number and they were doing poorly" (Griffiths 1901).

Changes in vegetative cover and composition are well documented especially on the Jornada Experimental Range in Southern New Mexico (Buffington and Herbel 1965; Hennessy et al. 1983; York and Dick-Peddie 1969) (Figure 1). These changes in vegetation were accompanied by soil erosion, arroyo cutting and changes in soil structure (Gibbens et al. 1983; Grover and Musick 1990). Similar changes have been recorded in the desert grassland ecosystems of southeastern Arizona (Glendening 1952) and in Texas (Archer 1990). There is a lack of agreement among scientists and land managers about the causes of these changes. Overgrazing and other land use practices are frequently cited as primary causes of desertification by some groups while climatic drought and competition between grasses and shrubs are invoked as causal agents by other groups.

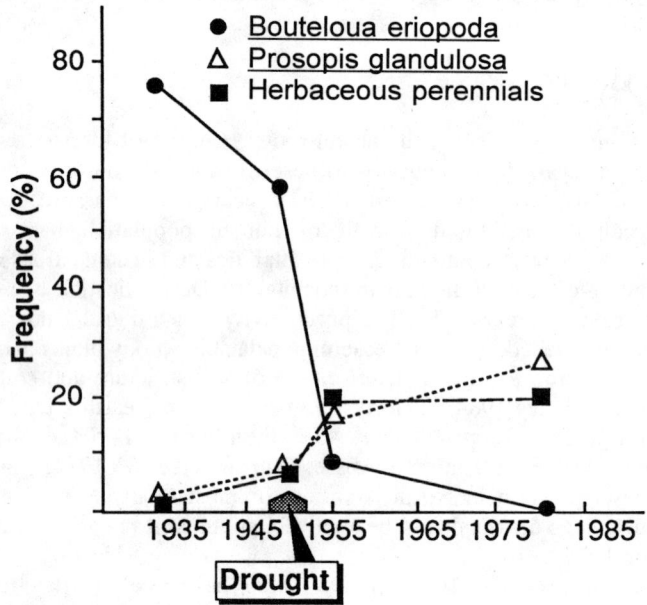

Figure 1. Changes in the percent frequency of black grama grass (*Bouteloua eriopoda*) and of mesquite shrubs (*Prosopis glandulosa*) in plots on the Jornada Experimental Range in New Mexico (Data from Hennessey et al. 1983).

There is evidence that the changes described as desertification occur even in the absence of grazing. There was no increase in perennial grass cover in a grazing exclosure established in 1933 on an area partially occupied by mesquite (*Prosopis glandulosa*) when re-measured in 1980 and there were no differences between the grazing exclosure and grazed area transects that were grassland in 1935 that degraded to mesquite dunes in 1980 (Hennessy et. al 1983). Similar results were obtained in grazing exclosure studies in Arizona where the densities of mesquite plants doubled between 1932 and 1949 in both cattle excluded and grazed areas (Glendening 1952). During that same period, there was a twenty-fold decrease in densities of clumps of perennial grasses (Glendening 1952). In a comparison of a grazed and an ungrazed paddock established in 1941, Smith and Schmutz (1975) recorded some recovery of grasses and reduction in sub-shrubs in the ungrazed paddock but a large increase in mesquite, (*Prosopis glandulosa*) was recorded in both paddocks. These authors also reported that an exotic, Lehmann's lovegrass was establishing on both sites. Brady et al. (1989) report a similar result in an area where grazing was excluded for 15 years and concluded that continued animal impact is not necessary for ecosystem deterioration to continue. The results of these and other studies suggests that once shrub establishment and grass cover reduction has passed some threshold, the degradation trajectory is followed even when grazing pressure is removed. The presence of some shrubs provides a seed source and animals such as rodents can act as dispersers and play important roles in the spread of shrubs in a grassland even in the absence of grazing (Cox et al. 1993). Another factor affecting continued change in exclosures even in the absence of grazing is that the surrounding landscape continues to be impacted by grazing which can lead to the development of erosion cells (Pickup 1985; Pickup and Chewings 1986). The growth of erosion cells eventually envelop an exclosure destroying grasses by burial in wind blown sand, by abrasion, and by loss of topsoil by sheetflow from up-slope disturbed areas (Figure 2). The end result is that with or without continued grazing the areas change to the point of being identified as degraded or desertified.

In this century, development of pastoral industries in semi-arid regions has centered on establishing reliable supplies of water for livestock. Established water points result in the development of what has been termed a piosphere. A piosphere is a zone of attenuating impacts of livestock radiating outward from the central water point. Piospheres reflect the concentric nature of animal activity around water (Andrew and Lange 1986a). Within 2.5 years after the establishment of a well and trough in a near pristine chenopod shrubland in South Australia, there were measurable changes due to the activity of sheep that were stocked at the lowest stocking levels in the Australian arid pastoral zone. Soil changes included marked reduction in surface lichen crust, increased soil compaction and increased bulk density especially within 40m of the trough. Vegetation changes included reduction in biomass of *Atriplex vesicaria*, a palatable shrub, and short-lived perennial grasses, and estalishment of an exotic species, *Marsubium vulgare* (hore hound), within a 13 metre radius from the trough. There was substantial mortality of *A. vesicaria* close to the trough after 8 years of grazing. Andrew and Lange (1986a) emphasize that these changes occurred despite the low stocking rate in the paddock. Water points tend to concentrate the activities of native herbivores thus exacerbating the grazing and browsing effects (Andrew and Lange 1986b).

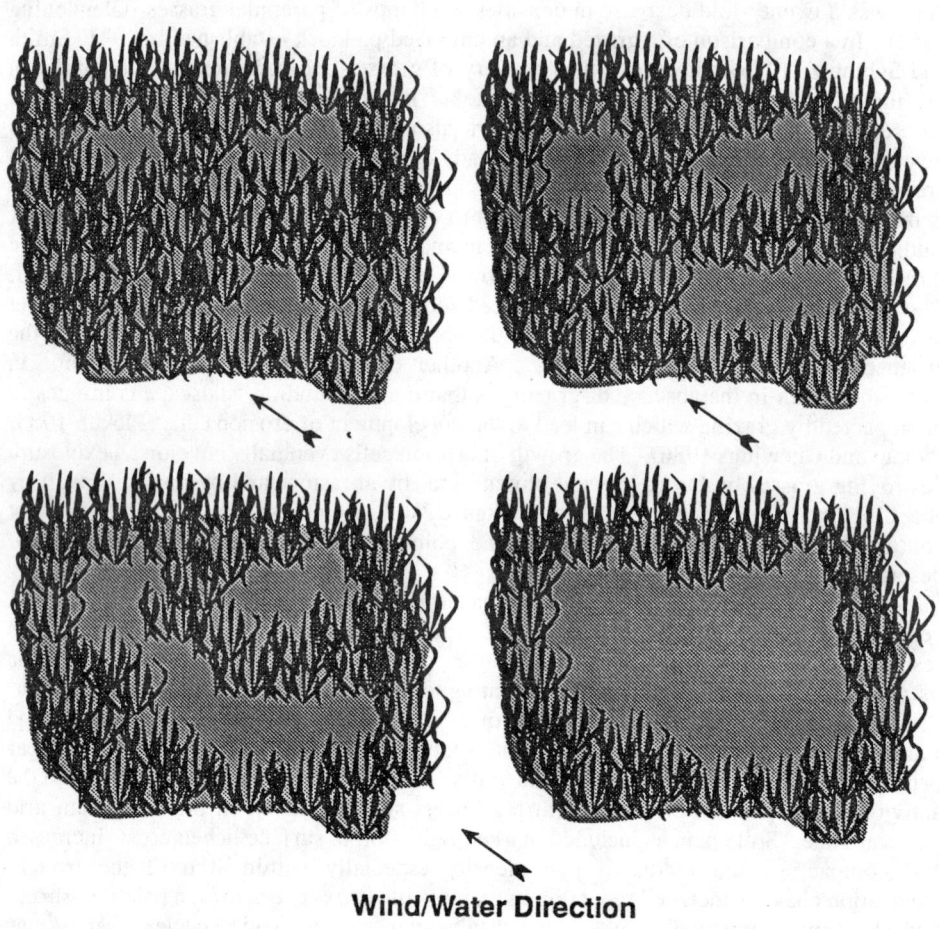

Figure 2. A diagrammatic model of the development and coalesence of erosion cells in a desert grassland.

These changes are consistent with the ecosystem distress syndrome described by Rapport et al. (1985). Andrew and Lange (1986b) go on to cite an extreme case where 6000 sheep denuded an area of saltbush up to 1000m from water in the first 3 months of stocking. Reductions of vegetative cover in the piospheres provide the initial erosion cells which expand by abrasion and burial of vegetation plus soil loss by erosion.

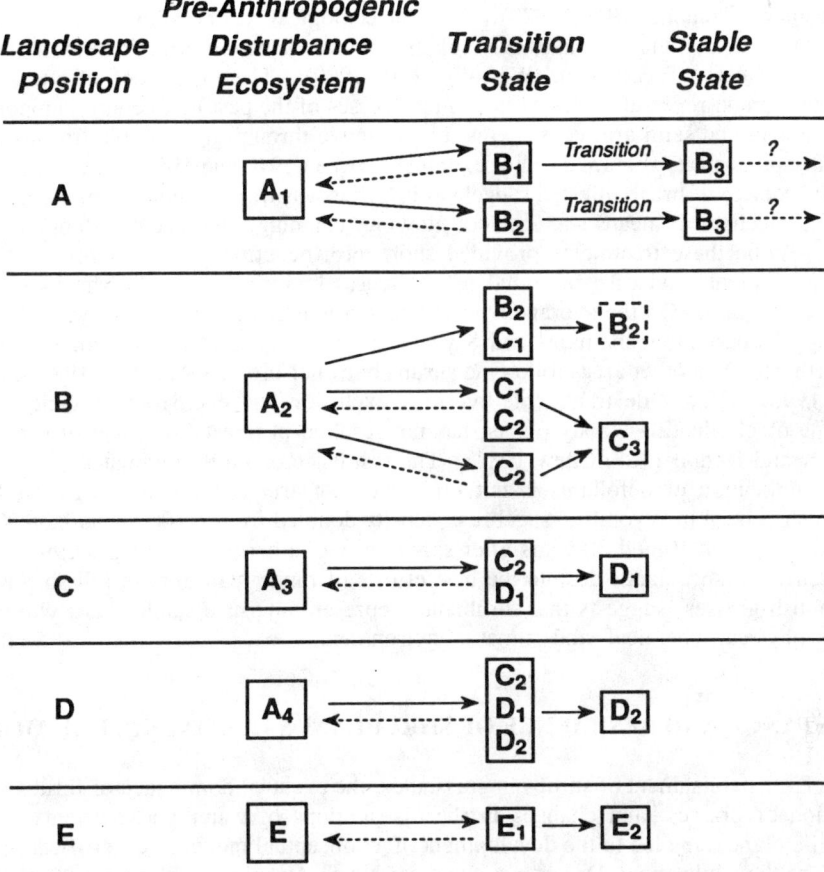

Figure 3. A conceptual model of the dynamics of ecosystems that make up and arid or semi-arid landscape in response to environmental stressors. Transition states may be similar or dissimilar to the eventual stable state which is generally different from the pre-stressed condition.

RECLAMATION OF DESERTIFIED ECOSYSTEMS

The concepts of succession and climax community structure have provided the theoretical base for most land management and revegetation efforts in areas subject to desertification in North America and in the rest of the world. Indeed, the Clementsian view of single equilibrium communities and deterministic successional stages in the process of return to the "climax" condition has been the basis for range management in this century (Clements 1916, 1928). Current ecological theory allows for alternative stable states, discontinuous and irreversible transitions, nonequilibrium communities and stochastic effects in succession (Westoby et al. 1989). The experience of attempts at revegetation and reversal of desertification processes of the past half century support the idea that arid and semi-arid ecosystems tend to move through irreversible transitions to alternate stable states (Figure 3). For example between 1940 and 1981 an average of 0.6 million hectares of brush infested rangeland in Texas was treated annually by herbicides, fire, and mechanical means such as root plowing, chaining, and discing (Rappole et al. 1986). While these treatments provided short term benefit in terms of production of herbaceous plants and ease of round up of livestock, regrowth of the shrubs quickly reversed the gain. On the average, re-treatment was necessary within 15 years of root plowing, 2 years after chaining, and 8 years after treatment with herbicides. Similar regrowth occurs in other areas following shrub control efforts (Herbel et al. 1983, Jacoby et al. 1990). Herbicide treatment, livestock exclusion and even root plowing, while reducing or eliminating woody plants, has not resulted in re-establishment of grassland (Roundy and Jordon 1988; Chew 1982). The resilience of shrub dominated ecosystems to the application of defoliants is not limited to hot arid ecosystems. In a study of sagebrush control in Wyoming, sagebrush density doubled from 2,100 plants/ha to 4,400 plants/ha between 10 and 20 years after spraying with a herbicide (Sturges 1993). The persistence of shrublands despite the best efforts of range managers to kill shrubs and re-establish grasses, suggests that shrublands represent alternate stable states within the current physical, chemical, and climatic environment.

RESISTANCE AND RESILIENCE OF SHRUBLANDS - A CONCEPTUAL MODEL

The establishment of shrubs in grasslands, the eventual dominance of shrubs to the exclusion of grasses, and changes in the distributions of water and nutrients on the desertified landscape led to the development of a conceptual model of desertification that applies to the Chihuahuan Desert grasslands of North America (Schlesinger et al. 1990, Figure 4). That model focuses on the redistribution of resources in shrub dominated ecosystems.

In order to understand the resistance of shrub dominated ecosystems to disturbance such as herbicide treatment, root plowing, chaining etc., we developed a conceptual model for the stability of the resource "island" produced by the establishment of shrubs. Seedling shrubs are probably not functionally different from grass clumps. Seedling and small shrubs do not affect water distribution in the soil profile and lack the structure to cause the accumulation of wind borne soil and particulate organic matter. Small shrubs within a grassland with relatively high cover of perennial grasses may actually be at a competitive disadvantage with respect to the grasses.

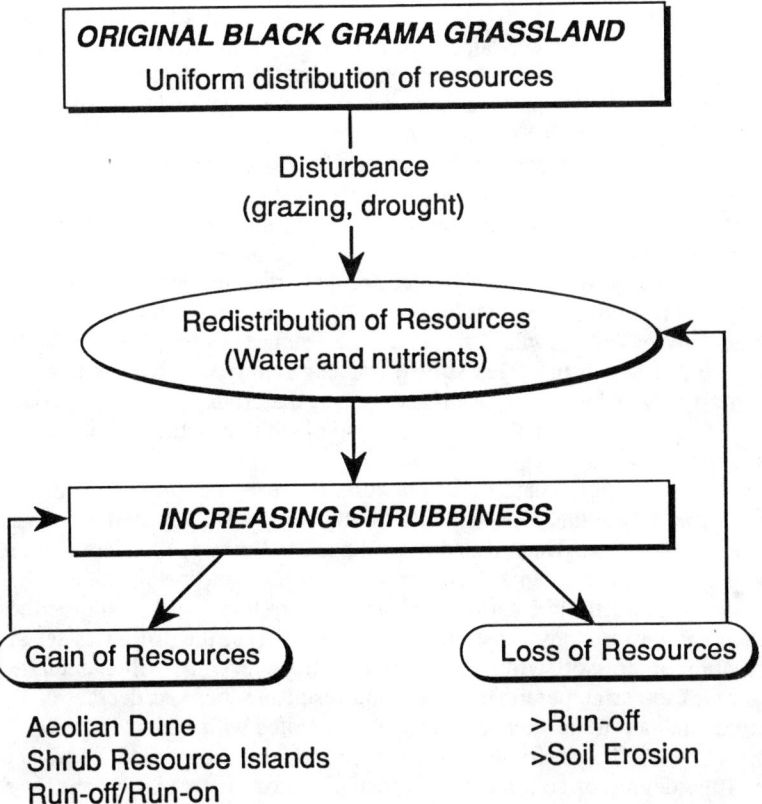

Figure 4. The Jornada model for desertification focusing on the redistribution of resources as the key process resulting in the establishment of alternate stable ecosystems.

However if drought and grazing in combination adversely affect the grasses, the shrubs are released from competition and can grow rapidly. As the shrubs increase in size, their morphological characteristics affect the distribution of rainfall in the soil and the distribution of organic matter and wind blown soil at the surface. Large shrubs in several deserts have been described as "islands of fertility" in a sea of relatively unfertile soil that makes up the intershrub spaces (Barth and Klemmedson 1982; Charley and West 1975; Garcia-Moya and McKell 1970, 1977; Parker et al. 1982; Virginia and Jarrell 1983; West and Klemmedson 1978). However rather than simply islands of fertility many desert shrubs and small trees may be more accurately describes as "islands of resources" which include water resources in addition to soil nutrients. Shrubs and small trees affect interception, infiltration and water storage (Elkins et al. 1986; Joffre and Rambal 1988; Nulsen et al. 1986; Pressland 1976). Rainfall intercepted by the canopy and funneled to the root crown by stem flow may follow preferential root pathways to deep soil layers thereby by-passing the resistance to percolation by the soil matrix (Nulsen et al. 1986; Pressland 1976) (Figure 5).

Many arid zone shrubs and trees have morphologies that produce relatively large amounts of stem flow which may be an important factor in the growth and survival of these species and a factor in their resistance to drought (Pilgrim et al. 1988; Pressland 1975; Slayter 1965). At some point in the growth of shrubs and trees, the plant becomes a self augmenting entity that is uncoupled from short-term climatic variation (Figure 6). From that point in time, the plant becomes increasingly resistant to disturbance and the immediate environs of the plant represent the patches where seedling establishment is most probable if the shrub is killed or physically removed. The areas between shrubs lose fine particulates by wind erosion and surface materials by water erosion. The result of these combined processes is a stable ecosystem with very different properties from the original system.

There are scattered data in the literature that provide support for these conceptual models. These conceptual models are currently being tested by studies of historical data and historical plots, experiments and modelling. Preliminary data from an experimental study is presented here as an example of a test of the conceptual model.

Not all desertified, shrub-dominated ecosystems will achieve the relatively resistant, resilient state described above. In some areas, the processes set in motion by a combination of stressors will continue along a trajectory that will eventually lead to the elimination of the large perennial shrubs and result in a marked decrease in net primary productivity and all of the attendant changes associated with such a decrease. The critical determinant of whether an arid ecosystem continues along a degradation trajectory toward stage 3 is the degree of continued soil erosion. Ecosystems that reach this stage in the desertification continuum have sparse cover of suffrutescent shrubs and most of the primary productivity is due to the growth of ephemeral plants. In such ecosystems there is close coupling of primary productivity to rainfall and net primary productivity is low in comparison to perennial grasslands or shrublands. Ecosystems at stage 3 provide the smallest number of options for land use by humans and the lowest probability of responding to efforts to force the systems to a more productive state within a time frame that is meaningful for human populations.

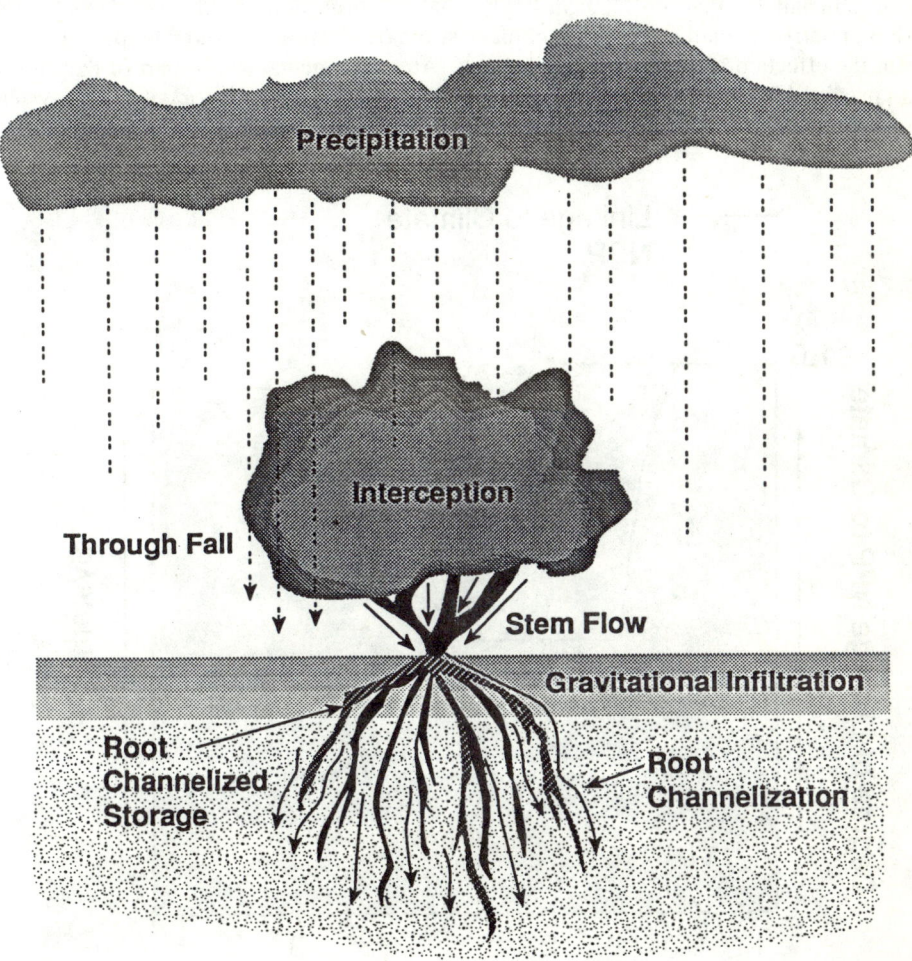

Figure 5. Water redistribution patterns as a result of small trees and shrubs in the landscape.

RESPONSES OF A DESERTIFIED ECOSYSTEM TO STRESSORS: DROUGHT AND ABOVE AVERAGE RAINFALL

In the conceptual model of desertification in the Chihuahuan Desert, establishment of shrubs results in the development of "resource islands" that uncouple the shrubs from environmental fluctuations. Accordingly, mature shrub dominated ecosystems should exhibit relatively small responses to climatic stress (resistance) and should rapidly recover from the effects of that stressor (resilience). An experimental test of part of that model was provided by data collected on a study of a creosotebush (*Larrea tridentata*) ecosystem in which plots had been subjected to repeated droughts or irrigation for three consecutive

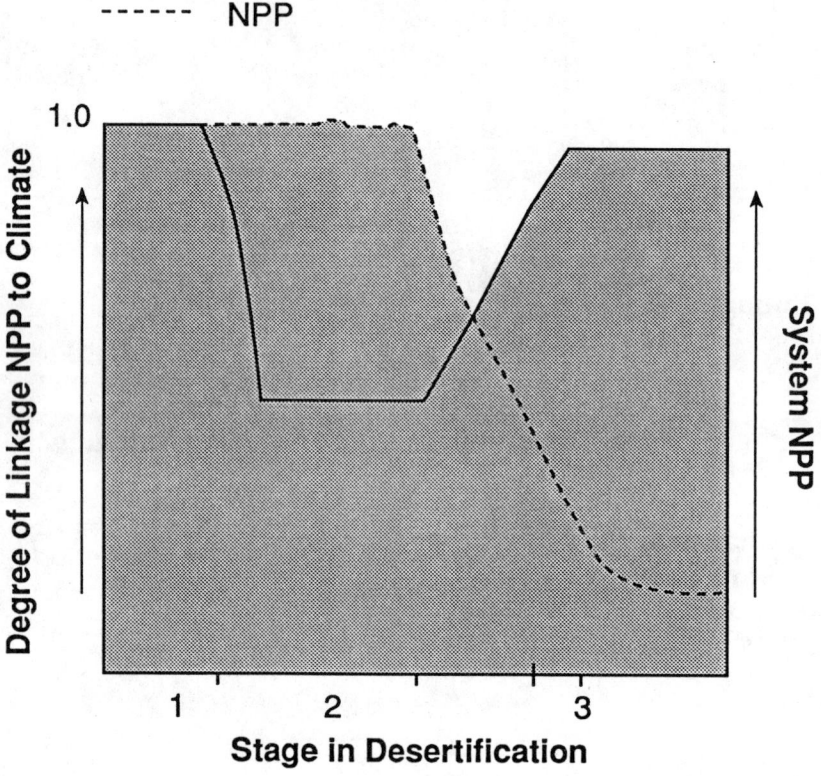

Figure 6. A diagram showing the hypothesized relationships of degree of desertification to linkage with climate, especially rainfall and net primary productivity.

summers in a study of soil nitrogen dynamics. In the subsequent years, the treatments were continued with "rain-out" shelters covered to eliminate summer rainfall (June - September) and irrigation to double the long-term average precipitation over the same period. Measurements were made of growth and productivity of vegetation, populations of microarthropods and decomposition of annual plant roots and creosotebush leaf litter. The data from those studies are briefly summarized here as the experimental test of the resistance-resilience hypothesis for established shrubs in desertified landscapes.

The experimental treatments produced soil water potentials that averaged - 2.5 MPa in the control plots, -1.9 MPa in the irrigated plots and - 7.9 MPa in the drought plots. For reference, permanent wilting point for agronomic crop plants is -1.5 MPa. Despite the large differences in soil water potential during the summer growing season for 5 consecutive years there were only marginal effects on soil microarthropod populations sampled in May following the last year of the treatments (Table 1).

The microarthropods that exhibited the largest response to the climate manipulation were the small fungivore-predator mites of the family Tydeidae and the predatory prostigmatid mites. The reductions of populations of these groups in the drought plots

Table 1. Average numbers of microarthropods ($\pm$ one standard deviation) extracted from soil cores taken at the edge of the canopy of *Larrea tridentata* shrubs in plots exposed to summer drought, irrigation (25mm every 2 weeks), and no treatment (control). Cores were collected 9 months after the last treatments were applied (October - June). Cores were 27 cm^2 to a depth of 15 cm.

	Irrigated	Drought	Control
Prostigmata			
Nanorchestidae			
Speleorchestes spp.	4.5±3.7	8.8±2.8	4.5±3.7
Nanorchestes spp.	6.3±1.4	10.5±7.0	5.0±1.6
Tydeidae	8.0±3.8	2.5±5.8	9.8±20.4
Fungivores	15.8±13.9	10.3±3.9	22.8±9.1
Predators	2.0±1.8	0.5±0.5	1.9±1.8
Cryptostigmata			
Oribatida	0	0	1.1±1.0
Mesostigmata	0.2±0.2	0.2±0.2	1.0±0.9
Collembola	2.5±2.0	2.5±1.7	4.2±3.5

may be attributable to the low numbers or absence of active nematodes in the drought plots during the reproductive season for these mites (Whitford 1989). The fungivorous nanorchestids occured at higher densities on the drought plots which is probably related to the increased food supply from fungi growing on the dead roots of annual plants. The biomass of annual plants on the plots in May, at the time that the soil samples for microarthropods were collected, averaged 3.1 g/m^2 on the control plots, 0.8 g/m^2 on the irrigated plots and 17.0 g/m^2 on the drought plots. This pattern of annual plant production appears to be a function of nitrogen availability. Drought periods result in increased rates of nitrogen mineralization when water becomes available at the end of the drought (Fisher et al. 1987). Repeated drought and irrigation had no effect on the growth of creosotebushes, *Larrea tridentata*, in May, nine months after the cessation of the treatments (Table 2). Measurements made in November, approximately one month after termination of the experimental treatments, revealed no differences in growth attributable to drought (Table 2). In these measurements, plants on the control plots exhibited a higher coefficient of variation in growth increment than either of the treatments (Table 2).

The experimental treatments had significant effects on the most abundant perennials, a herb *Bahia absinthifolia* and fluffgrass, *Erioneuron pulchellum*. After five years of treatments, densities of *B. absinthifolia* were: control - $0.34/m^2$, irrigated - $0.46/m^2$, and drought - $0.15/m^2$. Densities of fluffgrass were: control - $0.53/m^2$, irrigated - $1.13/m^2$, and drought $0.05/m^2$. These perennials were present on all of the irrigated plots, all but one of the control plots and only two of the drought plots. Obviously drought had an adverse effect on the survival of the non-woody perennials that are dependent upon summer rainfall for their growth and reproduction. The virtual absence of non-woody perennials on the drought plots also contributed to the abundance of spring annuals on the drought plots because of the absence of competition for critical resources. The loss of perennials and the abundance of annuals on the drought plots is consistent with the pattern of replacement of long-lived with short lived species in disturbed ecosystems (Rapport et al. 1985).

Irrigation had no effect on decomposition of roots during the final year of the experiment but drought resulted in a significant reduction in mass loss. Average mass losses of roots of a perennial forb were: control - $83.8 \pm 10.4\%$, irrigated - $87.6 \pm 7.8\%$, drought - $51.3 \pm 10.2\%$. There were no significant effect of treatments on decomposition of leaf litter on the soil surface during the last year of the study: control - $52.3 \pm 4.0\%$, irrigated - $47.6 \pm 3.0\%$, drought - $56.2 \pm 5.6\%$. These data are consistent with the hypothesis that in hot, arid ecosystems, decomposition of leaf litter on the soil surface is primarily a physical process and decomposition of roots is a biological process (Steinberger et al. 1990; Whitford et al. 1988).

When these data are considered together, it is obvious that the creosotebush dominated ecosystem is extremely resistant and resilient to a climatic stress (drought). There are some surprising effects of luxury amounts of water with respect to the responses of the ecosystem. Luxury amounts of water did not affect decomposition rates, tended to reduce the abundance of spring annual plants and had no long term beneficial effect on the growth of creosotebush.

Table 2. Responses of creosotebushes (*Larrea tridentata*) to drought and irrigation for three or more consecutive summers.

a) Average growth increments (cm) of terminal stems of creosotebushes in plots receiving drought or irrigation from June 1 through October 1. Numbers followed by the different letters are significantly different ($p < 0.05$). No letters indicate no significant difference.

	Nov '89-May '90	Jun '90-Nov '90
Control	11.2±0.6	12.9±1.6 a
Irrigated	11.3±1.2	13.8±2.0 b
Drought	11.4±2.0	13.2±2.4 b

b) Average biomass production (gms) of creosotebush branch tips in plots subject to drought or irrigation from June 1 through October 1. Statistical presentation the same as 2a). Number in parenthesis = coefficient of variation.

	Nov '90-May '91	Jun '91-Nov '91
Control	1.9±0.4	0.8±0.7 a,b (85%)
Irrigated	1.9±0.3	1.1±0.4 a (30%)
Drought	1.9±0.5	0.4±0.2 b (50%)

There were no obvious adverse effects of consecutive droughts on the population sizes and/or diversity of the soil microarthropod community nor any demonstrable effect of consecutive above average summer rainfall on these organisms. The documented resilience of this ecosystem in response to stress is one characteristic of a "healthy" ecosystem (Rapport 1992). Despite the obvious vigor of this desertified ecosystem, it is perceived as a degraded ecosystem by society because it represents a change from ecosystems that produced commodities (livestock) to an ecosystem that has little value as a producer of livestock.

IMPLICATIONS OF NON-EQUILIBRIUM ECOSYSTEMS AND ECOSYSTEM HEALTH

There now appears to be ample evidence that southwestern rangelands behave as non-equilibrium systems. Many of the ecosystems that currently make up the landscape of the southwestern United States appear to be functioning at some new operating points which are very different from the original operating points (Kay 1991). Since these systems appear to have been rapidly driven past one set of catastrophe thresholds, it is reasonable to expect that the current systems may change dramatically to yet other operating points (Kay 1991). Many of the ecosystems in the desertified areas of the southwest currently maintain their organization in the face of changing environmental conditions as demonstrated by the experimental studies reported here. Such ecosystems demonstrate resilience to stress hence can be said to be "healthy". The primary indicator of "poor health" in these ecosystems is the reduced carrying capacity for livestock and increased run-off and sediment yield from areas between shrubs when compared to grasslands on the same topography. However comparisons of the shrub systems to grasslands reveal that species diversity of vertebrate animals is higher in the shrub ecosystems than in the "original" grasslands Menke and Bradford 1992). Diversity of invertebrates may also be higher as demonstrated by studies on ants (unpublished data, Jornada Long Term Ecological Research Program, New Mexico State University, Las Cruces, N.M.). Depending upon the indicators selected, some of the recent shrublands can be classified as "healthier" than the reference grasslands. However, when the total array of management options and ecosystem services are considered, these systems are obviously less "healthy" than the reference grasslands. The characteristics of such disturbed ecosystems may best be understood by examining the manifestations of the disturbance syndrome as discussed below.

The characteristics of the desertification process described here raises some important questions about the ecological health metaphor and how it should be applied to ecosystems that may be undergoing desertification. In human health, the physiological basis of health and medicine is that of the relatively constant internal environment, ie. a homeostatic system. When physiological parameters are above or below the narrow range of natural variability in that characterizes those parameters, the individual is judged to be unhealthy or sick. For many of the ills that afflict humans, the deviations from the physiological norm is understood along with an understanding of the causal agent. However in human medicine, not all departures from the norm are understood but are described as a syndrome of symptoms that characterize named disease syndromes such as multiple sclerosis or Reyes Syndrome. Such syndromes vary considerably among individual humans with respect to their manifestation, progression, and endpoint. Even the array of symptoms varies among individual patients but a patient is judged to have the disease if a minimum number of symptoms are present. In the natural world, desertification is a syndrome exhibiting differences in manifestation, progression and endpoint but is clearly recognized as an unhealthy state. As with human disease syndromes, the diagnosis and treatment of desertification must be approached as a syndrome where cause is unknown and where effect is not necessarily predictable. The ecosystems that are subject to desertification are probably not in equilibrium or not homeostatic to use the human health terminology. As with human disease syndromes, there is much that needs to be learned about the ecosystems and their responses to stress in order to gain an understanding of cause and effect that will provide the basis for maintaining the ecological health of the ecosystems.

It is evident from the difference in species composition and the changes in the key species in ecosystem processes resulting from desertification that options for land use may be decreased or simply changed. Land managers may be faced with choices of options that limit or eliminate the possibility of additional options. This is illustrated by the case study of desertification in southeastern Arizona. Attempts to restore grasslands in the southwestern United States to support the livestock industry began in 1890 but most of the plantings of native species failed because the changes in soil and vegetation produced conditions that were unsuitable for most native perennial grass species (Cox et al. 1990). Run-off and sediment loads increased from the desertified watersheds and ground water re-charge was reduced or eliminated. Revegetation of watersheds was deemed essential to reduce damage from flooding and to restore some livestock production. Screening of exotic grasses in the 1930's led to the recommendation that an African perennial bunch grass, Lehmann's lovegrass (*Eragrostis lehmanniana*) that produces abundant seed, be used to rehabilitate rangeland ie. increase livestock fodder production and reduce erosion. Between 1940 and 1980 land managers in Arizona established Lehmann's lovegrass on 70,000 ha and that species spread to an additional 130,000 ha (Cox and Ruyle 1986, Anable et al. 1992). The spread of Lehmann's lovegrass has transformed large areas of shrubland to grassland with scattered shrubs. Lehmann's lovegrass continues to spread into shrub dominated areas and into native perennial grasslands where it appears to replace the native species (personal observations on the Jornada Basin in Southern New Mexico). The consequences of using this option is that other options have been lost. Lehmann's lovegrass is poor quality forage for livestock which use the grass growing in nitrogen enriched patches (under mesquite) and then only during the active growing season. The diversity of animals is severely reduced in areas dominated by Lehmann's lovegrass (Anable et al. 1992) and it is likely that few if any of the original perennial grassland species are able to maintain populations in a Lehmann's lovegrass grassland. By applying a restoration technique that provided some benefits (reduction of erosion and limited livestock production), the land managers have virtually eliminated the possibility of using other options that would provide a different suite of benefits.

Similar choices in the exercise of options are being made in many areas of the world that have been subjected to desertification (Alonso 1990; Ovalle et al. 1990). Since desertification leads to a limitation of land use options, should not land use options be criteria for assessing ecological health?

APPLYING THE ECOHEALTH METAPHOR

Considering that arid ecosystems are chaotic or at best non-equilibrium systems, the choice of indicators of "health" status must be carefully chosen in order to eliminate the possibility that an "alternate healthy state" be judged to be unhealthy. It must be recognized that there is a degradation continuum along which some ecosystems may move through time whilst others stop at some point and develop a resilience to external stressors. It may not be possible to predict which ecosystems in an arid landscape are in transition and which have attained some alternate stable state with reference to an "undisturbed original ecosystem". It is probable that all of the ecosystems of a landscape are changing but at different rates. The challenge is to identify those that are changing along a trajectory that continues to reduce the land use options. The lesson to be gained from application of restoration techniques that have limited goals is that in that application, land use options may be lost or severely compromised. Given the state of our knowledge about desertification and the ecosystems that make up the landscapes of the arid and semi-arid regions of the world, using a knowledge based model for assessing the status and probability of change in structure and processes of the component ecosystems may be a good first step in applying a "medical" model to arid regions.

ACKNOWLEDGEMENTS

Discussions with personnel of the Environmental Monitoring and Assessment Program (EMAP) stimulated the formulation of many of the ideas presented in this chapter.

REFERENCES

Abernathy G.H., Herbel C.H. (1963) Brush eradicating, basin pitting and seeding machine for arid to semiarid rangeland. J. Range Manag. 26:189-192

Alonso J.L. (1990) Reforestation of arid and semi-arid zones in Chile. Agric.Ecosystems Environ. 33:111-127

Anable M.E., McClaran M.P., Ruyle G.B. (1992) Spread of introduced Lehmann lovegrass *Eragrostis lehmanniana* Nees. in Southern Arizona, USA. Biol. Cons. 61:181-188

Andrew M.H., Lange R.T. (1986a) Development of a new piosphere in arid chenopod shrubland grazed by sheep. 1. Changes to the surface. Australian J. Ecology 11:395-409

Andrew M.H., Lange R.T. (1986b) Development of a new piosphere in arid chenopod shrubland grazed by sheep. 2. Changes to the vegetation. Australian J. Ecology 11:411-424

Archer S., Scifres C., Bassham C.R., Maggio R. (1988) Autogenic succession in a subtropical savanna: conversion of grassland to thorn woodland. Ecol. Monographs 58:111-127

Archer S. (1990) Development and stability of grass/woody mosaics in a subtropical savanna parkland, Texas, U.S.A. J. Biogeog. 7:453-462

Barth R.C., Klemmedson J.O. (1982) Amount and distribution of dry matter, nitrogen, and organic carbon in soil-plant systems of mesquite and palo verde. J. Range Manag. 35:412-418

Buffington L.C., Herbel C.H. (1965) Vegetational changes on a semi-desert grassland range from 1858 to 1963. Ecol. Monographs 35:139-164

Charley J.L., West N.E. (1975) Plant-induced soil chemical patterns in some shrub-dominated semi-desert ecosystems of Utah. J. Ecology 63:945-964

Charley J.L., West N.E. (1977) Micro-patterns of nitrogen mineralization activity in soils of some shrub dominated semi-desert ecosystems of Utah. Soil Biol. Biochem. 9:357-365

Clements F.E. (1916) Plant succession: an analysis of the development of vegetation. Carnegie Inst. Publ. 242. Washington DC

Clements F.E. (1928) Plant Succession and Indicators. Hafner Publ. Co., NY

Chew R.M. (1982) Changes in herbaceous and suffrutescent perennials in grazed and ungrazed desertified grassland in southeastern Arizona 1958-1978. Amer. Midl. Nat. 108:159-169

Cox J.R., DeAlba-Avila A., Rice R.W., Cox J.N. (1993) Biological and physical factors influencing *Acacia constricta* and *Prosopis velutina* establishment in the Sonoran Desert. J. Range Manag. 46:43-48

Cox J.R., Ruyle G.B., Roundy B.A. (1990) Lehmann lovegrass in southeastern Arizona: biomass production and disappearance. J. Range Manag. 43:367-372

Elkins N.Z., Sabol G.V., Ward T.J., Whitford W.G. (1986) The influence of subterranean termites on the hydrological characteristics of a Chihuahuan Desert ecosystem. Oecologia 68:521-528

Fisher F.M., Parker L.W., Anderson J.P., Whitford W.G .(1987) Nitrogen mineralization in a desert soil: interacting effects of soil moisture and nitrogen fertilizer. Soil Sci. Soc. Amer. J. 51:1033-1041

Garcia-Moya E., McKell C.M. (1970) Contribution of shrubs to the nitrogen economy of a desert-wash plant community. Ecology 51:81-88

Gibbens R.P., Tromble J.M., Hennessy J.C., Cardenas M. (1983) Soil movement in mesquite dunelands and former grasslands of Southern New Mexico from 1933 to 1980. J. Range Manag. 36:145-148

Glendening G.E. (1952) Some quantitative data on the increase of mesquite and cactus on a desert range in southern Arizona. Ecology 33:319-328

Griffiths D. (1901) Range Improvement in Arizona. U.S. Dept. Agric. Bureau Plant Industry Bull. 4:1-31

Grover H.D., Musick H.B. (1990) Shrubland encroachment in southern New Mexico, U.S.A.: An analysis of desertification processes in the American southwest. Climatic Change 17:305-330

Hastings J.R., Turner R.M. (1965) The Changing Mile. Univ. Ariz. Press. Tucson, Arizona

Hennessy J.T., Gibbens R.P., Tromble J.M., Cardenas M. (1983) Vegetation changes from 1935 to 1980 in mesquite dunelands and former grasslands of southern New Mexico. J. Range Manag. 36:370-374

Le Houerou H.N., Bingham R.L., Skerbek W. (1988) Relationship between the variability of primary production and the variability of annual precipitation in world arid lands. J. Arid Environ. 15:1-18

Joffre R., Rambal S. (1988) Soil water improvement by trees in the rangelands of southern Spain. Acta Oecologia Plantarum 9:405-442

Karrar G., Stiles D. (1984) The global status and trend of desertification. J. Arid Environ. 7:309-313

Kay J.J. (1991) A nonequilibrium thermodynamic framework for discussing ecosystem integrity. Environ. Manag. 15:483-495

Menke J., Bradford G.E .(1992) Rangelands. Agric., Ecosystems, Environ. 42:141-163

Milton, S.J., Dean W.R.J., duPlessis M.A., Siegfried W.R .(1994) A conceptual model of arid rangeland degradation. Bioscience 44:70-76

Nulsen R.A., Bligh K.J., Baxter I.N., Solin E.J., Imrie D.H. (1986) The fate of rainfall in a mallee and heath vegetated catchment in southern Western Australia. Australian J. Ecology 11:361-371

Ovalle C., Aronson J., Del Pozo A., Avendano J. (1990) The espinal: agroforestry systems of the mediterranean type climate region of Chile. Agroforestry Systems 10:213-239

Parker L.W., Fowler H.G., Ettershank G., Whitford W.G. (1982) The effects of subterranean termite removal on desert soil nitrogen and ephemeral flora. J. Arid Environ. 5:53-59

National Research Council (1994) Rangeland Health. National Academy Press. Washington DC

Pickup G. (1985) The erosion cell - a geomorphic approach to landscape classification in range assessment. Australian Rangelands J. 7:114-121

Pickup G., Chewings V.H. (1986) Mapping and forecasting soil erosion patterns from Landsat on a microcomputer-based image processing facility. Australian Rangelands J. 8:57-62

Pilgrim D.H., Chapman T.G., Doran D.G. (1988) Problems of rainfall-runoff modelling in arid and semi-arid regions. Hydrol. Sci. J. 33:379-400

Pressland A.J. (1975). Rainfall partitioning by an arid woodland (*Acacia aneura* F. Meull.) in southwestern Queensland. Australian J. Bot. 21:235-245

Pyke D.A., Archer S. (1991) Plant-plant interactions affecting plant establishment and persistence on revegetated rangeland. J.Range Manag. 44:550-557

Rappole J.H., Russell C.E., Norwine J.R., Fulbright T.E. (1986) Anthropogenic pressures and impacts on marginal, neotropical, semiarid ecosystems: the case of South Texas. Sci. Total Environ. 55:91-99

Rapport D.J. (1992) Evaluating ecosystem health. J. Aquatic Ecosystem Health 1:15-24.

Rapport D.J., Regier H.A., Hutchinson T.C. (1985) Ecosystem behaviour under stress. Am. Nat. 125:617-640

Roundy B.A., Jordan G.L. (1988) Vegetation changes in relation to livestock exclusion and rootplowing in southeastern Arizona. Southwest. Nat. 33:425-436

Schlesinger W.H., Reynolds J.R., Cunningham G.L., Huenneke L.F., Jarrell W.M., Virginia R.A., Whitford W.G. (1990) Biological feedbacks in global desertification. Science 247:1043-1048

Slayter R.O. (1965) Measurements of precipitation interception by an arid zone plant community (*Acacia anuera* F. Muell.) UNESCO Arid Zone Research 25:181-192

Sturges D.L. (1993) Soil-water and vegetation dynamics through 20 years after big sagebrush control. J. Range Manag. 46:161-169

Steinberger Y., Shmida A., Whitford W.G. (1990) Decomposition along a rainfall gradient in the Judean Desert, Israel. Oecologia 82:322-324

Verstraete M.M. (1986) Defining desertification: a review. Climate Change 9:5-18

Virginia R.A., Jarrell W.M. (1983) Soil properties in a mesquite dominated Sonoran Desert ecosystem. Soil Sci. Soc. Amer. J. 47:138-144

West N.E., Klemmedson J.O. (1978) Structural distribution of nitrogen in desert ecosystems. In West N.E., Skujins J.J. (eds.) Nitrogen in Desert Ecosystems. Dowden, Hutchinson, and Ross. Stroudsburg, Pa. pp:1-16

Westoby M., Walker B., Noy-Meir I. (1989) Opportunistic management for rangelands not at equilibrium. J. Range Manag. 42:266-274

Whitford W.G., Steinberger Y. (1989) The long-term effect of habitat modification on a desert rodent community. In Morris D.W., Abramsky Z., Fox B.J., Willig M.R. (eds.) Pattern in the Structure of Mammalian Communities. Special Publ. No. 28, The Museum, Texas Tech. Univ., Lubbock, Texas pp:33-43

Whitford W.G., Stinnett K., Anderson J. (1988) Decomposition of roots in a Chihuahuan Desert ecosystem. Oecologia 75:8-11

Whitford W.G. (1989) Abiotic controls on the functional structure of soil food webs. Biol. Fert. Soils. 8:1-6

York J.C., Dick-Peddie W.A. (1969) Vegetation changes in southern New Mexico during the past hundred years. In McGinnies W.G., Goldman B.J. (eds.) Arid Lands in Perspective. Univ. Arizona Press, Tucson, Arizona pp:157-166

RAPPORTEUR'S REPORT

Tom Forbes
Department of Marine Ecology and Microbiology
National Environmental Research Institute
Fredersborgvej 399
P.O. Box 358, DK-4000 Roskilde
Denmark

DISCUSSION

The discussion focused on several key issues. It was generally felt that detection and measurement of undesirable changes in ecosystems was not a major problem. Techniques are currently well developed. However, the choice of indicator parameters may often be difficult. Questions raised included: what should be protected or restored? and how important is the selection of key species?
Further discussion focused on questions that were considered relative to case studies:

Question 1: What information was used to reach a conclusion of system degradation?

Most frequently the initial observation of ecosystem degradation has tended to be due primarily to public perception of the problem (e.g. often people reacting to a catastrophe). Initial perception was generally not science driven and usually involved an economic resource (e.g. fisheries) or an obvious change in community structure. Only later was scientific information brought to bear to determine cause and effect.

Question 2: What information was missing?

In nearly every case a clear understanding of cause and effect was missing. Baseline information was also typically missing so that system state before disturbance was unknown. What information was available often tended to be scattered and in need of synthesis.

Question 3: Were there misdiagnoses? Why?

Misdiagnoses often occurred due to lack of understanding of the relative importance of anthropogenic and natural influences contributing to the observed changes. When new phenomena are encountered, the wrong tools were sometimes brought to bear. This suggested a need for broadly trained trouble-shooting groups that could respond rapidly to environmental problems. Incorrect paradigms sometimes misdirected study of the system. Multiple causation has created problems for cause-effect determination and interpretation.

Question 4: How would one validate predictions or assessments?

Relatively little detailed discussion occurred on this topic. However, it was mentioned that validating assessments requires a strategy that incorporates both flexibility and feedback of information about the system. Particularly in problems involving interdisciplinary or large-scale problems a greater synthesis of available data is required.

Session IV

RECOVERY AND REHABILITATION OF LARGE-SCALE ECOSYSTEMS

INTRODUCTION

Chair: David Rapport

This session will examine the processes of ecosystem recovery from stress in the context of several case studies. These case histories should stimulate general discussion around the following core issues:

1) Can the general pattern of recovery, as measured by a variety of indicators, be characterized as a reversal of trends exhibited by stressed ecosystems? Or alternatively is the recovery trajectory qualitatively different than that of ecosystem breakdown?
2) Which indicators are most effective in signalling that ecosystem recovery is underway? Are these the same indicators the most sensitive in signalling the onset of the breakdown process?
3) How can one differentiate between a "healthy" recovery and an "unhealthy" recovery? What ultimately are the criteria for ecosystem health in the context of recovery? Historic states? A new "domain" characterized by specific structural and functional properties?
4) Is the speed of recovery from perturbation, itself a measure of the underlying health of the ecosystem?
5) Can one distinguish between the dynamics of a system that has once been stressed but recovered and one that has never been subject to stress; ie., in a recovered system, is there still evidence of impairment, or a stress history?
6) What mechanisms (processes) are involved in ecosystem recovery from stress and do they differ from those which are part of mechanisms which enable ecosystems to cope with normal perturbations?
7) Do theses and other case studies suggest that the quest for "leading" and "lagging" indicators will be useful in spotting the early onset of recovery?

The aim of this session will be to draw upon particular case studies to search for possible general properties of ecosystem recovery from stress.

16. PALEOLIMNOLOGICAL APPROACHES TO THE EVALUATION AND MONITORING OF ECOSYSTEM HEALTH: PROVIDING A HISTORY FOR ENVIRONMENTAL DAMAGE AND RECOVERY

John P. Smol
Paleoecological Environmental Assessment and Research Lab (PEARL)
Dept. Biology, Queen's University
Kingston, Ontario K7L 3N6
Canada

INTRODUCTION

A recent workshop on ecosystem health concluded that:

> *An ecological system is healthy and free from "distress syndrome" if it is stable and sustainable -- that is, if it is active and maintains its organization and autonomy over time and is resilient to stress.*
>
> (Haskell et al. 1992; p. 9)

Implicit in this conclusion is the requirement for time-series data of sufficient length and quality to determine if the above criteria have been met. Unfortunately, ecologists and environmental scientists rarely have such data, and most environmental decisions are based on information from "the invisible present" (Magnuson 1990).

The shortage of long-term data was exemplified by Tilman's (1989) library study of papers published between 1977-1987 in the journal *Ecology*. He found that about 40% of the studies were of less than one year duration (and generally less than one field season), and 86% lasted three field seasons or less. Less than 2% of field study experiments lasted at least 5 years.

Aquatic studies also often suffer because data sets are too short-term. I attempted a similar survey to Tilman's by searching 111 papers in a recent volume of *Limnology and Oceanography*, and found that less than 30% of the contributions were based on more than one year's data. Short-time series are often entirely appropriate for some types of research. In general though, this may not be the case for ecological and environmental studies, for without data of sufficient quality and time span one cannot distinguish environmental noise (i.e. natural variability) from signals of environmental degradation and recovery. Not surprisingly, ecologists often invoke "unusual events" to explain the outcomes of their short-term observations (Weatherhead 1986). Moreover, since almost all environmental assessments are done "after-the-fact", managers can rarely determine the trajectory or the causes of degradation, nor can they accurately estimate likely "target" or reference conditions for their mitigation efforts. In short, without long-term data, it is difficult to assess ecosystem health (Smol 1992).

In this essay, I hope to demonstrate that longer time-series data are required and are often attainable, at least in an indirect manner, from paleoecological studies. Most researchers and managers stress "looking forward" when discussing ecosystem health. I agree we should look forward, but it is also very helpful and often essential to look backwards - for back in history lies much of the information that we need. The answers may not always be clearly recorded, but our ability to interpret these data is improving steadily and dramatically, and literature continues to accumulate showing how these approaches can be used to provide meaningful information for ecosystem managers. This paper describes how paleolimnological data can be used as proxies for the data sets we never measured directly, but we now realize we need for effective ecosystem management.

THE PALEOLIMNOLOGICAL APPROACH

Stratigraphic deposits, such as the sediments accumulating at the bottom of lake basins, contain a wealth of paleoecological information. This discussion will generally be restricted to studies of lake histories, or the multidisciplinary science of paleolimnology, which uses the physical, chemical, and biological information preserved in sedimentary profiles to reconstruct past environmental conditions. Biological studies will be highlighted. A number of review articles and commentaries have recently been published and so this introduction will be brief. These include general reviews (e.g. Anderson 1993; Binford et al. 1987; Smol 1990ab; Smol 1992; Smol and Glew 1992), as well as those specific to, for example, the approaches used to "calibrate" indicators with respect to environmental variables (Charles and Smol 1994), the ways that paleolimnological approaches can be melded with biomonitoring programs (Dixit et al. 1992ab; Charles et al. 1994), and how paleolimnology can be used to study climatic change (Smol et al. 1991). In addition, a number of books (e.g. Gray 1988; Haworth and Lund 1984; Warner 1990) have been published on methodologies, assumptions, and case studies. The overall approach can be challenged by a variety of potential problems, but in general these can be assessed and quantified (reviewed in Charles et al. 1994). Below, I briefly summarize some of the general approaches.

Although paleolimnological protocols have become much more sophisticated over recent years, the basic approach is still relatively straightforward. Paleolimnologists must first remove sediment cores from near the centre of the lake basin, which are of appropriate length (i.e. cover the appropriate time span of sediment accumulation) and quality to study the problem at hand. For most North American work, lake managers will be most interested in the last 200 years of history, and so relatively short sediment cores (e.g. about 50 cm) will usually suffice.

Smol and Glew (1992) recently summarized and illustrated some of the most commonly used equipment for paleolimnological studies; much of this equipment has been modified to collect and section high quality cores of a lake's recent sediments. Moreover, if the cores are undisturbed by bioturbation (apparently a surprisingly less serious problem than most people may expect in many lacustrine environments) or other such processes, close-interval sectioning of the core (e.g. Glew 1988) can often provide a temporal resolution of a few years or less per sediment section (e.g. at 0.25 cm intervals). The sediment stratigraphy can then be dated using a variety of methods (also reviewed in Smol and Glew 1992; and Charles et al. 1994), among which ^{210}Pb geochronologies are the most widely used.

Lake sediments contain a large selection of indicators that can be used to reconstruct past lake environments. Past lake biota are represented by both morphological (i.e. the hard parts of organisms) and biogeochemical (i.e. some chemical signature that can be related to past populations) remains. A good example of the latter are fossil pigments, many of which are specific to certain algal groups, which can be used to trace, for example, past blue-green algal or bacterial populations, which do not usually leave reliable morphological fossils. Gray (1988) and Warner (1990) have recently edited and collated a number of review articles summarizing most of the major indicators that can be used in paleolimnological reconstructions.

The most widely used bio-indicators are algal (such as diatom valves, chrysophyte scales and cysts) and invertebrate (such as the chitinous exoskeletal parts of zooplankton and aquatic insect) fossils. Pollen grains and spores (usually treated under the separate science of palynology) can also be studied in sedimentary profiles to reconstruct past terrestrial vegetation. Geochemical analyses of sediments can be used to, for example, trace the history of airborne contaminants (e.g. PAHs, PCBs), or other variables of environmental importance (e.g. heavy metals). Paleolimnologists continue to describe new indicators, and the amount of information contained in sedimentary profiles is often surprising.

The precision with which paleolimnologists can infer environmental variables has improved greatly over the last few years, and this is especially true of inferences using biological data. Quantification of the environmental optima and tolerances of many indicator groups has been done to a high degree of certainty, with known variance, for many indicators. Many ecologically realistic and statistically robust transfer-function equations have been developed that relate species distributions (e.g. diatoms) to particular environmental variables (e.g. pH, total lakewater phosphorus, temperature, salinity). These transfer functions can then be used to reconstruct past environmental conditions in dated sediment cores so that long- term trends can be inferred.

Charles and Smol (1994) describe how transfer functions are usually constructed. In brief, a suite of perhaps 100 calibration or "training set" lakes is chosen that represents a range of present-day environmental variables. The contemporary limnological conditions in these lakes should be fairly well known, preferably over the last few years. The paleolimnologist then samples the surface sediments of each of these lakes, usually the top 0.5 to 1 cm depth, which contain indicators (such as diatom valves, insect parts) that lived in the lake over the recent past. Sampling equipment specifically developed for these types of studies is available (Wright 1990; Glew 1991). In effect, the surface sediments contain a temporally and spatially integrated sample of the biota living in the lake over the last few years.

The above analyses yield two matrices of data: one is a listing of, in this hypothetical example, 100 lakes with their present-day limnological data, and a second with the percentages of diatom species from the surface sediments of these same lakes. The next step is to use statistical treatments to construct quantitative transfer functions, by which environmental variables can be inferred from the fossil biota. These transfer functions incorporate both ecological realism and statistical robustness and have been subjected to a large amount of quality control considerations. The techniques are strong and defendable both theoretically and empirically. A few examples of such transfer functions include those for lakewater pH (Birks et al. 1990; Dixit et al. 1993), trophic variables (Hall and Smol 1992; Fritz et al. 1993), lakewater salinity (Fritz 1990; Cumming and Smol 1993), and temperature and other climatic related variables (Walker et al. 1991; Pienitz and Smol 1993; Pienitz et al. 1994).

Although most detailed paleolimnological studies attempt to achieve temporal resolutions of a few years, some studies can actually infer changes at the sub-annual level (some examples shown below). Moreover, what paleolimnology may lack in temporal resolution can be compensated for by extending the time scale back in time, and back in time is where many of our answers lie. Below, I provide a few examples of this approach.

DEFINING ECOSYSTEM HEALTH

This essay began with a quote defining ecosystem health within the context of stability and sustainability. Below I explore some of these concepts using paleoecological data.

One of the supposed symptoms of a stressed ecosystem is the "lack of stability". Most scientists who claim they are studying "stable" systems are basing their observations on perhaps only 2 or 3 years of data, and often even less. A paleolimnological perspective may temper some of these observations. For example, Renberg (1990, Figure 1) analysed 700 contiguous sediment samples for diatoms from a 12,000 year long sediment core from a lake in southwestern Sweden. He found that there were considerable short-term variations in the diatom assemblages and pH. For example, over a few hundred years, pH could change several times with an amplitude of up to 0.5 pH units. Marked acidification, of much greater magnitude of change than natural variability, was recorded in the lake's most recent sediments, coinciding with the period of atmospheric deposition of strong acids. This example illustrates the "natural variability" or noise that can be present in a system.

Similarly, researchers studying the recent limnology of Little Round Lake, Ontario, might easily conclude that this presently meromictic lake is a stable system, with some predictable changes observed in its characteristics. If one looked at the paleoecological record of this lake (Smol and Boucherle 1985), the lake's stability may be less apparent. Some of these pre-cultural changes are difficult to interpret, but may have been caused by past changes in the lake's chemical stratification; other changes appeared to be the result of a catastrophic decimation of hemlock trees in the lake's drainage, presumably due to a pathogen, about 4800 years ago (Boucherle et al. 1986). Hall and Smol (1993) completed paleolimnological assessments of the effects of this decimation on other Ontario lakes. The paleolimnological literature contains many other such examples, which may challenge some ideas of "stability" and "equilibrium", especially as they relate to anthropogenic causes. "Natural" systems have changed dramatically in the past, without any human interference.

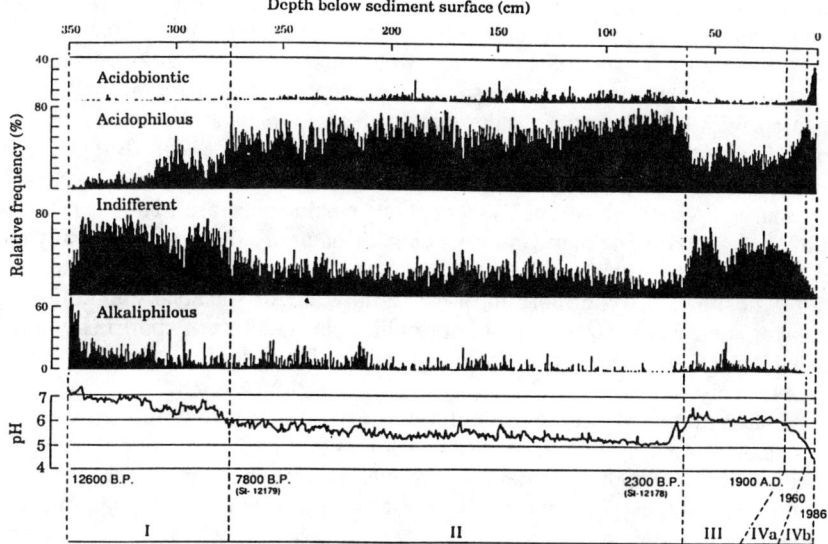

Figure 1. Distributions of fossil diatom taxa, grouped into Hustedt's four pH categories, for the 12,600 year history of Lilla Öresjön, southwest Sweden (Renberg 1990). Diatoms were identified and enumerated from 700 contiguous sediment samples (0.5. cm thick) from this 3.5 m long core. The pH inferences are based on weighted averaging regression and calibration. The sediment core was dated using ^{14}C and ^{210}Pb geochronologies. The lake began as an alkaline system (12,600-7,800 BP) after deglaciation (I). This was followed by a naturally acidic period (7,800-2,300 BP) when diatom-inferred pH decreased from 6.0 to 5.2 (II). Between 2,300 BP and AD 1900, lakewater pH increased as a result of land-use changes in the drainage (III). The recent declines in pH in this century (IV) are largely the result of atmospheric deposition of strong acids. Note the natural variability in the assemblages; it would be difficult to discern these trends without long-term data sets. Reproduced from Renberg (1990); used with permission.

Ecosystem health is also often defined with the term "sustainability"; the latter "...implies the system's ability to maintain its structure (organization) and function (vigour) over time in the face of external stress (resilience)." (Costanza 1992). Paleolimnological studies show that lakes undergo successional changes over time that have little, if anything, to do with "stress" as it is often defined. Once a system is subjected to "stress", I would argue that the changes that take place in an ecosystem, and hence a system's "resilience", is far more stressor-dependent than some inherent quality of the system itself. For example, if some lakes have had a low acid neutralizing capacity for thousands of years (e.g. Whitehead et al. 1989), and so are especially susceptible to anthropogenic acidification, I am not certain that these lakes are any less "healthy" than those that naturally have a high buffering capacity. Using North American lakes as an example, both systems have supported, over about the last 12,000 years (since deglaciation), thriving but distinct food webs adapted to the ambient environmental conditions. It was only once a specific pollutant or stressor, in this case acidic deposition, was placed on the system that "deterioration" was realized in the lakes with a naturally low acid neutralizing capacity (e.g. Charles et al. 1990). But this is stressor-dependent, not system-dependent. Had the stressor been, for example, a contaminant that is especially toxic in higher pH waters, then would the more alkaline lakes be "less healthy"? I would argue that naturally acidic or naturally fishless or naturally eutrophic lakes are just as "healthy" as their counterparts. The difference in opinions as to what is more "healthy" may lie more in our culturally and/or financially modified views of what lakes should be: non-acidic, oligotrophic, but with lots of sport fish. Rather than focusing on stressor-dependent "symptoms", we should look at the effects of specific stressors themselves.

I have previously argued (Smol 1992) that an alternate definition of a "healthy" ecosystem is the ecosystem that existed before significant anthropogenic impact. The parameters defining such ecosystems can be gleaned from paleoecological records. I fully acknowledge that these are not entirely realistic goals for mitigation programs, as humans are now part of the systems and they are here to stay. Nonetheless, this information is vital to provide some realistic constraints to our definitions. This is discussed in more detail below.

DETERMINING THE EXTENT OF THE PROBLEM

One of the first requirements of a manager is to determine if in fact a problem exists. To answer such a question, we need to know background or pre-impact conditions, and the trajectory of environmental change. For many environmental problems, there are no historical data, but paleolimnology can provide some of the missing data.

Recently, controversy raged in parts of North America and Europe over the problem (or some would claim "supposed problem") of lake acidification. Limnologists, working with present-day systems, could record that a large percentage of lakes in, for example, the Adirondack Mountains of New York, were presently acidic and many were presently fishless. A major scientific and policy question was: Are they naturally acidic, or were they acidified due to atmospheric deposition or some other cause? In short, is there a problem?

Because the distribution of many paleolimnological indicators, like diatoms (Dixit et al. 1992a, 1993) and chrysophytes (Cumming et al. 1992a), are closely related to lakewater pH and pH-related variables (e.g. monomeric aluminum, dissolved organic carbon), statistically robust transfer functions were developed to relate the distribution of taxa to these environmental variables, using surface-sediment calibration or training sets, as described above. Once these quantitative transfer functions were developed, paleolimnologists could use the distribution of these indicators in dated sediment cores to reconstruct past trends in lakewater variables.

To use the example from the Adirondacks, this was done by sampling, using a statistical design, 38 Adirondack lakes with a range in present pH from 7.8 to 4.4. In order to determine if these lakes had acidified since pre-industrial times, the fossil algal biomonitors were analysed in the lakes' surface sediments (i.e. the "tops", to represent conditions over the last 3 or so years), and from sediments deposited before AD 1850 (i.e. sediments that were deposited in pre-industrial times, or the "bottoms" of short cores, from about 30 cm depth in the core). A summary of the diatom results is presented in Figure 2. The 38 study lakes are presented in order of increasing pH, and the differences between pre-1850 pH and present-day inferred pH is shown in the histograms; other pH related variables are also illustrated. Cumming et al. (1992b) showed that only those lakes with present-day pH values < 6.5 show evidence of chronic long-term acidification. On a regional basis, this represents approximately 25-30% of 675 statistically representative Adirondack target lakes (target lakes had acid neutralizing capacities of less than 400 μeq $-L^{-1}$). In short, presently acidic lakes did, in fact, acidify since AD 1850, and acidic precipitation was the most likely cause (see Cumming et al. 1992b for full discussion).

Such a "top/bottom" approach was also used in other acid affected regions, such as Sudbury (Dixit et al. 1992c) and northern Sweden (Korsman 1993). As perhaps expected, the Sudbury example showed marked acidification and metal contamination as a result of the local smelting operations. Interestingly, although a number of northern Swedish lakes are presently acidic, Korsman's data suggested that many may have been naturally acidic. This clearly has important implications for liming programs.

Often related to the issue of acidification is the problem of fisheries loss. Similar to the acidification questions posed above, it is relatively easy for a manager to determine if a lake is presently fishless. However, without long-term data, one cannot be certain if the lake ever supported fish populations, or if it is naturally fishless (lakes of the latter type are not necessarily good candidates for fish-stocking programs!). Newly developed paleolimnological techniques can be used to infer the past status of fish populations.

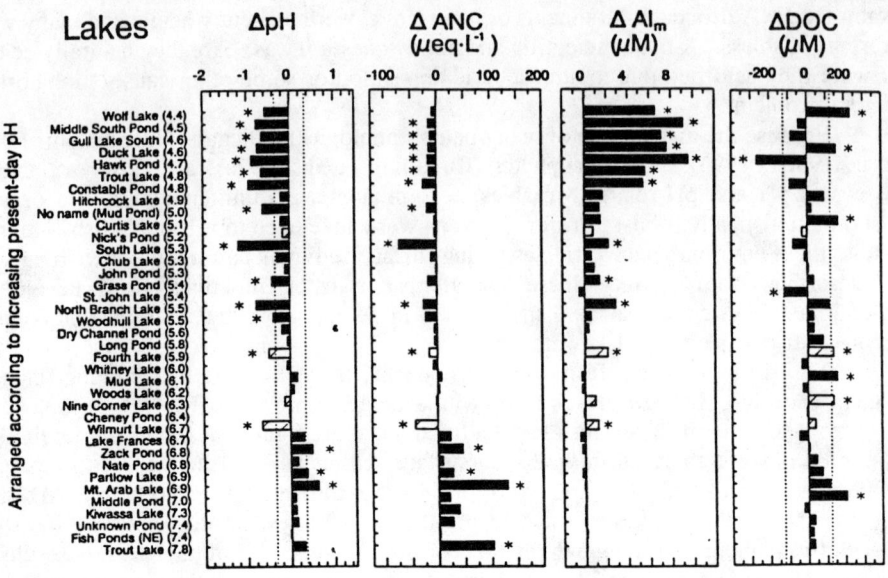

Figure 2. Diatom-based estimates of historical change (pre-industrial to the present) in pH, acid neutralizing capacity (ANC), total monomeric aluminum (Al_m), and dissolved organic carbon (DOC) from 37 Adirondack study lakes (Cumming et al., 1992b). The estimates are presented as the differences in inferred water chemistry between the "top" (0-1cm, representing recent limnological conditions) and the "bottom" (usually >25cm, representing pre-industrial time periods) sediment core intervals. Asterisks denote changes that are greater than the bootstrapped root mean squared error for the various inference models. Reproduced from Cumming et al. (1992b); used with permission.

Uutala (1990) and Johnson et al. (1990) have shown that the preserved mandibles of phantom midge (*Chaoborus*) larvae can be used to assess past fish populations. The key indicator is *C. americanus*, a species that does not usually co-occur with fish. Such studies are especially effective when they are done in conjunction with stratigraphic analyses of siliceous algal fossils (diatoms and chrysophytes), which can be used to infer past lakewater chemistry changes (Kingston et al. 1992; Uutala et al. 1994). These approaches can also be used to study, after-the-fact, the effects of other stressors on lake systems. For example, Miskimmin and Schindler (1994) analysed sediment cores from two toxaphene-treated and one untreated lake basin to demonstrate both the short-term effect of toxaphene applied in 1961/62 and the longer-term effect of subsequent trout stocking on invertebrates.

Paleolimnological approaches can be used to study many other types of problems. For example, a serious problem in many regions is eutrophication. Brugam (1988), Christie (1993), and Dixit and Smol (1994) used "top-bottom" approaches to assess the extent of cultural eutrophication in Washington, Ontario, and the north-eastern USA, respectively. Many lakes showed marked eutrophication with cultural disturbances, but some lakes were naturally productive, and some lakes revealed other background conditions that would be of interest to aquatic managers.

SETTING REALISTIC TARGETS FOR ECOSYSTEM HEALTH

The above discussion can also apply to setting realistic targets for mitigation efforts. Cairns (1991; p.186) notes that "restoration means recreating both the structural and functional attributes of a damaged ecosystem". But without long-term environmental data, we cannot estimate what background or pre-impact conditions were, and so how can we know what attributes we should be recreating? For example, if a lake is naturally acidic or naturally fishless, attempts to "return" this system to a circumneutral and fish-stocked system are not ecologically sound, and in fact it can be quite damaging if "exotics" are introduced, in this case fish, or changing the water chemistry to levels that the lake had never experienced. Similarly, a lake that is naturally eutrophic will not respond to treatments, and so forth. As acknowledged above, attempting to return lakes entirely to pre-impact conditions is not practical in many systems; humans are now part of the system. Nonetheless, paleoecological data can provide some realism to mitigation programs (as noted above, one should not immediately lime a naturally acid lake or stock a naturally fishless system), as well as provide information on how past mitigations have affected the system (Smol 1992).

PROVIDING INFORMATION ON A SYSTEM'S LIKELY RESPONSE TO SPECIFIC STRESSORS

The above can be re-phrased into a question, such as "Does knowing background conditions help determine if a lake will be adversely affected by a specific stressor?". Although our results may be preliminary, there are some indications that knowing background levels can be used to help predict how a system will respond to a stressor. My example is again from the Adirondacks of New York, although it builds on considerable previous paleolimnological work on acidification (e.g. Battarbee et al. 1990; Charles et al. 1990; Davis et al. 1990; Sweets et al. 1990; Kingston et al. 1990; and reviewed in Charles et al. 1989 and Davis 1987). As a continuation of the "top/bottom" study described earlier, we asked: If lakes did acidify, then when did this acidification begin? Cumming et al. (1994) showed that four distinct patterns of post-1850 to present-day pH trajectories were identified from the 20 lakes cored for this study. These included: 1) lakes that had background pH values around 5 that continued to acidify to the present-day, (2) lakes that had background pH values between 5.5 and 6 and acidified quickly *ca*. 1900, (3) lakes that had background pH values between 5.5 and 6.5 and acidified rapidly *ca*. 1940, and (4) those lakes that showed little or no evidence of lake acidification. Canonical variates analyses (multiple discriminant analyses) showed that these lake groupings were significantly discriminated from each other based on their background pH values [i.e. the lakes with lower pH values (little buffering capacity) acidified *ca*. 1900, whereas those that acidified *ca*. 1940 had intermediate pH values, and those that didn't acidify had the highest background pH values]. The lakes that did not acidify were also located outside the area of highest sulfate deposition. From these data, it would appear that knowing background conditions can help determine how a lake system will change if it is subjected to specific stressors.

SETTING CRITICAL LOADS FOR POLLUTANTS

A critical load can be defined as the load of a pollutant that causes a change in the structure or function of an ecosystem. Comparison of the above paleo-results with historical estimates of sulfate loadings for the Adirondacks (Husar et al. 1991) indicates that Adirondack lakes that acidified *ca*. 1900 and *ca*. 1940 lakes did so under sulfate loadings of approximately 5-10 kg/ha^2/yr and 20-25 kg/ha^2/yr, respectively (Cumming et al. 1994). Since the early 1970's, deposition in the Adirondacks has declined significantly. Nonetheless, inferred post-1970 pH trends in these lakes showed little change as would be expected from the Cumming et al. (1994) interpretations. This observation is in concordance with the small amount of water-monitoring data that exists for Adirondack lakes (Driscoll and van Dreason 1993).

Similar studies have been completed for the acid and metal stressed lakes of the Sudbury region (summarized in Dixit et al. 1994), and could be undertaken for many other regions receiving different types of pollutants. For example, Schmidt (1991) used high-resolution techniques (Figure 3) to study the diatom succession from a laminated core from an Austrian lake to reconstruct the effects of nutrient reductions as a result of sewage diversion in Mondsee. Anderson et al. (1990) documented the eutrophication and the partial recovery of a lake in Northern Ireland as a result of effluent from a local creamery, and Wolin et al. (1991) tracked recent changes in Lake Ontario's development as a result of phosphorus reductions.

BIODIVERSITY

A major global concern now appears to be to protect "biodiversity", and clearly "diversity" (however measured) is a popular metric when it comes to studies of ecosystem health (see chapters in Constanza et al. 1992). But how can we maintain biodiversity or use these measures without knowing what past diversity has been? Many conclusions are based on space-for-time substitutions by, for example, comparing a presently acidic lake to a presently circumneutral lake.

Few conclusions seem to be from actual studies of a single lake over the time period of degradation. We are again missing the long-term data sets. Perhaps the system has always had low diversity, and so in this case a less diverse system might be "healthier" than a more diverse system because it is closer to its "natural" state?

Paleoecologists may question some of the commonly held beliefs on diversity, even though quantitative estimates of diversity using paleoecological data may be problematic (Smol 1981). Paleolimnological studies often clearly show that diversity has changed markedly in many lake systems in the past. These changes are often unrelated to human activity, and even when they are, there may be some surprises. To return to the example of acidification, in many affected lakes chrysophyte diversity actually increased in lakes as a result of declines in lakewater pH. Moreover, although diatom plankton may have declined, diversity in the diatom periphyton may have increased.

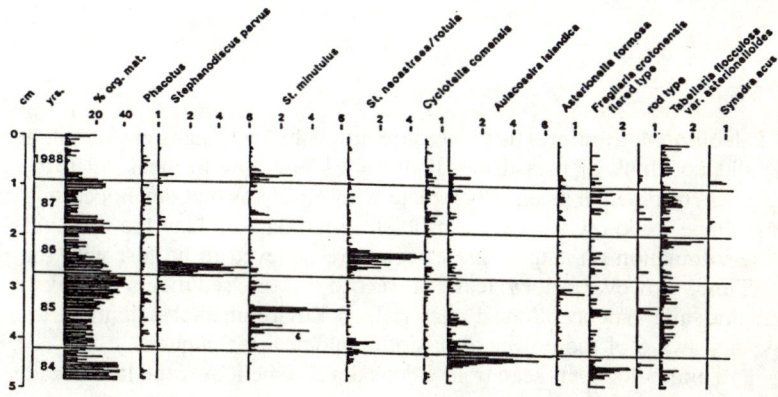

Figure 3. Microstratigraphy of planktonic diatoms of an *in situ* frozen sediment core from Mondsee, Austria (Schmidt 1991). Note the high resolution one can achieve in some paleolimnological studies. Reproduced from Schmidt (1991); used with permission.

THE SOURCE AND FATE OF CONTAMINANTS

Major issues in contaminant research are to define the sources and fates of contaminants in the ecosystem. In short: where did it come from, who is responsible, and where did it end up?

As with the previous examples, much of this information cannot be traced directly, but can be studied using a variety of paleoecological approaches. Many examples are available (e.g. see review by Autenrieth et al. 1991, and chapters in Monitoring Assessment and Research Centre 1985). Figure 4 is an example of how polycyclic aromatic hydrocarbons (PAHs) were studied in sedimentary profiles from the Saguenay Fjord. Smith and Levy (1990) showed that changes in the chemical assemblage of PAHs tracked the region's aluminum industry. PAH inputs increased dramatically after the 1940s following a major expansion and changes in the processing facilities.

In addition to aquatic sediments, contaminants can be traced in other stratigraphic material. Bourton et al. (1991) used accumulated snow stratigraphy in Greenland to track changes in anthropogenic lead, cadmium and zinc deposition over the last 5,500 years. They showed a marked increase in deposition over the last few centuries, but then a decline in concentrations since the late 1960s, presumably in response to remedial actions. In a follow-up study, Rosman et al. (1993) used the isotopic composition of lead in the snow to trace sources. By studying the ratio of $^{206}Pb/^{207}Pb$, they concluded that the United States was a significant source of lead in the 1970s, but it had since declined in relative proportion to sources from Eurasia. The use of these and other isotopes to track the sources of pollutants will certainly be an important research area in the future.

CONCLUSIONS

Paleolimnology allows us to measure, after-the-fact, many environmental variables that we did not think of measuring or did not know how to measure in the past. The history of environmental science is replete with problems that developed without anyone noticing. Once we know we have a problem, vital data on how the problem developed, including infomation on causes and effects, are believed to be lost with the passage of time. Time, however, does leave a record in the sediments of lakes and other stratigraphic and paleoecological materials. Environmental scientists are becoming increasingly aware of the power of paleolimnological techniques.

Paleolimnology has seen tremendous development over the last decade, and many of these advances can be used by lake managers to assess aquatic resources. In fact, for many lake management issues, there are no substitutes for paleolimnological data. Although most of these procedures have been tailored to lake studies, paleolimnological approaches can be adapted to environmental assessments of, for example, wetlands (Stevenson and Flower 1991; Flower et al. 1992), rivers (Klink 1989), peatlands (Stevenson et al. 1990; Gorham and Janssens 1992), estuaries (Brush and Davis 1984), and deltas (Trefry et al. 1985). New techniques continue to be developed (Korsman et al. 1992), and new problems are emerging rapidly.

These are exciting and challenging times to be doing paleolimnology. I have just described a few examples of the approach and just a handful of case studies. In the coming years, there is little doubt that paleoecological techniques will continue to be integrated, at an accelerated rate, into the assessment and the study of ecosystem health. The challenge for paleolimnologists is to continue developing techniques and approaches that can provide these data in a form that is meaningful to other scientists and managers.

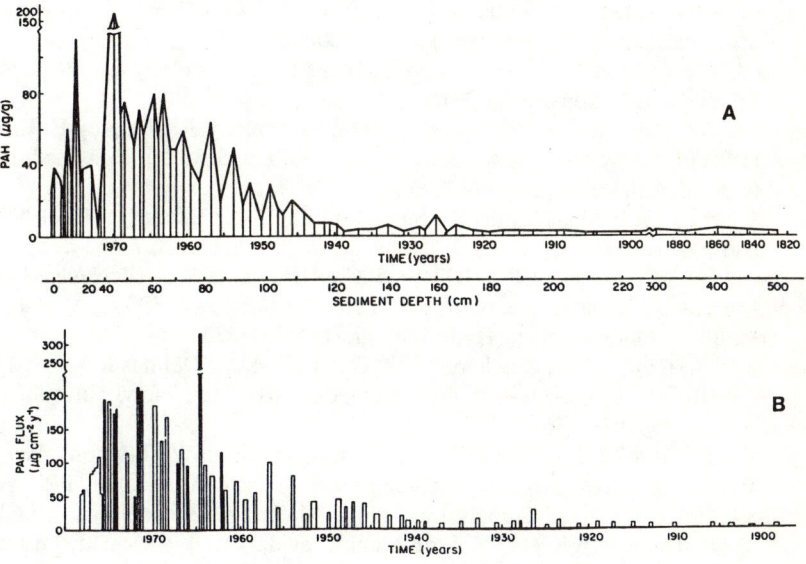

Figure 4. PAH concentration (A) and PAH flux (B) as a function of sediment deposition date from a core from the Saguenay Fjord (Smith and Levy 1990). Reproduced from Smith and Levy (1990); used with permission.

ACKNOWLEDGEMENTS

I am grateful to colleagues at PEARL, as well as W. Last, H.J. Birks, D.F. Charles, B. Cumming, P. Leavitt, and V. Forbes for helpful comments on the manuscript.

REFERENCES

Anderson N.J. (1993) Natural versus anthropogenic change in lakes: The role of the sediment record. Trends in Ecology and Evolution 8:356-361

Anderson N.J., Rippey B., Stevenson A.C. (1990) Change to a diatom assemblage in a eutrophic lake following point source nutrient re-direction: a palaeolimnological approach. Freshwat. Biol. 23:205-217

Autenrieth R., Bonner J., Schreiber L. (1991) Aquatic sediments. Res. J. Water. Poll. Control. Fed. 63:709-725

Battarbee R.W., Mason J., Renberg I., Talling J.F. (eds.) (1990) Palaeolimnology and lake acidification. The Royal Society, London

Berglund B.E. (ed.) (1986) Handbook of Holocene Palaeoecology and Palaeohydrology. John Wiley & Sons, New York

Binford M.W., Brenner M., Whitmore T., Higuera-Gundy A., Deevey E.S., Leyden B. (1987) Ecosystems, paleoecology and human disturbance in subtropical and tropical America. Quat. Sci. Rev. 6:115-128

Birks H.J.B., Line J.M., Juggins S., Stevenson A.C., ter Braak C.J.F. (1990) Diatoms and pH reconstruction. Phil. Trans. R. Soc. Lond. B 327:263-278

Boucherle M.M., Smol J.P., Oliver T.C., Brown S.R., McNeely R.N. (1986) Limnologic consequences of the decline in hemlock 4800 years ago in three Southern Ontario lakes. Hydrobiologia 143:217-225

Boutron C.F., Görlach U., Candelone J.P., Bolshov M.A., Delmas R.J. (1991) Decrease in anthropogenic lead, cadmium and zinc in Greenland snows since the late 1960s. Nature 353:153-156

Brugam R.B. (1988) Long-term history of eutrophication in Washington lakes. In Adams W.J. et al. (eds.) Aquatic Toxicology and Hazard Assessment:10th Vol, ASTM STP 971, American Society for Testing Materials, Philadelphia pp:63-70

Brush G.S., Davis F.W. (1984) Stratigraphic evidence of human disturbances in an estuary. Quat. Res. 22:91-108

Cairns J. (1991) The status of theoretical and applied restoration ecology. The Environmental Professional 13:186-194

Charles D.F., Battarbee R.W., Renberg I., van Dam H., Smol J.P. (1989) Paleoecological analysis of lake acidification trends in North America and Europe using diatoms and chrysophytes. In Norton S.A. et al. (eds.) Acidic Precipitation: Vol. 4. Springer-Verlag, New York pp:207-276

Charles D.F., Binford M.W., Furlong E.T., Hites R.A., Mitchell M.J., Norton S.A., Oldfield F., Paterson M.J., Smol J.P., Uutala A.J., White J.R., Whitehead D.R., Wise R.J. (1990) Paleoecological investigation of recent lake acidification in the Adirondack Mountains, N.Y. J. Paleolim. 3:195-241

Charles D.F., Smol J.P. (1994) Long-term chemical changes in lakes: Quantitative inferences using biotic remains in the sediment record. In Baker L. (ed.) Environmental Chemistry of Lakes and Reservoirs, American Chemical Society, Washington DC pp:3-31

Charles D.F., Smol J.P., Engstrom D.R. (1994) Paleolimnological approaches to biomonitoring. In Loeb S., Spacie A. (eds.) Biological Monitoring of Aquatic Systems. Lewis Press, Ann Arbor pp:233-293

Christie C.E. (1993) Paleoecological reconstruction of lake trophic status: the effects of human activity on lake conditions in southeastern Ontario in the recent (ca. 200 years) past. PhD Thesis, Queen's University, Canada

Costanza R. (1992) Toward an operational definition of ecosystem health. In Costanza et al. (eds.) Ecosystem Health - New Goals for Environmental Management. Island Press, Washington DC pp:239-256

Costanza R., Norton B.G., Haskell B.D. (eds.) (1992) Ecosystem Health - New Goals for Environmental Management. Island Press, Washington DC

Cumming B.F., Smol J.P. (1993) Development of diatom-based salinity models for paleoclimatic research from lakes in British Columbia (Canada). Hydrobiologia 269/270:179-196

Cumming B.F., Smol J.P., Birks H.J.B. (1992a) Scaled chrysophytes (*Chrysophyceae* and *Synurophyceae*) from Adirondack (N.Y., USA) drainage lakes and their relationship to measured environmental variables, with special reference to lakewater pH and labile monomeric aluminum. J. Phycol. 28:162-178

Cumming B.F., Smol J.P., Kingston J.C., Charles A.F., Birks H.J.B., Camburn K.E., Dixit S.S., Uutala A.J., Selle A.R. (1992b) How much acidification has occurred in Adirondack region (New York, USA) lakes since pre-industrial times? Can. J. Fish. Aquat. Sci. 49:128-141

Cumming B.F., Davey K., Smol J.P., Birks H.J.B. (1994) When did Adirondack Mountain lakes begin to acidify and are they still acidifying? Can. J. Fish. Aquat. Sci. 51 (in press)

Davis R.B., Anderson D.S., Whiting M.C., Smol J.P., Dixit S.S. (1990) Alkalinity and pH of 3 lakes in northern New England, USA., over the past 3000 years. Phil. Trans. R. Soc. Lond. B 327:413-421

Davis R.B. (1987) Paleolimnological diatom studies of acidification of lakes by acid rain: an application of Quaternary science. Quat. Sci. Rev. 6:147-163

Dixit S.S., Smol J.P. (1994) Diatoms as environmental indicators in the Environmental Monitoring and Assessment - Surface Waters (EMAP-SW) program. Env. Monitoring and Assessment (in press)

Dixit S.S., Smol J.P., Kingston J.C., Charles D.F. (1992a) Diatoms: Powerful indicators of environmental change. Environ. Sci. Tech. 26:22-33

Dixit S.S., Cumming B.F., Smol J.P., Kingston J.C. (1992b) Monitoring environmental changes in lakes using algal microfossils. In McKenzie D.H., Hyatt D.E., MacDonald V.J. (eds.) Ecological Indicators: Volume 2. Elsevier Applied Sciences, Amsterdam pp:1135-1155

Dixit S.S., Dixit A.S., Smol J.P. (1992c) Assessment of changes in lake water chemistry in Sudbury area lakes since preindustrial times. Can. J. Fish. Aquat. Sci. 49(Supp. 1):8-16

Dixit S.S., Cumming B.F., Kingston J.C., Smol J.P., Birks H.J.B., Uutala A.J., Charles

D.F., Camburn K. (1993) Diatom assemblages from Adirondack lakes (N.Y., USA) and the development of inference models for retrospective environmental assessment. J. Paleolim. 8:27-47

Dixit S.S., Dixit A.S., Smol J.P. (1994) History and extent of industrial damage to lakes: Reading the record stored in Sudbury lake sediments. In Gunn J (ed.) Long Term Ecological Research in Sudbury (in press)

Driscoll C.T., van Dreason R. (1993) Seasonal and long-term temporal patterns in the chemistry of Adirondack lakes. Water Air Soil Poll. 67:301-307

Glew J.R. (1988) A portable extruding device for close interval sectioning of unconsolidated core samples. J. Paleolim. 1:235-239

Glew J.R. (1991) Miniature gravity corer for recovering short sediment cores. J. Paleolim. 5:285-287

Flower R.J., Dearing J.D., Rose N., Patrick S.J. (1992) A palaeoecological assessment of recent environmental change in Moroccan wetlands. Würzb. Geogr. Arb. 84:17-44

Fritz S.C. (1990) Twentieth-century salinity and water level fluctuations in Devils Lake, N. Dakota: a test of a diatom-based transfer function. Limnol. Oceanogr. 35:1771-1781

Fritz S.C., Kingston J.C., Engstrom D.R. (1993) Quantitative trophic reconstruction from sedimentary diatom assemblages: a cautionary tale. Freshwat. Biol. 30:1-23

Gorham E., Janssens J. (1992) The paleorecord of geochemistry and hydrology in northern peatlands and its relation to global change. Suo 43:117-126

Gray J. (ed.) (1988) Paleolimnology. Elsevier, Amsterdam

Hall R.I., Smol J.P. (1992) A weighted-averaging regression and calibration model for inferring total phosphorus concentration from diatoms in British Columbia (Canada) lakes. Freshwat. Biol. 27:417-434

Hall R.I., Smol J.P. (1993) The influence of catchment size on lake trophic status during the hemlock decline (4,800 to 3,500 BP) in southern Ontario lakes. Hydrobiologia 269/270:371-390

Haskell B.D., Norton B.G., Costanza R. (1992) Introduction: What is ecosystem health and why should we worry about it? In Costanza R. et al. (eds.) Ecosystem Health - New Goals for Environmental Mangement. Island Press, Washington DC pp:3-20

Haworth E.Y., Lund J.W.G. (eds.) (1984) Lake sediments and environmental history. University of Minnesota Press, Minneapolis

Husar R.B., Sullivan T.J., Charles D.F. (1991) Methods for assessing long-term trends in atmospheric deposition and surface water chemistry. In Charles.D.F. (ed.) Acidic Deposition and Aquatic Ecosystems: Regional Case Studies. Springer-Verlag, New York pp:65-82

Johnson M.G., Kelso J.R.M., McNeil O.C., Morton W.B. (1990) Fossil midge associations and the historical status of fish in acidified lakes. J. Paleolim. 3:113-127

Kingston J.C., Cook R.B., Kreis R.G. Jr, Camburn K.E., Norton S.A., Sweets P.R,

Binford M.W., Mitchell M.J., Schindler S.C., Shane L.C.K., King G.A. (1990) Paleoecological investigation of recent acidification in the northern Great Lakes states. J. Paleolim. 4:153-201

Kingston J.C., Birks H.J.B., Uutala A.J., Cumming B.F., Smol J.P. (1992) Assessing trends in fishery resources and lake water aluminum for paleolimnological analyses of siliceous algae. Can. J. Fish. Aquat. Sci. 49:116-127

Klink A. (1989) The Lower Rhine: palaeoecological analysis. In Petts G.E. (ed.) Historical Change of Large Alluvial Rivers: Western Europe. John Wiley & Sons, Chichester pp:183-201

Korsman T. (1993) Acidification trends in Swedish lakes: an assessment of past water chemistry conditions using lake sediments. PhD Thesis, University of Umeå, Sweden

Korsman T., Nilsson M., Öhman J., Renberg I. (1992) Near-infrared reflectance spectroscopy of sediments: A potential method to infer the past pH of lakes. Environ. Sci. Tech. 26:2122-2126

Magnuson J. (1990) Long-term ecological research and the invisible present. BioScience 40:495-501

Miskimmin B.M., Schindler D.W. (1994) Long-term invertebrate community response to toxaphene treatment in two lakes: 50-year records reconstructed from lake sediments. Can. J. Fish. Aquat. Sci. 51: (in press)

Monitoring and Assessment Research Centre (1985) Historical Monitoring. MARC Report No. 31, University of London, London

Pienitz R., Smol J.P. (1993) Diatom assemblages and their relationship to environmental variables in lakes near Yellowknife (N.W.T., Canada). Hydrobiologia 269/270:391-404

Pienitz R., Smol J.P., Birks H.J. (1994) Assessment of freshwater diatoms as quantitative indicators of past climatic change in the Yukon and Northwest Territories, Canada, (under review)

Renberg I. (1990) A 12,600 year perspective of the acidification of Lilla Öresjön, southwest Sweden. Phil. Trans. R. Soc. Lond. B 327:357-361

Rosman K.J.R., Chisholm W., Boutron C.F., Candelone J.P., Görlach U. (1993) Isotopic evidence for the source of lead in Greenland snows since the late 1960s. Nature 362:333-334

Schmidt R. (1991) Recent re-oligotrophication in Mondsee (Austria) as indicated by sediment diatom and chemical stratigraphy. Verh. Internat. Verein. Limnol. 24:963-967

Smith J.N., Levy E.M. (1990) Geochronology for polycyclic aromatic hydrocarbon contamination in sediments of the Saguenay Fjord. Environ. Sci. Tech. 2:874-879

Smol J.P. (1981) Problems associated with the use of "species diversity" in paleolimnological studies. Quat. Res. 15:209-212

Smol J.P. (1990) Are we building enough bridges between paleolimnology and aquatic ecology? Hydrobiologia 214:201-206

Smol J.P. (1990) Paleolimnology - Recent advances and future challenges. Mem. Ist.

Ital. Idrobiol. 47:253-276
Smol J.P. (1992) Paleolimnology: An important tool for effective ecosystem management. J. Aquat. Ecos. Health 1:49-58
Smol J.P., Boucherle MM (1985) Postglacial changes in algal and cladoceran assemblages in Little Round Lake, Ontario. Arch. Hydrobiol. 103:25-49
Smol J.P., Glew J.R. (1992) Paleolimnology. In Nierenberg WA (ed) Encyclopedia of Earth System Science, vol. 3. Academic Press, Inc., San Diego, CA, pp 551-564
Smol J.P., Walker I.R., Leavitt P.R. (1991) Paleolimnology and hindcasting climatic trends. Verh. Internat. Verein. Limnol. 24:1240-1246
Stevenson A.C., Flower R.J. (1991) A palaeoecological evaluation of environmental degradation in Lake Mikri Prespa, NW Greece. Biol. Cons. 57:89-109
Stevenson A.C., Jones V.J., Battarbee R.W. (1990) The cause of peat erosion: a palaeolimnological approach. New Phytol. 114:727-735
Sweets P.R., Bienert R.W., Crisman T.L., Binford M.W. (1990) Paleoecological investigations of recent lake acidification in northern Florida. J. Paleolim. 4:103-137
Tilman D. (1989) Ecological experimentation: Strengths and conceptual problems. In Likens G.E. (ed.) Long Term Studies in Ecology: Approaches and Alternatives. Springer-Verlag, New York, NY pp:136-157
Trefry J.H., Metz S., Trocine R.P., Nelsen T.A. (1985) A decline in lead transport by the Mississippi River. Science 230:439-441
Uutala A.J. (1990) *Chaoborus* (Diptera: Chaoboridae) mandibles - paleolimnological indicators of the historical status of fish populations in acid-sensitive lakes. J. Paleolim. 4:139-151
Uutala A.J., Yan N., Dixit A.S., Dixit S.S., Smol J.P. (1994) Paleolimnological assessment of declines in fish communities in three acidic, Canadian Shield lakes. Fish. Res. 19:157-177
Walker I.R., Smol J.P., Engstrom D.R., Birks H.J.B. (1991) An assessment of Chironomidae as quantitative indicators of past climatic change. Can. J. Fish. Aquat. Sci. 48:975-987
Warner B.G. (ed.) (1990) Methods in Quaternary Ecology. Geoscience Canada, Reprint Series 5, Geological Association of Canada, St. John's, Newfoundland
Weatherhead P.J. (1986) How unusual are unusual events? Am. Nat. 128:150-154
Whitehead D.R., Charles D.F., Jackson S.T., Smol J.P., Engstrom D.R. (1989) The developmental history of Adirondack (N.Y.) lakes. J. Paleolim. 2:185-206
Wolin J.A., Stoermer E.F., Schelske C. (1991) Recent changes in Lake Ontario 1981-1987: Microfossils evidence of phosphorus reduction. J. Great Lakes Res. 17:229-240
Wright H.E. Jr (1990) An improved Hongve sampler for surface sediments. J. Paleolim. 4:91-92

17. RECOVERY AND REHABILITATION OF MEDITERRANEAN TYPE ECOSYSTEM: A CASE STUDY FROM TURKISH MAQUIS

Munir Ahmet Ozturk
Ege University
Science Faculty, Centre for Environmental Studies
35100 Bornova-Izmir
Turkey

INTRODUCTION

The Mediterranean type ecosystem shows a restricted distribution as compared to other major ecosystem types of the world. Two of the typical Mediterranean ecosystems are maquis and phrygana, both being shrub formations. This vegetation type in general shows indisputable similarities in Mediterranean regions. Similarities are visible in forest formations as well as their principal stages of degradation. Studies on these ecosystems have attracted the attention of many investigators such as Akman (1982), Akman and Ketenoglu (1986), Akman et al. (1978,1979), Aschmann (1973), DiCastri and Mooney (1973), DiCastri et al. (1981), Flauhalt (1937), Kilickiran (1991), Mooney and Dunn (1970b), Mooney et al. (1970,1974b), Nahal (1981), Naveh (1971,1973,1975), Pons (1981), Specht (1969ab), Tomaselli (1974,1976), Trabaud (1981) and Zohary (1973). In these investigations, three types of studies have been followed; phyto-ecological, energy-budget ecology and an evolutionary approach. However, a complete synthesis of the results is rather difficult due to their multilingual publication as well as different approaches used in studying these heterogenous ecosystems.

Sixty percent of the shrub formations of these ecosystems exist in the Mediterranean basin. The basin is characterised by lands in 3 continents embodying countries with widely different economic development, 40,000 km of shore, numerous islands with calcareous substrates, and high plant species diversity. The area is accepted as the gene centre of many crops. It has influenced other Mediterranean regions through an introduction of some ephemerals and dispersion of grasses as well as weeds. Highest plant cover is found in Spain followed by Turkey, Morocco and Italy. This chapter will focus on maquis in Turkey, in particular their distribution, evaluation, recovery and rehabilitation.

DISTRIBUTION OF MAQUIS IN TURKEY

Turkey is included in the eastern zone of Mediterranean Basin, extending from the eastern half of Italy up to Lebanon. It is difficult to draw a clearcut border line of the Mediterranean phytogeographical region in Turkey because terrestrial plant formations easily intermix with each other; in particular Irano-Turanian and Mediterranean show a great similarity. The region is dominated by typical plant cover of *Pinus brutia* forests and shrub formations. These generally cover hill tops and low altitudes of mountains, rarely smooth plains, giving an evergreen look to these areas. In some places maquis extends horizontally hundreds of kilometres inwards from the coast as in Bucakkisla around Isparta, where a major role is played by coastal winds along the Goksu, Dalaman and Manavgat Rivers. A general altitudinal distribution lies around 500-600 m or 600-800 m; however some members like *Quercus coccifera*, *Q. ilex*, and *Arbutus unedo* grow at 1420 m as in Cameli. *Q. coccifera* exists even at 1600 m in Gursu village around Yaylacik.

The first report concerning total area of maquis in Turkey is that of Yigitoglu (1941) who estimated that there was around 216,660 ha. This was followed by studies undertaken during 1950-1956. According to this report there exist 7,169,720 ha of maquis in Turkey, out of which 241,218 ha are accepted as forest and 475,754 ha as non-forest. The latter mainly include 55,921 ha from Antalya, 51,795 ha from Balikesir, 6,521 ha from Bursa, 80,415 ha from Istanbul, 33,210 ha from Izmir, 247,323 ha from Mugla, in addition to 380 ha from Kastamonu on the Black Sea coast. Changes in the forest evaluation rules during the last 3 decades and surveys conducted in the Mediterranean climatic zone of Turkey showed that maquis covers a total area of around 1,015,375 ha up to an altitude of 500 m. This does not include the maquis vegetation present in the form of patches on the Black Sea coast which is termed as "Pseudomaqui" by Turrill (1929) and Ansin (1980,1983).

Generally maquis is accepted as a climatic climax (Philippson 1922; Rikli 1943) but there is no general agreement on this subject because not a single criterion but many factors - natural, technical, economic and social - are involved. Two main views put forth in this connection are:

1) Maquis are a climax or primary formation under certain conditions.
2) Maquis are a sub-climax left behind from old Mediterranean forests, i.e. these areas are secondary or anthropogenic formations created by biotic factors.

It is more plausible to accept the second view, because it is not possible to find maquis as a climax formation (Figure 1). According to Polunin and Huxley (1972) climax plant formations developing naturally under Mediterranean climatic conditions are leathery-leaved, evergreen oaks and *Pinus brutia* forests, which can exist for years together if left untouched. In Turkey, maquis elements also exist as a second story in 2,115,157 ha of forests in the Mediterranean zone. They resemble each other physionomically; forming impenetrable thickets with a dense cover of climbers in some places, thus hindering the passage through these areas. A general vegetational set up of maquis is summarized below:

Sarcopoterietalia spinosi generally appears after the ephemeral stage in the

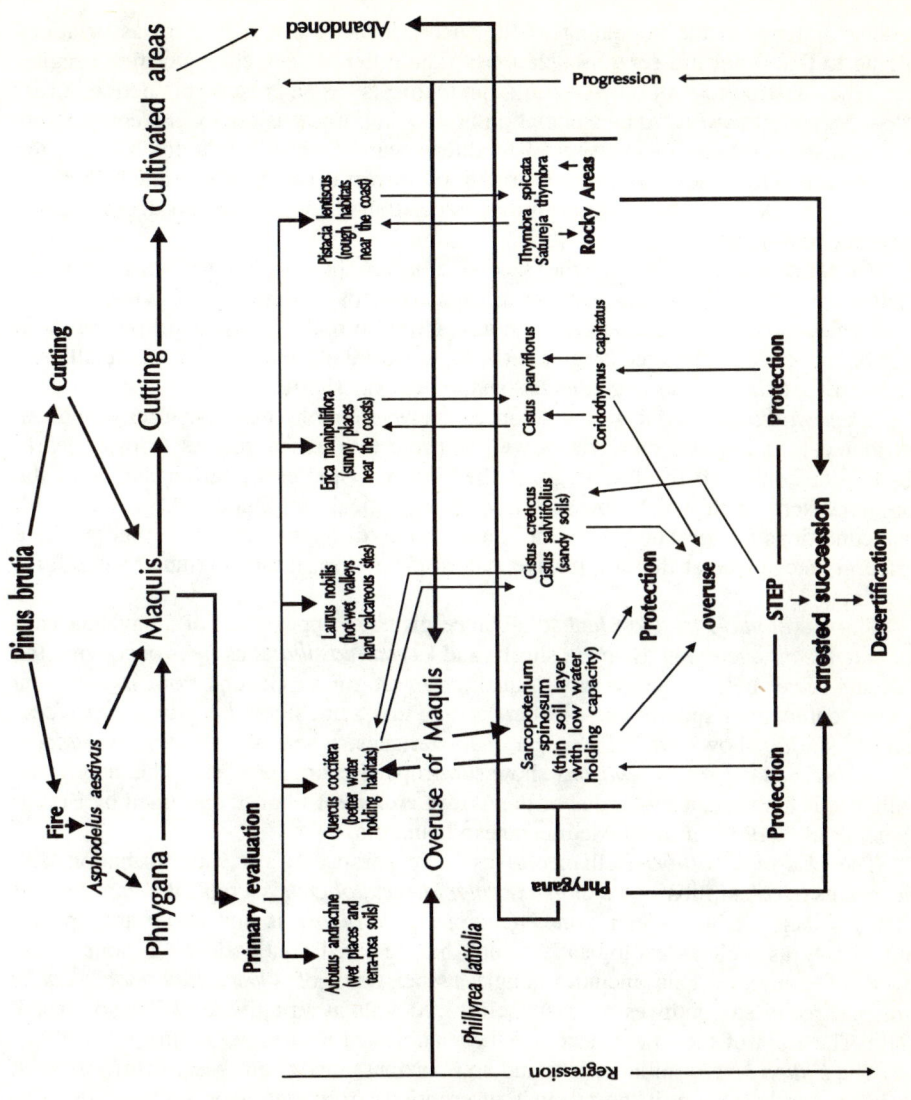

Figure 1. Role of biotic factors in the development of maquis (modified from Gemici and Seçmen 1986).

succession and lies at the beginning of the successional sere and thus all associations belonging to this order are seral associations. The order is very common in the region due to the destruction of the maquis and forests as a result of active fires. *Sarcopoterietum spinosi typicum* association is very important as one of the components of the successional stages under typical Mediterranean climatic conditions towards the redevelopment of the maquis and forest formations, on terra-rosa, rendzina and calcareous soil types. The floristic composition of this association changes with soil depth, human interference and exposure.

Coridothymion capitati is the second alliance showing typical Mediterranean characteristics from both an ecological and composition point of view. The *Coridothymetum capitati typicum* association is found in rocky areas, marn or chalky in nature, being naked or covered by a shallow layer of soil. The *Cistion creticae* alliance includes an *Cistetum cretici aegeaum* association (Bekat 1980).

Quercetalia calliprini, with 9 alliances, shows the richest composition which can be attributed to human interference as well as protection of the species from the early stages of succession till the last stage of the climax. Species in earlier stages of the successional sere in this order possess a wide ecological amplitude. At places with normal conditions for seral development, these species disappear, but at the places where destruction takes place at the progressive stages of the sere, these continue to live for a longer time.

The *Ceratonio-Pistacion lentisci* alliance gives an appearance of a savanna with *Pistacia lentiscus* occurring as single shrubs and *Ceratonia siliqua* as trees occupying the hottest and lower belt of the evergreen mediterranean zone. Ground flora is very rich and other companion species are *Olea europeae*, and sometimes *Juniperus phoenice*a. *Ceratonia siliqua* shows typical tropical characteristics, whereas *Juniperus phoenicea*, *Olea europeae* and *Pistacia lentiscus* show sub-tropical characteristics. The ecology of this alliance is homologous with that of the *Olea-Ceratonion* alliance discussed by Braun-Blanquet et al. (1952) in west Mediterranean countries.

The *Quercion calliprini* alliance gives an appearance of a maquis formation with dense groups of oak shrubs. *Quercion macrolepidis anatolicum* is a typical alliance found in the Turkish Mediterranean zone. *Quercus macrolepis* is an important species taxonomically as well as ecologically which has suffered continuously in spite of its richness of tannin. This includes single associations of *Quercetum macrolepidis anatolicum* recorded southwest of Demirci on red soils at an altitude of 750 m (Gork 1982). The flora of the lower strata has been subjected to heavy grazing.

The *Pinion brutiae* alliance is found on rendzinas, sandy soil, basaltic tuff, as well as terra-rosa soils and on different kinds of eruptions from magmatic rocks at altitudes of 0-1200 m, rarely going up to 1800 m. The alliance can occupy the associations of the *Quercion calliprini* alliance if destroyed. The elements of the latter thus form the constituents of the *Pinion brutiae* alliance. Two main associations of this alliance are *Pinetum brutiae aegeaum* and *Pinetum brutiae-Quercus infectoria*. Like many other workers (Adamovic 1909; Quezel 1976ab; Perelman 1981; Gemici and Secmen 1986) it appears to me as a climax plant formation of the Mediterranean zone. The *Pinion pineae* alliance too is placed in the typical Mediterranean plant formation but is rare in the east Mediterranean region.

SOCIO-ECONOMIC CONDITIONS

The demographic explosion in the Mediterranean basin took place 4000 years earlier in the eastern part as compared to the west (Le Houerou 1981). This increase in population is still continuing in the Turkish Mediterranean zone (2.88%). As a result, maquis and phrygana which cover about 8500 km^2 in the area are under an immense demographic pressure. These are either burnt down (563,000 ha up to 1991) and used for different purposes, opened up for cultivation (73,000 ha up to 1991) and for urbanisation (10,000 ha up to 1991), or subjected to intensive grazing of about 25,000 domestic animals.

The educational standard in the area varies. Out of a population of 5 million above 10 years age - 25% are illiterate; 55% are in primary schools; 16% are in middle schools and 4% are in higher educational institutes. This affects the economic development in the area which mainly depends on agriculture, stock-farming and forestry. Technological developments in the region like the use of tractors, improved seed, fertilizers, pesticides and good irrigation facilities are helping about 8% of the active population in dealing with agricultural production, and about 10% involved in industrial sectors like mining and construction. In spite of all this, primitive agricultural methods are still practiced in the region and this leads to negative effects on maquis. Due to misguided political decisions as well as misguided rules concerning the use of land, maquis are accepted as spare areas for short-term agricultural practices.

The methods used in livestock raising are primitive and the nomadic way of livestock breeding, particularly with goats (6,000), proves harmful to the maquis pastures. Due to climatic conditions, livestock breeders migrate earlier towards the pastures at higher elevations. These inappropriate practices of livestock husbandry introduced into the maquis result in an inhibition of full growth of pastoral species through indiscriminate grazing, browsing, as well as fodder removal leading towards a rapid site deterioration and ultimately desertification (Figure 1).

Although Mediterranean forests and shrub formations with an area of 65,000 km^2 apparently represent a considerable value, 60% of it is of degraded type, with a low productive potential due to short-circuiting of elements important to nutrient cycling.

The potential continues to decrease due to social pressures as well as financial stringency. Previously maquis was accepted as a potential sector for increasing the living standards of the public in the Mediterranean area and preventive measures were imposed easily. However this situation has now changed. This problem is of immense importance for a proper recovery and rehabilitation of maquis. For this purpose its historical background needs to be dealt with in detail.

EVALUATION OF MAQUIS

The first use of Turkish maquis is recorded on government papers dated back to 1772 and 1796. These mention that bundles of brushwood should be used for the construction of ships together with wooden pegs of holly (an oak species). This was followed by a law in 1858 indicating how shrub formations (without mentioning maquis) and degraded forest can be cleared for crop cultivation. It also mentioned that any person grafting shrubby species in the vicinity of his dwellings or field can own the area without payment. A law passed in 1937 did not mention the use of maquis directly but indirectly referred to these by stating that any type of spiny species and brushwood areas will not be considered as a forest. This actually referred to *Erica* species. In 1939 another law was passed concerning the improvement of wild olives, carobs, *Pistacia* species and other maquis elements through grafting, but again the term maquis was not mentioned. The same is the case with a law of 1945 which stated that brushwoods should be excluded from forest areas.

For the first time in 1950, laws governing the use of maquis were put into force and during 1950-1956 a large number of these areas along the coast were opened up for the cultivation of cotton, wheat, oranges and banana, whereas areas at higher altitudes, maquis areas were used for grazing as well as production of firewood and secondary products. Such areas had economically valuable trees, shrubs, herbs and grape plantations forming multi-storied systems resembling the natural ecosystem in structure as well as function. Thus there was a harmony existing with environment.

In 1956 areas dominated by mastic tree too were brought under the regulation of grafting. Out of 14 thousand hectares of wild olive areas, 9 thousand ha were distributed, 7 thousand ha grafted, and 2 thousand ha taken back in the states of Adana, Antalya, Isparta, Icel and Mugla. Six thousand hectares of these still exist as fallow areas. All these activities were affected by land speculators and led to a destruction of maquis, with the result that most of these areas lost the productive capacity through a disruption of teh delicate balance in nutrient cycling. This prevented the ability of maquis ecosystems to recover.

An abandonment of these habitats has added to the erosion problem of Turkey. Today 45 million ha (58.7%) of our land area is facing intense erosion, and 21.5 million ha (27.24%) are facing medium erosion. Every year 450-500 million tons of our soils are eroded (RAFEF 1992). This is 20 times greater than that in the European continent as a whole. However, there are still many possibilities for a recovery and rehabilitation of maquis and these can add to the economy of our country, if a proper land use planning on a wider basis is undertaken in the light of social, natural, economic, technical and political conditions. These recovery and rehabilitation patterns are outlined below.

Recovery and Rehabilitation as a Forest

For centuries maquis in the Turkish Mediterranean have served as a potential source of firewood, coalwood and grazing lands for cattle. They have also been used for the production of poles and stakes. This has led to a great loss in the productive potential of maquis (Schwarz 1936; Ozturk 1971; Ozturk et al. 1983, 1989, 1990, 1992, 1993ab, Quezel 198; Secmen 1974, 1977; Secmen et al. 1986; Vardar et al. 1980). Today environmental awareness, socio-economic developments and advances in the wood technology have brought the possibilities of more rational use of maquis to the forefront. In many places fast growing plants like *Eucalyptus* species together with pine species are planted in degraded maquis to overcome this loss of soil due to erosion as well as to add to the countries economy. This plant helps to increase the income per capita without causing land degradation (Geray and Gorcelioglu 1983). Such mixed plantation systems are supported by UNEP too. Eucalyptus is accepted as a proper plant for "agro-silvicultural" practices (Hoddy 1988).

Under our climatic conditions a 3 year old plantation of *E. camaldulensis* produces 97 m of wood per hectare (Avcioglu 1989). They can be used for the production of *Eucalyptus* honey too as in Australia (Hoddy 1988). Some of the plants are used as raw material for cellulose production. At some places small reserved firewood forests are created. The latter have proved successful in areas which have not lost their productivity characteristics. At some places these have been designated as natural reserves or national parks. Here a healthy recovery is seen with a lush green cover of typical indicator species of our maquis, such as *Ceratonia siliqua, Pistacia spp., Arbutus spp.* This emphasizes that a stressed maquis ecosystem can regain its potential if protected from fires, severe cutting and grazing. The above species appear to be significant in indicating that ecosystem recovery is underway and their destruction is a signal for the onset of breakdown. In fact the monoculture type plantation mentioned above is in my opinion an unhealthy recovery, because natural destruction and cutting for economic purposes is continuous.

Recovery and Rehabilitation as Agricultural Lands

During the last four decades a major trend in Turkey has been to change maquis into agricultural lands, in particular the heavily grazed areas. First wheat cultivation is practiced and when production decreases these areas are abandoned and sold at low prices for housing. In these areas mainly oranges, vine and banana orchards, and olive plantations are planted and there is some vegetable production in greenhouses. Although very little data are available about the grassland value of maquis they still are used and can be evaluated as such even if the species composition is changed. However, this too in my opinion is an unhealthy recovery. In addition to the recoveries mentioned above, maquis areas along the coast or around archaeological sites are cleared and used for the construction of hotels for tourist purposes, again an unhealthy recovery.

Limitations for the Recovery and Rehabilitation of Maquis

All the areas covered by maquis elements in Turkey have not been separated rationally. Each one of the Departments (i.e. forestry, agriculture and treasury) try to

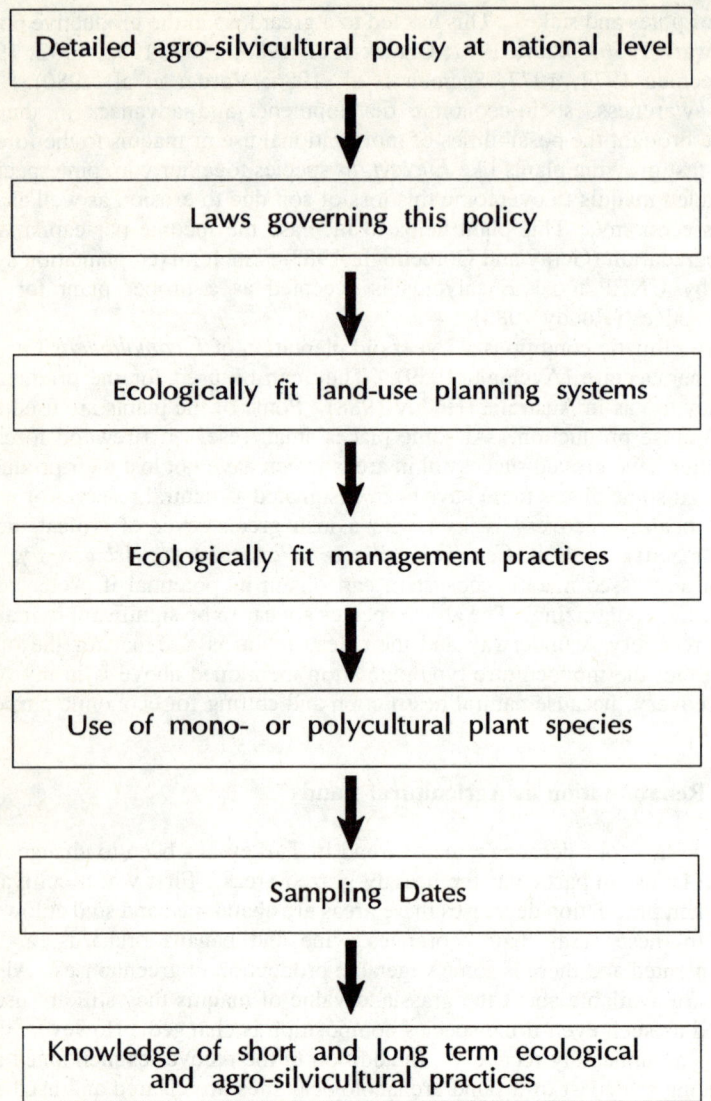

Figure 2. What different officials dealing with the evaluation of maquis need to know.

interfere in their evaluation, creating departmental frictions, as well as encouraging illegal use of these areas, leading to unhealthy developments (Figure 2). This points to the fact that, for a proper recovery and rehabilitation of maquis there is a need for long-term planning with a firm base (Constanza et al. 1992). The responsibilities of all departments involved need to be well defined. Thus there is an urgency for an interdisciplinary research (Figure 3) on the classification of soils supporting maquis depending on land use capability and also the socio-economic situation of the area vis-a-vis other related factors. This will permit us to use these areas in a more profitable way and work out the models for their healthy recovery.

The socio-economic situation is playing a greater role in this connection due to high illiteracy, high population increases, and lack of land for cultivation purposes. These produce a social pressure on maquis as well as mediterranean forests. Most areas are evaluated as pastures or as firewood stock. Social and economic measures are needed at national level to overcome this problem. The criteria for adequateness of rules have not been specified fully anywhere. Sometimes some of the maquis plantations are excluded from forested areas, at the same time a part of these is included with forests. Evaluation of maquis is thus not done on sound basis.

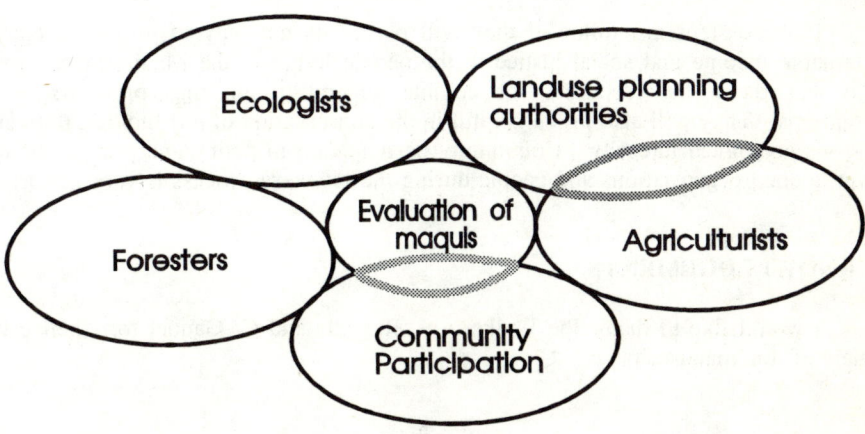

Figure 3. Possibilities for interdisciplinary actions.

CONCLUSIONS

For a healthy recovery the following steps need be taken to improve the system:

- If it is not imperative, allotment of maquis should be stopped for the time being.
- A major part of the maquis should be left for natural recovery and propagation methods should be employed.
- Part of these should be reforested with *Ceratonia siliqua, Pistacia spp., Laurus nobilis, Arbutus spp.* as well as medicinal and aromatic plants together with local climax species such as *Pinus brutia*. Attention should be paid towards fast growing native shrubs and trees.
- Nitrogen-fixing plant species should be included in the recovery processes in order to improve nitrogen economy.
- For a recovery of maquis, the time span of succession needs to be condensed through a study on tree growth strategies as well as adjustment of species polyculture in time and space.
- At lowest altitudes orchards, vineyards and olive plantations should be created after proper soil conservation methods are employed. An incorporation of ecological knowledge on canopy structure will help in the rearrangement and strengthening of agro-silvicultural practices.
- The impact of grazing should be minimized by reducing the number of animals and by extending the rotation time.

If these steps are followed they will permit us not only to provide a constant sustainable income and social justice to the people living in the Mediterranean region today but retain the resources for coming generation too thus preventing social disruptions. They will also prove helpful in the conservation of our biological diversity vis-a-vis ecological integrity. Community participation in policy and practice between governmental organisations and people during the recovery process is very important.

ACKNOWLEDGEMENTS

I would like to thank Joe Walker, F. Ramade and C. Gaudet for their critical review of this manuscript.

REFERENCES

Adamovic L. (1909) Die vegetations verhaltnisse der Balkanlander. In: Engler (ed) Die vegetation der Erde. II.W.Fischer Jena pp: 379-460

Akman Y. (1982) Climats et bioclimats mediterraneens en Turquie. Ecologia Mediterranea 8:73-87

Akman Y., Ketenoglu O. (1986) The climate and vegetation of Turkey. Proceedings of the Royal Society of Edinburgh 89:123-134

Akman Y., Barbero M., and Quezel P. (1978) Contribution a l'etude de la vegetation forestiere de l'Anatolie, mediterraneen. Phytocoenologia 5:1-79

Akman Y., Barbero M., and Quezel. P (1979) Contribution a l'etude de la vegetation forestiere d'Anatolie mediterraneene. Phytocoenologia 5:189-346

Ansin R. (1980) Floral composition of main vegetation types in east Black Sea region. D.Sc. Thesis, Karadeniz Univ., Trabzon 305 pp.

Ansin R. (1983) Floristic regions and major vegetation types of Turkey. Journal of Forestry Fac. Karadeniz Univ. 6:318-339

Aschmann H. (1973) Distribution and peculiarity of Mediterranean ecosystems. In DiCastri F., Mooney H.A. (eds). Mediterranean Type Ecosystems, Origin and Structure. Springer, Berlin 7:11-19

Avcioglu E. (1989) Importance and place of *Eucalyptus* in east mediterranean forestry. Forest Engineering Bulletin 15:6-19

Bekat L. (1980) Flora and vegetation of Karaburun Akdagi from aegean region. M.Sc Thesis. Ege University Science Faculty 126 pp.

Braun-Blanquet J., Roussine N., Negre R. (1952) Les Groupements vegetaux de la France Mediterraneen. CNRS, Paris 297 pp.

Constanza R., Norton B.G., Haskell B.D. (eds.) (1992) Ecosystem Health: New Goals for Environmental Management. Island Press, Washington DC 269 pp.

DiCastri F., Goodall D.W., Specht R.L. (eds.) (1981) Mediterranean-Type Shrublands. Elsevier, Amsterdam 11:643 pp.

DiCastri F., Mooney H.A. (eds.) (1973) Mediterranean Type Ecosystems, Origin and Structure. Springer, Berlin 7: 405 pp.

Flauhalt C. (1937) La distribution geographique des vegetaux dans la Region mediterraeen Francaise. Lechevalier, Paris 178 pp.

Gemici Y., Secmen O. (1983) Etude phytosociologique et phytoecologique de la vegetation de la montagne Yamanlar-Izmir. Journal of Science Faculty, Ege University 8:51-65

Geray U., Gorcelioglu E. (1983) Mixed systems in the use of agricultural and forest lands. Istanbul University Forest Faculty Journal 33:10-26.

Gork G. (1982) Vegetation and floral aspects of Emet Egrigoz Mountain in Kutahya. PhD Thesis, Ege University Science Faculty 119 pp.

Hoddy E. (1988) Eucalyptus: Boon for forestry or danger for environment. Development Corporation 4

Kilickiran S. (1991) Les possibilites d'utilisation des maquis dans la region mediterraneen de la Turquie. Journal of the Turkish Forest Research Institute 37:61-84

Le Houreau H.N (1981) Impact of man and his animals on mediterranean vegetation. In DiCastri F. et al. (eds.) Mediterranean Type Shrublands. Elsevier, Amsterdam 11:479-521

Mooney H.A., Dunn E.L. (1970b) Convergent evolution of mediterranean-climate evergreen sclerophyll shrubs. Evolution 24:292-303

Mooney H.A., Dunn E.L., Shorpshire F., Song L. (1970) Vegetation comparisons between the mediterranean climatic areas of California and Chile. Flora 159:480-496

Mooney H.A., Parsons D.J., Kummerow J. (1974) Plant development in mediterranean climates. In Leith H. (ed.). Phenology and Seasonality Modelling. Springer, Berlin pp:255-267

Nahal I. (1981) The mediterranean climate from a biological view-point. In DiCastri F. et al. (eds.) Mediterranean Type Shrublands. Elsevier, Amsterdam 11:63-86

Naveh Z. (1971) The conservation of ecological diversity of mediterranean ecosystems. In Duffey E., Watt A.S. (eds.) The Scientific Management of Animal and Plant Communities for Conservation. Blackwell, Oxford pp:605-622

Naveh Z. (1975) The evolutionary significance of fire in the mediterranean region. Vegetatio 29:199-208

Naveh Z., Dan J. (1973) The human degradation of mediterranean landscapes in Israel. In DiCastri F., Mooney H.A. (eds.) Mediterranean Type Ecosystems, Origin and Structure. Springer, Berlin 7:373-399

Ozturk M.A. (1971) Ecology of *Myrtus communis* in Turkish mediterranean phytogeographical region. PhD Thesis, Ege University, Science Faculty, Izmir 220 pp.

Ozturk M.A., Secmen O., Gork G., Kondo K, Segawa H. (1983) Ecological studies on maquis elements in aegean region of Turkey. Memoirs Fac. Integ. Arts and Science. Hiroshima Univ. 8:51-86

Ozturk M.A., Turkan I., Yurekli K. (1989) Studies on the water relations of some maquis elements. Journal of Science Faculty, Ege University, Izmir 11:17-24

Ozturk M.A., Secmen O., Gemici Y., Gork G. (1990) Plants and Landscape in the Aegean Region of Turkey. Sentez Press 176 pp.

Ozturk M.A., Gemici Y., Secmen O., Gork G. (1992) High mountain vegetation of mediterranean part of Turkey. Journal of Science Faculty, Ege University, Izmir 13:51-59

Ozturk M.A., Pirdal M, Turkan I, Ay G. (1993a) Ecology of Aegean Grasslands. Ege University Press, Izmir 250 pp.

Ozturk M.A., Erdem U., Sukatar A., Secmen O., Guner H. (1993b) Coastal mediterranean plant cover and pollutants - a case study from Izmir. Okeanos'93 Colloquium, Montpellier, France. pp:1-4.

Perelman R. (1981) Perception of mediterranean landscapes, particulary of maquis landscapes. In Di Castri F. et al. (eds). Mediterranean Type Shrublands. Elsevier, Amsterdam 11:539-553

Philippson A. (1922) Das mittelmeergebiet. Druck and Verlag Leipzig-Berlin

Polunin O., Huxley A. (1972) Flowers of the Mediterranean. Chatto and Windus, London 260 pp.

Pons A. (1981) The history of the mediterranean shrublands. In DiCastri F. et al. (eds.) Mediterranean Type Shrublands. Elsevier, Amsterdam 11:132-138

Quezel P. (1976a) Les forets du pourtour de la mediterraneen. In Forets et Maquis

Mediterraneens: Ecologie, Conservation et Amenagement. Notes techniques du MAB 2 UNESCO Paris 9-35

Quezel P. (1976b) Les chenes sclerophylles en region mediterraneene. Options Mediterr. 35:25-29

Quezel P. (1981) Floristic composition and phytosociological structure of sclerophyllous matorral around the mediterranean. In DiCastri F. et al. (eds.) Mediterranean Type Shrublands. Elsevier, Amsterdam 11:107-121

RAFEF (1992) Let Us Save Our Forests. Research Association of Rural Environment and Forestry. Kavaklidere Ankara, Turkey 12 pp.

Rikli M. (1943) Das pflanzenkleid der Mittelmeerlander. Verlag H. Huber Bern.

Schwarz O. (1936) Die Vegetations Verhaltnisse Westanatolien. Engler

Secmen O. (1974) Ecology of *Ceratonia siliqua*. PhD Thesis. Ege University, Science Faculty, Izmir 202 pp.

Secmen O. (1977) Studies on the flora and vegetation of Nif mountain-Izmir. D.Sc.Thesis. Ege University, Science Faculty 172 pp.

Secmen O., Gemici Y., Bekat L., Gork G. (1986) *Phrygana* vegetation around Izmir. Doga Bilim Dergisi 10:197-206

Specht R.L. (1969a) A comparison of the sclerophyllous vegetation characteristic of mediterranean type climates in France, California and Southern Australia. I. Structure, morphology and succession. Aust. J. Bot.17:277-292

Specht R.L. (1969b) A comparison of the sclerophyllous vegetation characteristic of mediterranean type climates in France, California and Southern Australia. II. Dry matter, energy and nutrient accumulation. Aust. J. Bot. 17:293-308

Tomaselli R. (1974) Etude sur la degradation du maquis mediterraneen. Conseil de l'Europe, Strasbourg CE/NA (74) Doc. 26 (rev): 68 pp.

Tomaselli R. (1976) La degradation du maquis mediterraneen. In Forets et Maquis Mediterraneens: Ecologie, Conservation et Amenagement. Notes Techniques du MAB 2. UNESCO, Paris pp:35-76.

Trabaud L. (1981) Man and fire: impacts on mediterranean vegetation. In DiCastri F. et al. (eds.) Mediterranean Type Shrublands. Elsevier, Amsterdam 11:523-537.

Turrill W.B. (1929) The Plant Life of the Balkan Peninsula. Oxford Press

Vardar Y., Secmen O., Ozturk M.A. (1980) Some distributional problems and biological characteristics of Ceratonia in Turkey. Portugaliae Acta Biologica 16:75-86

Yigitoglu A.K. (1941) Importance and Place of Forestry in Turkish Economy. Publication of Agricultural Institute, Ankara 110 pp.

Zohary M. (1973) Geobotanical foundations of the middle east. Geo Bot. Selecta, Stuttgart 3:379 pp.

18. FOREST RECOVERY FOLLOWING PASTURE ABANDONMENT IN AMAZONIA: CANOPY SEASONALITY, FIRE RESISTANCE AND ANTS

Daniel C. Nepstad[1], Peter Jipp[1,2], Paulo Moutinho[3], Gustavo Negreiros[1], Simone Vieira[1]

[1]Woods Hole Research Center, P.O. Box 296, Woods Hole, MA 02543, USA
[2]Duke University, School of Environmental Studies, USA
[3]DPE/Centro de Filosofia e Ciencias Humanas, Universidade Federal do Pará, Campus Guama, 66075,Belém, PA, Brasil

INTRODUCTION

Tropical forests are important regulators of the flux and storage of carbon, water, and energy in the Biosphere, and they are the habitat of more than three-fourths of the world's plant and animal species. These ecosystems are also undergoing rapid conversion through pasture formation, shifting cultivation and timber highgrading as the people of tropical nations turn to forestlands for sustenance and wealth.

In Brazilian Amazonia, deforestation has claimed 10% of the 4-million-km^2 formation of closed forest (Fearnside 1993). Roughly half of this forestland was converted to cattle pasture, much of which was subsequently abandoned as resprouting trees, invading weeds and declining soil fertility reduced forage quality (Serrão and Toledo, 1990). Fire has been superimposed upon the Amazonian deforestation frontier, burning country-sized areas of deforested land and logged forest each year (Setzer and Pereira 1991, Uhl and Buschbacher 1985), severely limiting the potential for forest recovery. The large area of moist tropical forests converted to ephemeral cattle pasture in Amazonia, in Mexico, and elsewhere (WRI 1990) calls into question the long-term effects of this temporary form of land-use on forest health. What elements of forest health are recovered on abandoned pasturelands, and at what rate?

Questions concerning forest recovery on abandoned land cannot be understood without knowledge of the intact, healthy forest. In this chapter, we argue that the most important feature of healthy forests in eastern Amazonia is an evergreen leaf canopy that helps to maintain regional rainfall patterns and that protects the forest from fire by preventing significant drying of fine fuels (organic debris) on the forest floor. Other parameters of healthy forests include the amount of biomass and the composition of plant and animal species assemblages.

Many of the ideas and data presented in this chapter are the product of research conducted on the *Fazenda Vitoria*, a farm near the town of Paragominas (Figure 1). At

30 years, this regional centre of cattle ranching and logging is a relatively old agricultural frontier with land-use patterns that may be the destiny of younger Amazonian frontiers. Like most of the "arc of deforestation" along the eastern and southern margins of Amazonia (Figure 1), where forest clearing has been greatest, the Paragominas region has deeply-weathered, kaolinite-dominated soils (Oxisols) and a pronounced seasonal drought. An average of only 250 mm of rainfall, of a 1750 mm annual total, fall during the months of July through November (Nepstad et al. 1991).

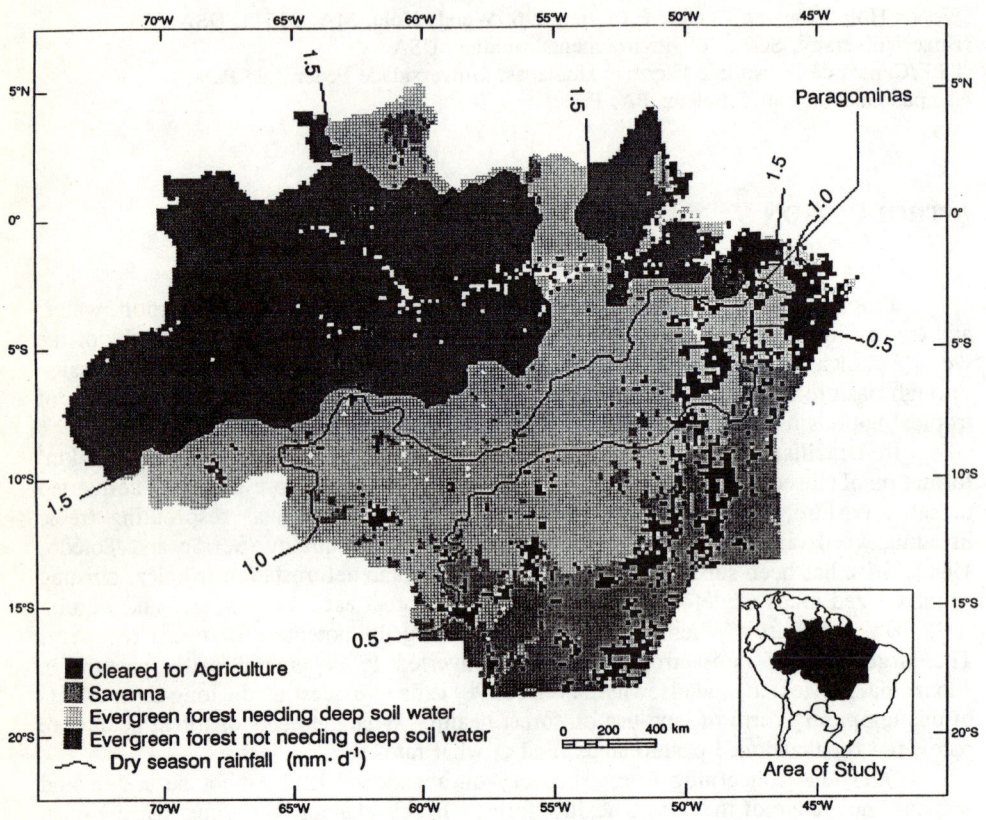

Figure 1. The Brazilian Amazon Basin, showing zones of drought intensity (isolines), areas of deforestation, the forests that maintain evergreen canopies despite pronounced seasonal drought, and the location of Paragominas. The southern and eastern edges of the Basin have the most deforestation, the severest seasonal drought, and support a native vegetation of evergreen forest and savanna (after Nepstad et al. in press).

THE HEALTHY FOREST

An important objective of this book is to identify biologically meaningful measures of health for some of the major terrestrial and aquatic ecosystems of the world (Rapport, this volume). Forest ecosystems can be studied at the scale of continents, watersheds, organisms, or chloroplasts, but no single level of analysis fully represents forest health. Part of the difficulty lies in the numerous emergent properties of the combined organisms of a forest, which are virtually impossible to measure through study of individual organisms.

The task of identifying measures of forest health is simplified, however, by the hierarchical organization of these ecosystems. The very existence of many forests depends on certain structural features, and these features are the logical choice as first-order indicators of forest health, i.e., as forest vital signs. In this section, we argue that the existence of forests in eastern and southern Amazonia depends on an evergreen leaf canopy. Without this fundamental feature, the other functional, structural and compositional attributes of the forest are subject to radical change. A more complete understanding of forest health is obtained through additional levels of analyses, as we illustrate with data for biomass and ant communities.

The Importance of Evergreen Canopies

Through a compilation of satellite imagery and rainfall data for the Amazon Basin, we find that deforested land is concentrated in areas of forests that maintain their leaf canopies throughout the year despite pronounced seasonal drought (Figure 1). In roughly half of Brazilian Amazonia, including most of the area of deforested land, the average daily precipitation during the driest three-month period of the year is <1.5 mm, which is about 2.5 mm day^{-1} lower than the potential rate of water vapor release to the atmosphere via evapotranspiration (ET) by the region's forests. Despite this relatively severe seasonal moisture deficit, the "greenness" of the forest leaf canopy, measured as the Normalized Difference Vegetation Index (NDVI), does not exhibit a gradual decline as the dry season progresses (Nepstad et al. in press; Figure 1). Unlike the seasonal patterns of NDVI observed for the savannas of Amazonia, most forests of eastern and southern Amazonia maintain high NDVI throughout the year.

The phenomenon of leaf canopy maintenance during prolonged dry periods is a crucial feature of forest health in eastern and southern Amazonia. In the absence of drought-induced leaf shedding, water vapour flux to the atmosphere (evapotranspiration, ET) continues during the dry season (Nepstad et al. in press). ET influences the region's climate by supplying moisture to the atmosphere that falls as rain downwind, and by absorbing a large amount of radiant energy through evaporation. Isotopic studies of rainwater (Salati et al. 1979; Victoria et al. 1991) and global circulation models (Shukla et al. 1990; Lean and Warrilow 1989) indicate a tight coupling between the Amazonian rainfall regime and forest ET, such that extensive conversion of forest to pasture might provoke a reduction in rainfall and an increase in air temperatures.

Forests with evergreen canopies and little seasonal depression of ET also exert a profound influence on soil water run-off, stream-flow and other hydrologic features. In Paragominas, water flux to the atmosphere via ET exceeds water input to the vegetation via rainfall during the six-month dry season, and is supplied by roots that absorb water from soil layers up to 18 m deep (Nepstad et al. in press). By the end of the dry season, the water content of Paragominas soils is 300-800 mm below field capacity, the maximum water content at which soils can hold water against gravity. During the first two or three months of the rainy season, rain that enters the forest is retained by the soil, and deep seepage to the water table, or lateral flow to streams, is very low. Evergreen forests in seasonally dry Amazonia therefore reduce run-off and stream flow relative to ecosystems that shed leaves during the dry season.

In addition to their important role in forest hydrology, evergreen leaf canopies also reduce the fire susceptibility of Amazonian forests. The principal determinants of forest flammability in the moist tropics are the mass and moisture content of fine fuels on the forest floor. If seasonal drought provokes substantial leaf shedding, the conditions for fire are met as leaf litter on the forest floor accumulates and dries through the accompanying increase in sunlight reaching the ground surface (Figure 2). In the forests of Paragominas, the litter layer of the forest floor remains too damp to be ignited during average dry seasons because of the high relative humidity of the understorey air (Uhl and Kauffman 1990). Subsequent years of low rainfall may deplete soil moisture reserves and provoke leaf shedding, rendering the forest floor flammable (Figure 2). Evergreen forests of eastern and southern Amazonia currently act as extensive fire breaks in the landscape, preventing the spread of thousands of fires, both intentional and accidental, that sweep across agricultural plots, cattle pastures, secondary forests and logged forests (Setzer and Pereira 1992).

Forest leaf canopies can remain active during dry periods (a) if roots absorb water stored in the soil or in the water table, or (b) by remaining physiologically active even at very low xylem pressure potentials. We studied the forest at the *Fazenda Vitoria* to determine which of these mechanisms permitted evergreen behaviour and found evidence that both are operative. In this forest, there are no canopy tree species that remain leafless for more than 2 weeks during the 5-month dry season, although several tree species replace their leaf crops (e.g. *Tabebuia serratifolia*, *Parkia pendula*) and others reduce their leaf area as the dry season progresses (e.g. *Tetragastris altissima*). During the record drought of 1992, when an El Niño Southern Oscillation provoked a reduction in annual rainfall to only 1100 mm (vs. 1700 mm average), the average leaf area of ten tree and liana species declined by only 12%, as measured through leaf counts of marked branches. During this same period, the water stress experienced by the leaf canopies of these species was very high, with xylem pressure potentials at pre-dawn ranging from -3.6 to -1.8 MPa. Measurements of water absorption from the upper 8 m of the soil indicated that ET declined to ca. 3 mm day^{-1} by the end of the dry season, only 25% below the rate measured at the beginning of the dry season.

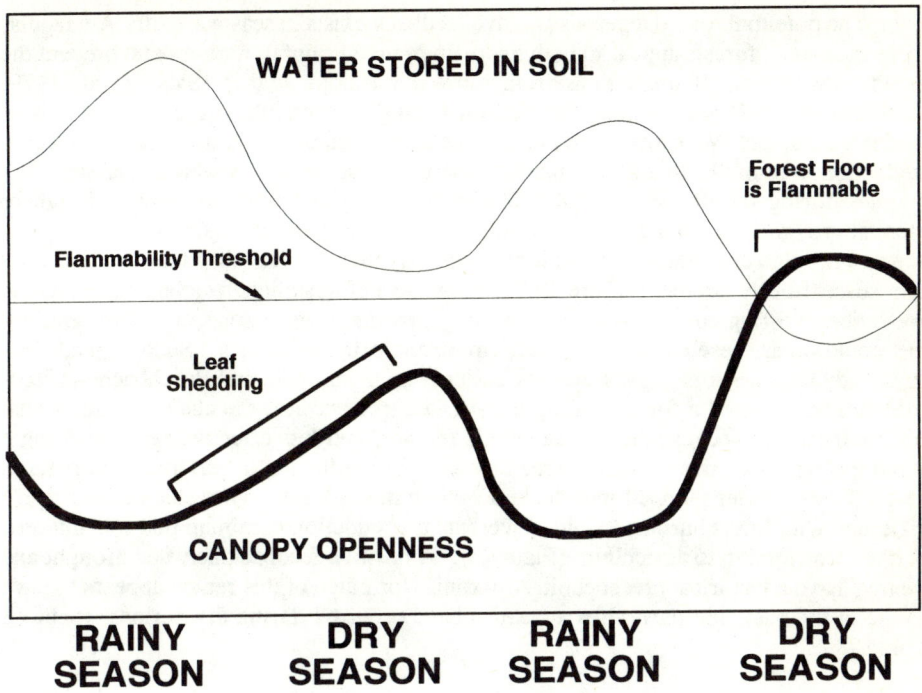

Figure 2. A model of primary forest resistance to flammability. Because of its access to a very large soil volume (through deep-rooting), successive years of low precipitation are necessary to provoke enough leaf shedding for the forest floor to become flammable. The flammability of the forest floor is limited most years by the dense leaf canopy which maintains a cool understory with high relative humidity.

Hence, it appears that the Paragominas forest avoids drought by tapping the moisture stored deep in the soil profile and it tolerates drought through characteristics of the leaves that allow them to maintain gas exchange despite low xylem pressure potential. In this sense, tree and liana species with deeply-penetrating root systems and drought tolerant leaves are necessary for evergreen canopy behaviour and, hence, for the health of these forests.

A Positive Feedback: Deforestation, Drought and Fire

The potential for a dangerous positive feedback exists in seasonally-dry Amazonia, where evergreen forests appear to both maintain regional rainfall regimes and prevent the spread of wildfire. If predictions from rainwater isotope studies (Salati et al. 1979, Victoria et al. 1991) and initial predictions from global circulation models (Shukla et al. 1990, Lean and Warrilow, 1989) are correct, deforestation in Amazonia is leading to a reduction in rainfall. In a drier Amazonia, some forests may be unable to maintain leaf canopies during the dry season, particularly in eastern and southern Amazonia where seasonal droughts are already very severe (Figure 1). Forests that exhaust available soil moisture through dry-season evapotranspiration may undergo leaf shedding that increases their susceptibility to fire (Figure 2). Along the deforestation frontier, where cattle production, shifting cultivation and timber highgrading provide sources of fire ignition, fires could enter these drought-stressed, fire-prone primary forests. Such ground fires are already common in logged forests of eastern Amazonia (Uhl and Buschbacher 1985). Once fire has entered a forest, killing thin-barked tree species, the chance of additional burning increases because the leaf canopy is thinner, and less effective at maintaining a humid understorey microclimate. Tree mortality and reduction in leaf area also reduces the quantity of water pumped into the atmosphere through ET. Thus, the decline in ET associated with forest burning would exacerbate the reduction in rainfall that was initiated by forest conversion to agriculture (Figure 3). A positive feedback between drought and fire may have a historical precedent in Amazonia, for much of this region appears to have burned in the past, and these fires appear to have occurred during dry periods (Sanford et al. 1985).

Other Criteria of Forest Health

There are numerous other parameters that could be employed as vital signs of Amazonian forest health and its recovery, although most are difficult to measure or interpret. The quantity of biomass contained in forests is fairly well documented and is easy to measure relative to such ecosystem features as gross primary productivity. The amount of biomass accumulation in a recovering forest is an integrative measure of health because it is influenced by processes of both tree establishment (reflecting seed and seedling availability, herbivory and competition) and tree growth (reflecting herbivory and edaphic conditions).

Species-based measures of forest health, similar to the Index of Biotic Integrity developed for stream ecosystems (Karr this volume), have not been developed for tropical forests and are complicated by the sheer number of species, taxonomic uncertainties, and the lack of specific information on the ecological roles and sensitivities of different taxa. For example, in a 200-ha mosaic of primary forest, secondary forest and pasture at the *Fazenda Vitoria*, there are 300 species of trees, almost 400 species of birds, and nearly 100 ant species.

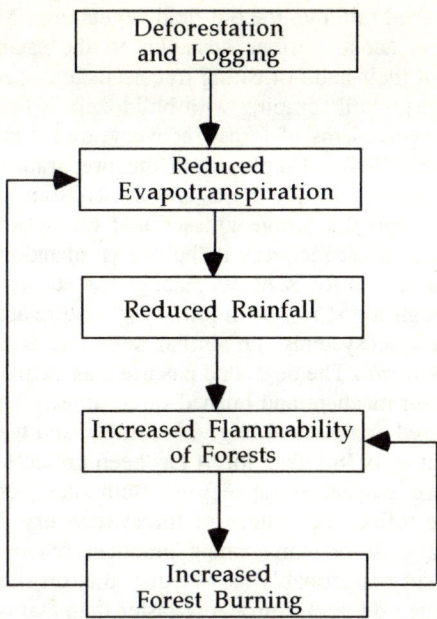

Figure 3. Diagram of a potential positive feedback between Amazonian drought and forest fire.

THE RECOVERY OF FOREST HEALTH

Forest recovery following clearing and burning begins when the first rains extinguish the smoldering boles of felled trees. Within days of the rain event, seeds germinate and new stems sprout from root systems and stumps. By the end of the rainy season, young trees and lianas cover the site and recovery is underway.

The rate of forest recovery following land abandonment declines as the duration or intensity of land-use increases (Uhl et al. 1981, 1990). Agricultural practices such as cutting woody stems with machete and burning vegetation destroy tree seedlings and sprouts, depleting the seed bank buried in the soil and killing residual root systems of trees. Woody shrubs and treelets that emerge above the vegetation attract seed-carrying birds and bats from nearby forests, and their removal further delays tree seedling establishment by reducing seed rain into the old field vegetation. Moreover, rodents and ants of old field vegetation act as further obstacles to the establishment of trees in abandoned fields because of their habit of eating tree seeds and seedlings (Nepstad et al. 1991; Moutinho et al. 1993). Soil scraping with bulldozers is the agricultural practice that is most damaging to mechanisms of forest recovery, and it is expanding rapidly in Amazonia (Mattos and Uhl 1994). Employed in the preparation of land for pasture reformation, bulldozing removes woody roots and seeds from the site, reducing the incursion of woody shoots into the forage grasses that are subsequently planted, and diminishing the potential for forest recovery if the site is abandoned.

We examine here the loss of forest health through forest conversion to pasture, and the recovery of health through forest regrowth following pasture abandonment. The data presented are from adjacent ecosystems, on similar soils and with similar topographic positions, at the *Fazenda Vitoria*. The degraded pasture was cleared and planted to grass in 1969, then "cleaned" with machete and burned three times. The secondary forest is on a site that was also cleared and planted to grass in 1969, and was cleaned and burned twice prior to abandonment in 1976. This forest has been protected from accidental fire since 1984, and is therefore unusual in the region. Both sites were grazed intensively. The results presented here reflect the pattern of forest recovery following a history of relatively intensive manual pasture management practices (cleaning and burning) and grazing. This forest recovery is probably slower than that on sites that were used less intensively or for shorter periods, and is probably faster than that on sites that have been bulldozed.

Canopy Seasonality

One of the most important measures of the loss and recovery of forest health in eastern Amazonia is the seasonality of the leaf canopy. Cattle pastures have highly seasonal leaf canopies compared to the primary forests that they replace. Forage grasses often lose their green leaf area completely during the first two to three months of the dry season (Nepstad 1989). In degraded pastures, the only patches of green during the dry season are the foliage of sprouting forest trees (e.g. *Stryphnodendron pulcherrimum*) and invading pioneer treelets (e.g. *Solanum crinitum*). On the Fazenda Vitoria, the leaf area of the degraded pasture declined by 60% during the severe dry season of 1992. During this same period, leaf area of the adjacent primary forest declined by only 12% (Figure 4). Sixteen years after pasture abandonment, the secondary forest at the Fazenda Vitoria had not fully recovered the evergreen canopy behaviour of the primary forest. Leaf area of this forest declined by 40% during the 1992 dry season and was more seasonal than the primary forest (Figure 4) because of complete leaf-shedding by the dominant tree species (*Banara guianensis*) and partial leaf-shedding by several other common tree species.

There are two possible explanations for the observed differences in leaf canopy seasonality. First, pasture vegetation may shed more leaves because pasture plants are more sensitive to drought stress. Second, pasture vegetation may shed more leaves because pasture plants have less access to water stored in the soil, i.e., the root systems of pasture plants are relatively shallow and occupy a smaller soil volume than the roots systems of forest trees.

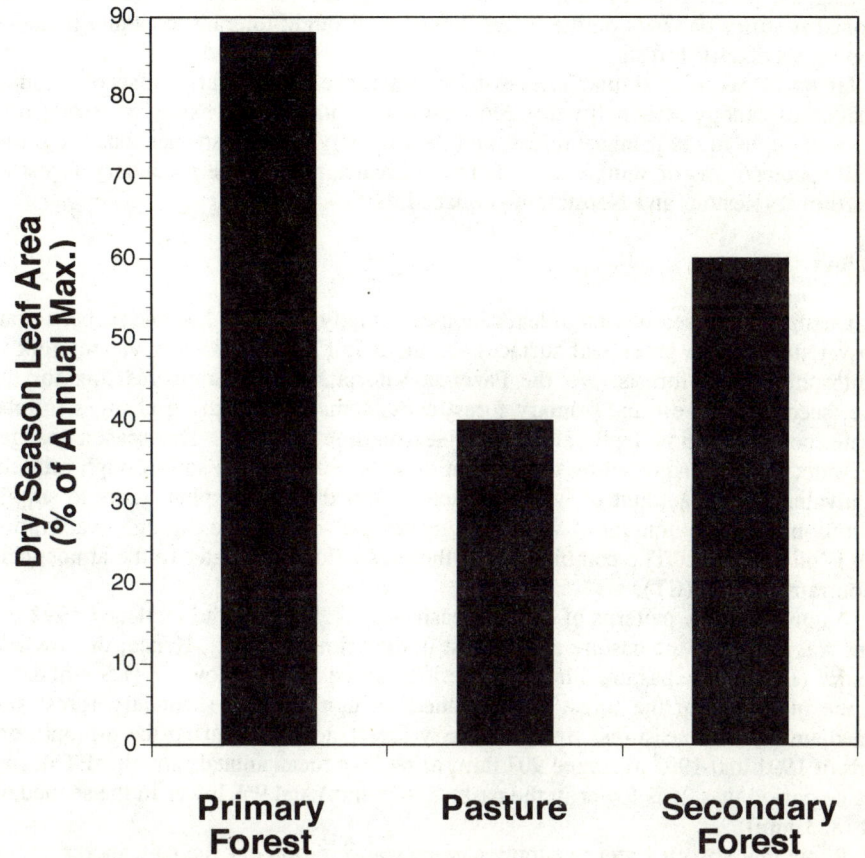

Figure 4. Dry season leaf area of Paragominas ecosystems, expressed as a percentage of annual maximum in 1992. These data are based on monthly observations of marked branches representing the ten most common plant species in each ecosystem.

We tested these two explanations at the Fazenda Vitoria by comparing seasonal patterns of leaf-shedding and drought stress (measured as xylem pressure potential, XPP) in pasture, secondary forest and primary forest. The first explanation was supported because leaf-shedding was negatively related to drought stress; the pasture exhibited greatest leaf-shedding and lowest drought stress, while the primary forest exhibited the least leaf-shedding and the greatest drought stress. Hence, primary forest trees and lianas maintain evergreen leaf canopies by tolerating the large tensions that develop in xylem sap during the prolonged dry season. In this sense, the recovery of forest health on abandoned pastures depends on the establishment of drought-tolerant trees and lianas in developing secondary forests.

Given plants with similar levels of leaf tolerance to drought, the second causal explanation of canopy seasonality differences might be important. Water is extracted to at least 8 m depth in the primary forest, and the diversity of plant species that reach this depth (9 species/6 m^{-2} of sample area) is much greater than in the secondary forest (2 species/6 m^{-2}) (Restom and Nepstad unpublished data).

Hydrology

Changes in the seasonality of leaf canopies strongly affect the hydrologic behaviour of ecosystems because green leaf surfaces are the major source of water vapour flux to the atmosphere from forests. At the Fazenda Vitoria, we compared this flux for the pasture, secondary forest and primary forest ecosystems by measuring changes in total soil water content (to 8 m depth) along a dry-season time sequence. Dry-season changes in soil water content, adjusted by the amount of water entering the soil through rainfall, are equivalent to the amount of water extracted from the soil by plant roots to supply transpiration, plus the amount of water that leaves the soil surface through evaporation (which is quite small). The combination of these two fluxes of water to the atmosphere is evapotranspiration (ET).

As predicted by patterns of canopy seasonality, ET during the 1991 and 1992 dry seasons was lowest in the pasture and highest in the primary forest. Hence, the low leaf area index (LAI) of the pasture during this period corresponded to low ET (1.9 mm day^{-1}) while the high LAI of the forest corresponded to high ET; the secondary forest was intermediate in both measures. In the primary forest, total ET during the 6-month dry seasons of 1991 and 1992 averaged 907 mm, or 64% of mean annual rainfall. ET during this same period was 26% lower in the pasture (675 mm) and 9% lower in the secondary forest (823 mm).

Roots that absorb water to supply canopy transpiration dry the soil, increasing its sponge-like capacity to retain moisture. By the end of the 1992 dry season, the upper 8 m of soil was sufficiently dry to store 265 ± 48 mm of rain in the pasture, 627 ± 31 mm in the secondary forest, and 706 ± 23 mm in the primary forest. The large water storage capacity that develops under primary forest is an indirect effect of canopy evergreenness that reduces wet season run-off and stream flow, and therefore may lead to lower levels of soil erosion and down-stream flooding. This feature of "healthy" forests recovers substantially within the first 15 years of fire-free forest regrowth.

The reduction in annual ET associated with forest conversion to pasture at the Fazenda Vitoria is greater than other estimates for the region that do not take into account extraction of deep soil moisture by primary forest. Moreover, the loss of deep soil water uptake that accompanies this conversion is not included in the recent global circulation models that predict a regional decline in rainfall through Amazonia deforestation (Shukla et al. 1990; Lean and Warrilow 1990). The phenomenon of deep soil water uptake would exacerbate these predictions. Counteracting the influence of deforestation on regional climate, however, is the considerable recovery of hydrologic function in secondary forests.

Fire Resistance

Resistance to fire is an important feature of healthy forests in seasonally-dry Amazonia. Without fire resistance, the forest's hydrologic functions and its population assemblages are at risk of periodic disruption or permanent alteration. All aspects of forest recovery on abandoned pastures hinge upon whether or not the young forest burns before it recovers fire resistance.

Pastures are highly flammable ecosystems. When grass leaves brown up, they remain attached to the plant for several weeks, exposed to the full drying effect of sun and wind. Since the entire fine fuel load of the pasture is located within a contiguous, wind-ventilated, 1.5 m-high profile, points of ignition can become roaring blazes within seconds. During the dry season, pastures meet the fuel and microclimatic requirements of fire whenever there is a drought of two days or more (Uhl and Kauffman 1990). Pasture flammability declines following abandonment, particularly when invading trees shade out grasses. As the seasonality of the recovering forest's leaf canopy declines, the number of rain-free days required to dry the forest floor to a flammable state increases. Based on the data of Uhl and Kauffman (1990), the 16 year-old secondary forest at Fazenda Vitoria was flammable for only 71 days in 1992, vs. 207 days in the pasture. The primary forest was not flammable this year but may become flammable during a sequence of dry years (Figure 2).

Biomass

Biomass accumulation in recovering forests depends on successful tree establishment and adequate edaphic conditions for tree growth. The speed of accumulation varies greatly on abandoned pastures near Paragominas and is best predicted by the intensity of pasture use prior to abandonment (Uhl et al. 1988). Soil fertility is not correlated with the trajectory of biomass accumulation on these abandoned sites (Buschbacher et al. 1988) and may have little utility as a measure of forest health on the deep, kaolinite-dominated oxisols of Amazonia.

At the Fazenda Vitoria, pasture contained <5% of the aboveground biomass of the primary forest, and about 12% of the original root biomass. After 16 years of recovery, the secondary forest contained 12% and 25% of these biomass pools (Figure 5) and was accumulating aboveground biomass at the rate of 5 Mg ha^{-1} yr^{-1} (C. Uhl, unpublished data). The largest pool of biomass in these ecosystems is soil organic matter, and is the topic of research currently underway. Preliminary data suggest that up to 15% of the soil organic matter pool may be dynamic, turning over on annual or decadal time scales, and therefore subject to loss through deforestation (E. Davidson and S. Trumbore, unpublished data).

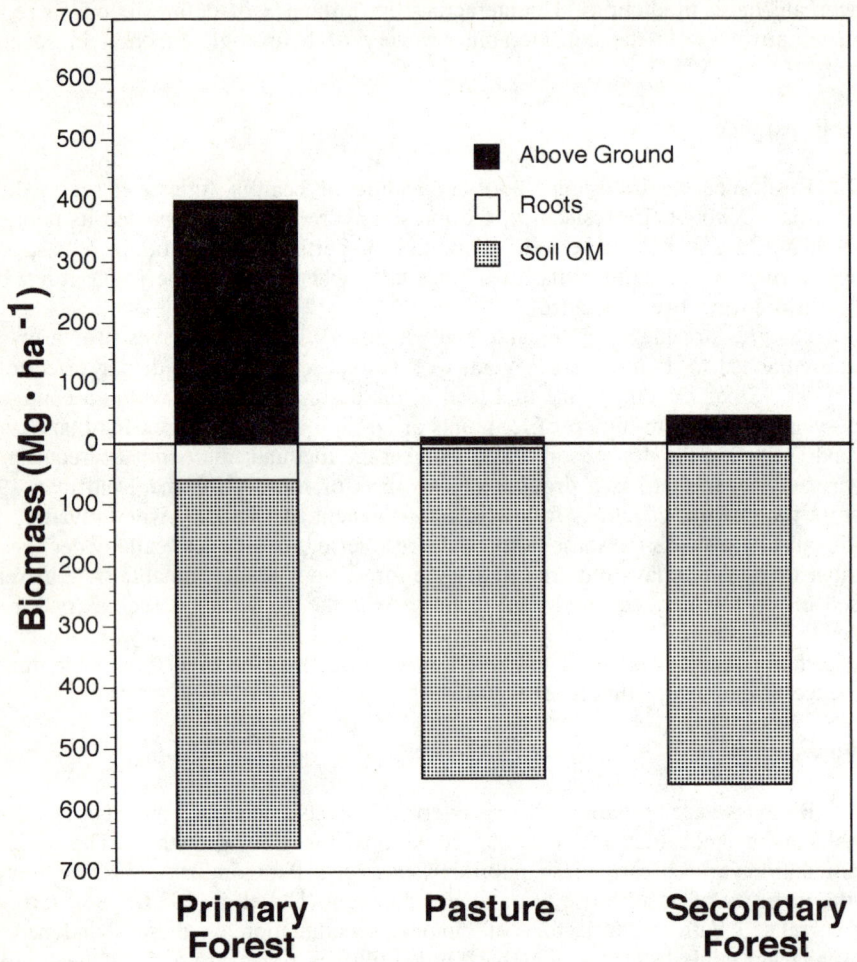

Figure 5. Biomass of the ecosystems of Paragominas. Aboveground data are from Nepstad (1989) and C. Uhl (unpublished data). Root data are from Nepstad (unpublished data).

Biomass accumulation in recovering forest is slow relative to the recovery of ET and fire resistance. More than 100 years may be required to achieve the biomass of the original forest. This rate is a direct measure of the ability of recovering forests to withdraw carbon from the atmosphere. Approximately 200 Mg ha^{-1} of carbon ($\sim$50% of above- and below-ground biomass) are released to the atmosphere as CO_2 through forest conversion to pasture and this carbon is reabsorbed at the rate of approximately 3 Mg ha^{-1} yr^{-1} during the first few decades, and may be reabsorbed at a slower rate thereafter. When multiplied by the area of secondary tropical forests globally, this carbon accumulation function becomes a significant component of the global carbon cycle (Houghton et al. 1993).

Ants

A full description of the shifts in faunal and floral populations that accompany forest conversion to pasture and subsequent forest recovery is beyond the scope of this chapter. However, we present data on shifts in ant species as a means of illustrating the potential that certain organismal groups hold as indicators of forest health and its recovery (Majer 1987). Ants are important in flower pollination, seed dispersal, seed predation and herbivory, and their role in altering soil structure and chemistry is potentially large but poorly understood (Haines 1978; Petal 1978). Moreover, ants comprise a large portion of animal biomass in Amazonian forests (Fittkau and Klinge 1973).

A simple comparison of ant fauna conducted at the Fazenda Vitoria using pitfall traps reveals that twice as many ant species occur in primary forest (53) and secondary forest (57) as in pasture (24) (Moutinho et al. 1992). These numbers do not reveal, however, a much greater reduction in <u>forest</u> ant species that accompanies forest conversion to pasture. Only five ant species that occur in the pasture also occur in the primary forest and only seven of the secondary forest species are common to the primary forest (Figure 6).

Comparisons of numbers of species do not provide information on population densities or ecological roles. The pasture ant fauna, for example, is dominated by four ant species, none of which are found in the primary forest. More than 90% of the 2324 ants collected in pasture pitfall traps belong to four species of the Subfamily Myrmicinae: *Wasmannia auropunctata* (68%), *Pheidole* sp. (18%), *Solenopsis* sp. (7%) and *Atta sexdens* (1%). Ant population shifts leading to the dominance of these four species have profound implications for the recovering forest. The first three species are generalist seed-eaters that move large numbers of small tree seeds to underground nests, potentially limiting the rate of forest recovery (Nepstad et al. 1991, Moutinho et al. 1993). The fourth species is a cutter ant, capable of moving seeds as large as corn grains and of cutting open larger seeds. *Atta sexdens* also defoliates and clips tree seedlings, preferring them over grass and shrub seedlings (Nepstad et al. 1991). Against these activities that reduce tree seedling establishment in abandoned pastures, this voracious species also constructs nests that extend >5 meters deep into the soil.

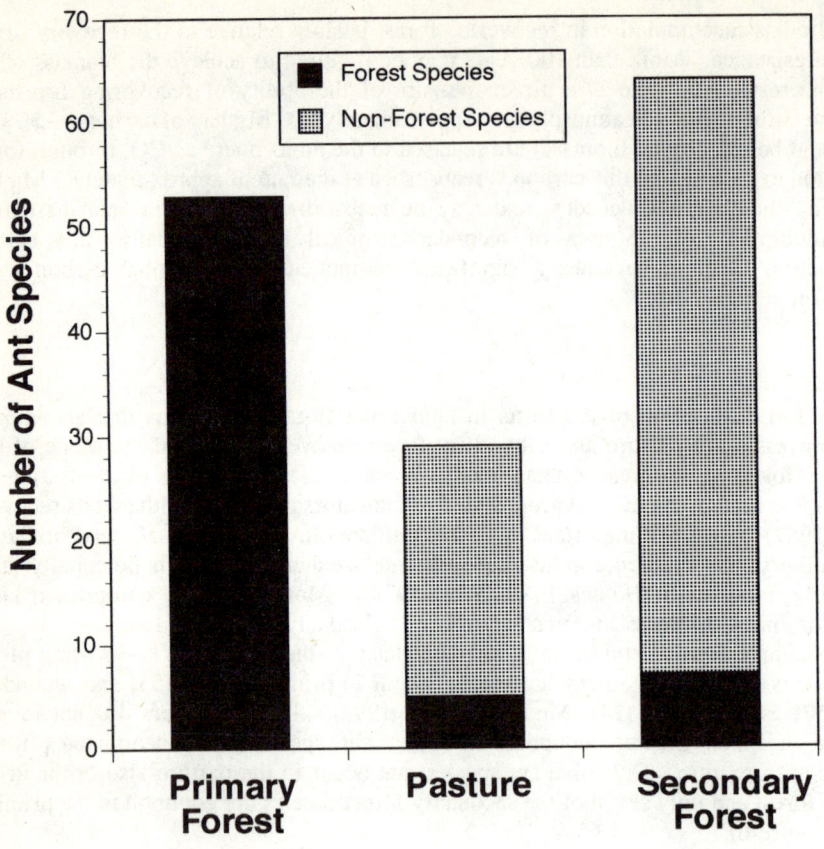

Figure 6. Number of ant species collected in pitfall traps placed in each of the ecosystems in Paragominas. (unpublished data of P. Moutinho).

These nests provide microsites rich in organic matter refuse in which roots proliferate and may thereby increase the rate of tree growth. Nest density of *Atta sexdens* in secondary forest is more than 20 times greater than nest density of cutter ants in the primary forest (Moutinho et al. in press).

Faunal and floristic species assemblages have potential as indicators of forest health, and the progress of forest recovery, and have been proposed as indicators of forest recovery on degraded tropical forestland. Development of indices of forest health based on such assemblages will depend on time-consuming species surveys, population censuses, and knowledge of the ecological roles of these species.

CONCLUSION

While much of the literature describing the ecology of moist tropical forest emphasizes nutrient cycles and the intricate interdependencies that exist between some species populations, neither of these provide particularly useful parameters of ecosystem health and its recovery. Canopy seasonality is the single most important measure of forest health in seasonally-dry eastern and southern Amazonia, for it determines forest resistance to fire and, therefore, the probability that the forest will continue to exist in the midst of human settlement. Canopy seasonality is determined by drought duration, rooting depth and the drought-tolerance of the leaf canopy, and may be monitored using measures of canopy greenness currently available through weather satellites. Biomass is also a measurable index of forest health that integrates the effects of plant regeneration processes and soil fertility. The use of species assemblages in assessing forest health holds considerable potential, as illustrated by studies of ants, but will require new knowledge of community composition and the ecological roles of individual species.

ACKNOWLEDGMENTS

The research presented in this paper was supported with grants from the US National Science Foundation, NASA, the US Agency for International Development, and the John D. and Catherine T. MacArthur Foundation. We are grateful to C. Uhl and F. Ramade for helpful comments on the manuscript. Graphics were prepared by P. Lefebvre and K. Schwalbe.

REFERENCES

Buschbacher R., Uhl C., Serrão E.A.S. (1988) Abandoned pastures in eastern Amazonia II. Nutrient stocks in the soil and vegetation. Journal of Ecology 76: 682-699

Fearnside P.M. (1993) Deforestation in Brazilian Amazonia: The effect of population and land tenure. Ambio 22: 537-545

Fittkau E.J., Klinge H. (1973) On biomass and trophic structure of the central Amazonian rain forest ecosystem. Biotropica 5: 2-14

Haines B.L. (1978) Element and energy flows through colonies of the leaf-cutting ants, *Atta colombica*, in Panama. Biotropica 10: 270-277

Houghton R.A., Unruh J.D., Lefebve P.A. (1993) Current land cover in the tropics and

its potential for sequestering carbon. Global Biogeochemical Cycles 7: 305-320
Lean J., Warrilow D.A. (1989) Simulation of the regional climatic impact of Amazon deforestation. Nature 342: 411-413
Majer J.D. (1987) Invertebrates as indicators for management. In Sunders D.A., Arnold G.W., Burbridget A.A., Hopkins A.J.M., (eds.) Nature Conservation: The Role of Remnants of Native Vegetation. pg. 353-355 Surry Beatty & Sons, Bentley, Australia
Mattos M., Uhl C. (1994) Economic and ecological perspectives on ranching in the eastern Amazon. World Development
Moutinho P.R.S., Nepstad D.C., Araujo K., Uhl C. (1993) Formigas e floresta [Ants and forests]. Ciencia Hoje 88: 59-60
Moutinho P., Nepstad D.C., Davidson E.A. (in press) Acabar com a saúva, mas nem tanto [Eliminate cutter ants, but not that much]. Ciencia Hoje
Nepstad D.C. (1989) Forest regrowth in abandoned pastures of eastern Amazonia: Limitations to tree seedling survival and growth. Ph.D. Dissertation Yale University, New Haven, CT
Nepstad D.C., Uhl C., Serrão E.A.S. (1991) Recuperation of a degraded Amazonian landscape: Forest recovery and agricultural restoration. Ambio 20: 248-255
Nepstad D.C., de Carvalho C.R., Davidson E.A., Jipp P., Lefebve P., Negreiros G.H., da Silva E.D., Stone T., Trumbore S., Vieira S. (in press) The deep-soil link between water and carbon cycles of Amazonian forests and pastures. Nature (in press).
Petal J. (1978) The role of ants in ecosystems. In Production Ecology of Ants and Termites. International Biology Programme, No. 13. Cambridge University Press New York pp:293-325
Salati E., Dall'Olio A., Gat J., Natsui E. (1979) Recycling of water in the Amazon Basin: An isotope study. Water Resour. Res. 15: 1250-1258
Sanford R.L., Saldarriaga J., Clark K., Uhl C., Herrera R. (1985) Amazon rain-forest fires. Science 227: 53-55
Serrão E.A.S., Toledo J.M. (1990) The search for sustainability In Anderson A.B. (ed.) Amazonian pastures. within Alternatives to Deforestation: Steps Toward Sustainable Utilization of Amazon Forests. Columbia University Press New York
Setzer A.W., Pereira M.C. (1991) Amazonia biomass burnings in 1987 and an estimate of their tropospheric emissions. Ambio 20: 19-22
Shukla J., Nobre C.A., Sellers P. (1990) Amazon deforestation and climate change. Science 247: 1322-1325
Uhl C., Buschbacher R. (1985) A disturbing synergism between cattle ranching burning practices and selective tree harvesting in the eastern Amazon. Biotropica 17: 265-68
Uhl C., Buschbacher R., Serrão E.A.S. (1988) Abandoned pastures in eastern Amazônia, I: Patterns of plant succession. Journal of Ecology 76: 663-681
Uhl C., Clark K., Clark H., Murphy P. (1981) Early plant succession after cutting and burning in the upper Rio Negro region of the Amazon Basin. J. Ecol. 69: 631-649

Uhl C., Kauffman J.B. (1990) Deforestation effects on fire susceptibility and the potential response of tree species to fire in the rain forests of the eastern Amazon. Ecology 71: 437-449

Uhl C, Kauffman JB, Silva ED (1990) Os caminhos do fogo na Amazônia. Ciência Hoje 65: 25-32

Victoria R.L., Martinelli L.A., Mortatti J., Richey J. (1991) Mechanisms of water recycling in the Amazon Basin: Isotopic Insights. Ambio 20: 384-387

World Resources Institute (1990) World Resources 1990-91: A Report by the World Resources Institute in Collaboration with the United Nations Environment Program and the United Nations Development Program. Hammond AL (ed.) Oxford University Press New York

RAPPORTEUR'S REPORT

Magda Havas
Environmental & Resource Studies
Trent University, Peterborough, ON K9J 7B8
Canada

The aim of this session is to:

(i) identify indicators of recovery
(ii) assess patterns and rates of recovery
(iii) determine general properties of ecosystem recovery
(iv) compare the process of recovery with that of degradation
(v) identify interventions needed to actualize the recovery process

SUMMARY OF PRESENTATIONS

Smol describes paleolimnological techniques that could be used to obtain historical information on species assemblages and water quality by examining diatom assemblages in lake sediment profiles. Hutchinson presents a case study of sulphur dioxide and metal pollution of terrestrial and freshwater ecosystems and their subsequent recovery following reduced emissions. Ozturk discusses effects of grazing pressure, urbanization, and fires on soil erosion and loss of biological richness and cover in the Turkish Maquis ecosystem and suggested a mechanism for co-operative efforts to protect this ecosystem. Nepstad discusses rate of recovery of abandoned pastures to secondary forests in the Brazilian Amazon and the factors that influenced this rate.

DISCUSSION

A series of questions were addressed in the discussion Session but only a few of these were attempted by the various groups. These questions and the responses to them are summarized below.

1. **Meaning of the word recovery:** One of the recurring themes was "what does recovery mean?" The definitions of recovery ranged from an improvement to a return to the original, pre-stressed state. Alternative words were used to describe the recovery process such as self-renewal, restoration, and rehabilitation. Obviously rates and patterns of recovery

will differ depending on your definition of the word. It was suggested that restoration should be used to mean a return to the original state and rehabilitation should be used to mean a deliberate management intervention to improve the system (peg leg). Also it is important to differentiate between structural and functional recovery of a system.

2. Another bottle neck is that there are so few good **examples of recovery** that it is difficult to formulate broad generalizations regarding the pattern and rate of recovery with any confidence.

3. A third bottle neck is that our **reference points** might be different and hence we have difficulty communicating with one another. For example European references are moderately disturbed and North American references are relatively pristine. Therefore we have to know where we are along a stress gradient or else we will misunderstand one another.

4. Also we have some qualifiers: Type of recovery, rate of recovery, and pattern of recovery depend on:

a) **the type of disturbance:**
   - physical disturbance and loss of habitat (e.g. erosion): for physical stressors recovery is different than the degradation process. If the resource base is affected (loss of soil or water resource), the rate of recovery will be slower and direction may be different than the original system.
   - chemical stressors (e.g. metals) may or may not have same recovery pattern.
   - biological stessors (e.g. over harvesting of fish populations) are dependent on hierarchical level affected (top predator vs. primary producer).

b) **the magnitude of the stress:** complete devastation vs. subtle stress

c) **the size of the area affected** (important as a seed source, larger areas may take longer than smaller areas)

d) **the distance of the affected site from nearby unaffected sites** (important seed source, inland sea slower to recover than coastal regions)

e) **the context of the disturbance** in terms of the evolutionary history of ecosystem, organisms, etc.

f) **the spatial and temporal scale**

g) **the type of system** (large lake, small lake, open ocean, bay, forest, grassland)

h) **the regional context** (e.g. Amazonian forest)

CONCLUSIONS

Is recovery a reversal of trends exhibited by stressed ecosystems?

Yes, but on a coarse level. If there are extinctions or if the soil is gone, system will not rebound in the same way. The direction of recovery depends on the level of destruction, if severe, it may not be a reversal of trends.

What is the value of management practices for rehabilitation?

With some exceptions, management practices were deemed to be undesirable for recovery of naturally sustainable systems (many examples were presented). For example, liming may enable fish to recover but will end up with a different system. Liming of acidic, metal-contaminated lakes will kill the few remaining species present because the metal becomes more toxic immediately after liming. Examples of poor rehabilitation practices include introduction of exotics (such as Eucalyptus in areas where Eucalyptus is not a native species, salmonids in Great Lakes which have been introduced, but are not reproducing).

Some key points were: it is important to have cost benefit analysis to evaluate any rehabilitation scheme; if a lake is naturally fishless you shouldn't stock it (paleolimnological evidence useful); naturally acidic lakes should not be limed (paleolimnological evidence); self-recovery might be more healthy than mitigation; stocking fish may be a placebo.

What are effective indicators of recovery?

Case studies are all different, perhaps a more useful approach might be to consider the indicators of recovery for a specific type of ecosystem (e.g.. freshwater, marine, terrestrial). This was not done in a comprehensive manner.

Are indicators for decline and recovery the same?

Some are (chemical changes, acidification, and metal concentrations in Sudbury lakes, for example); some are not (biological changes may or may not be the same; return of terrestrial biota at Sudbury is not the same). For lichens they are the same i.e. lichens are first killed and first to come back following disturbance by air pollution.

The production and re-establishment of the food web is an important indicator of recovery and decline. In aquatic systems, benthic organisms are critical.

How can one differentiate between a "healthy" recovery and an "unhealthy" recovery?

Most groups thought that "unhealthy" recovery was not possible and dismissed this question. An alternative term might be incomplete vs. complete recovery. Incomplete recovery might be deemed "good" in the sense that 90% recovery might be cheaper than complete recovery. Complete recovery might take too long for society; or, if extinctions occurred, 100% recovery might not be feasible.

What factors affect the rate of recovery?

Examples include: closed basin lakes may recover more slowly than flushing rivers (seed source); systems with r-species improve more quickly the systems with k-species; if physical structure destroyed, it will take a long time for recovery (e.g. coral reefs); western society wants instant gratification, but a long time-scale for recovery may be better than instant recovery; need for research on topics due to imperfect understanding; what we know about the rate of recovery for oil spills is that exposed sites take 3-5 years to recover, sheltered sites 10 years, and bays decades.

Science & Policy Issues: How can we promote the recovery process?

Prevention vs. cure: health maintenance philosophy is more important than fixing it after the fact. Prevention is cheaper in long run.

Education: education of the public important.

Communication: we have to learn to communicate our messages: keep it short and eliminate qualifiers; build right network from journalists to lobbyists. Communication with policy-makers is a problem. It is Important to reach lobbyist to influence decision-makers. We should learn how to use media contacts carefully; ecological training for media people is important

Need for generalist ecologists

Need a set of environmental ethics

Session V

**METHODOLOGICAL ISSUES IN DESIGN AND
ANALYSIS OF ECOSYSTEM HEALTH**

INTRODUCTION

Chair: Mikael Hilden

The aim of the workshop is to generate new insights on the evaluation and monitoring of large-scale ecosystems. The session on methodological issues will focus on how one can obtain data on the state of a system and how this data can be interpreted as information on the state of this system.

The starting point is large-scale ecosystems. Specific requirements can be put on the methodology aiming at measuring the state (health) of these systems. The methodology must be able to identify positive and negative changes in the state system, and in order to be able to do this a reference must be identified. In some cases the reference may represent a static level, but it may also represent a dynamic change. Preferably the methods to measure change are sensitive enough in order to become part of a feed back system related to the management of the system. Thus they should be able to provide early signals of changes. The methods must also be technically feasible on a large-scale. Further the data acquisition and analysis should be regarded together in order to create an efficient system for the assessment of the health of large systems. During the discussions the criteria for the methodologies are developed further. All authors are asked to identify which specific criteria their approaches meet.

When criteria for successful methodologies have been identified the issues of generality can be raised. Can a unifying theoretical foundation be identified for successful methods? Is it possible to identify general principles that guide the choice of methods or will the methods always be identified case by case? Which are the relevant boundary conditions for the application of a specific approach in different regions and types of systems?

As a consequence of the large-scale of the systems to be investigated, multidimensionality and the importance of spatial and temporal scales will be important in a discussion of methodological issues.

The multidimensional nature of the systems and hence their health raises the question on the identification of necessary and sufficient observations of the state of the system. To what extent can the information be aggregated to a limited set of dimensions? For example diversity can be aggregate measure of the state of a system, but is it sufficient? What is the correct scale for measuring diversity in relation to ecosystem health of large-scale systems? Population, species taxa, functional groups or other unit? Can other aggregate measures be identified?

Large-scale systems change on several different spatial and temporal scales. Some of the changes represent natural variability whereas others can be related to trends. To what extent are different methodologies able to separate these? Should focus be on some specific spatial and temporal scale and how can the focus be determined? Can the correct temporal scale be identified by examining the relationship between anthropogenic sources of stress (e.g. pollution, pollution control, rehabilitation) and the anticipated reaction time of the system? The focus on anthropogenic effects on the large-scale systems could justify such an approach.

Not all of the papers can answer all of the questions above, but by focusing the discussion on criteria for successful methodologies, the theoretical foundations of the methodologies and different ways of introducing spatial and temporal considerations the session aims at providing a significant input for the final analysis which draws on the main conclusions of all sessions.

19. TEMPORAL AND SPATIAL VARIABILITY AS NEGLECTED ECOSYSTEM PROPERTIES: LESSONS LEARNED FROM 12 NORTH AMERICAN ECOSYSTEMS

Timothy K. Kratz[1], John J. Magnuson[1], Peter Bayley[2], Barbara J. Benson[1], Cory W. Berish[3], Caroline S. Bledsoe[4], Elizabeth R. Blood[5], Carl J. Bowser[1], Steve R. Carpenter[1], Gary L. Cunningham[6], Randy A. Dahlgren[4], Thomas M. Frost[1], James C. Halfpenny[7], Jon D. Hansen[8], Dennis Heisey[1], Richard S. Inouye[9], Donald W. Kaufman[10], Arthur McKee[11], and John Yarie[12]

[1]Center for Limnology, University of Wisconsin, Madison WI 53706
[2]Illinois Natural History Survey, 607 E Peabody Dr., Champaign, IL 61820
[3]EPA Laboratory, Atlanta, GA 30322
[4]Department of Land, Air and Water Resources, University of California, Davis, California 95616
[5]J.W. Jones Ecological Research Center, ICHAUWAY, Newton, GA 31770
[6]Biology Department, New Mexico State University, Las Cruces, NM 88003
[7]P.O. Box 989, 300 Scott, Gardiner, MT 59030
[8]USDA ARS, P. O. Box E, 301 S. Howes, Fort Collins, CO 80522
[9]Department of Biological Science, Idaho State University, Pocatello, ID 83209-8007
[10]Division of Biology, Ackert Hall, Kansas State University, Manhattan, KS 66506
[11]Forest Science Department, Oregon State University, Corvalliss, OR 97331
[12]Forest Soils Lab, University of Alaska, Fairbanks, AK, 99775
USA

INTRODUCTION

Evaluating and monitoring the "health" of large-scale systems will require new and innovative approaches. One such approach is to look for ecological signals in the structure of ecological variability observed in space and time. Such variability is sometimes considered something to minimize by clever sampling design, but may in itself contain interesting ecological information (Kratz et al. 1991). In fact, much of ecology can be considered an attempt to understand the patterns of spatial and temporal variability that occur in nature and the processes that lead to these patterns. Despite widespread interest in patterns of variation there have been relatively few attempts to describe comprehensively the temporal and spatial variation exhibited by ecological parameters. As a result, we have no general laws that allow us to predict the relative magnitude of temporal and spatial variability of different types of parameters across the full diversity of ecological systems. Even within single ecosystems, understanding of the interplay between temporal and spatial variability is lacking. For example, Lewis (1978) noted that despite a large literature, the relation between temporal and spatial variability in plankton distribution within a lake is not well understood. Matthews (1990) makes a similar point regarding fish communities in streams.

In this paper we describe general patterns exhibited by ecological parameters across a wide variety of ecosystem types. We attempt to answer three basic questions regarding ecological variability: 1) do climatic, edaphic, and biological parameters differ systematically in variability, 2) how is variability partitioned between spatial vs temporal components, and 3) to what extent are ecological parameters spatially or temporally coherent (Magnuson et al. 1990)? By coherence we mean the tendency for different locations within a landscape to behave similarly in different years independent of the average for the locations (temporal coherence) or the tendency for locations within a landscape to be consistently different regardless of the year (spatial coherence). We use data collected at 12 diverse North American ecosystems represented in the Long Term Ecological Research (LTER) network. For each of the 12 LTER sites, data are available for several years at several locations. Therefore, we are able to analyze both the spatial and temporal aspects of variability at each site.

In addition to these general questions about patterns of variability and the effects of scale, we also used cross system comparisons of variability to test two smaller-scale ecological hypotheses: 1) that deserts are more variable than lakes temporally, but less variable spatially; and 2) that in predator-prey pairs, the smaller-shorter lived member of a pair is more temporally variable regardless of whether it is the prey or the predator. We posited that deserts are more variable among years than lakes because they are more sensitive to among year differences in weather than are physically buffered lakes. Conversely, we hypothesized that lakes would be more variable spatially because they are more isolated from one another than are areas of continuous desert. We also felt that variability in a population was more a function of its life history than its position in a food web.

Variability is highly dependent on the temporal and spatial scales of the data set and on the level of aggregation of the parameter of interest (e.g. species level vs community level) (Allen and Starr 1982, Frost et al. 1988, Frost et al. 1992, Wood et al. 1990) This scale dependence raises a complication in comparative studies, because it is possible to confound differences in patterns of variability with differences in scales of measurement at two or more systems. There are two different aspects of scale, grain and extent, and the effect of these on observations of variability need to be considered independently (Allen and Starr 1982; O'Neill et al. 1986; Turner 1989; Wiens 1989). Grain refers to the level of resolution of the study. Extent refers to the size of the study area or the duration of the study period. In this study we focus on a temporal grain of one year because a year is a physically and biologically meaningful unit of time for which data are available. The spatial grain was more difficult to fix, however, because the size of study units sampled was not standardized across LTER sites, and varied by a factor of about 64000 across the 12 LTER sites. Therefore, we were required to address two scale related questions: 1) what is the effect of sampling unit size on measured variability; 2) how sensitive is the observed variability to the temporal extent of the measurements. We also considered the degree to which observed variability is related to degree of aggregation of the parameter measured.

METHODS

The Study Sites

The 12 sites from the LTER network are listed in Table 1 and their locations are shown in Figure 1. Ecosystem types represented include desert, prairie, alpine tundra, forest, lake, estuary, and river. We intentionally chose diverse sites so that our comparison of variability would include a wide range of biotic and abiotic conditions. Descriptions of the sites are given in Van Cleve and Martin (1991).

Figure 1. Locations of 12 LTER sites participating in this study.

Table 1. Data set characteristics for each of the 12 LTER sites.

LTER Site	Abbreviation	Number of Locations	Number of Years	Size of Location (ha)	Total Number of Variables	Number of Climatic Variables	Number of Edaphic Variables	Number of Plant Variables	Number of Animal Variables
Cedar Creek Natural History Area	CDR	4-18	5-6	3	17	0	0	12	5
Hubbard Brook Experimental Forest	HBR	2-6	5-6	3.7	49	4	45	0	0
Illinois River	ILR	5	4-5	6400	19	5	6	0	8
Konza Prairie Research Natural Area	KNZ	6-9	5-7	50	52	3	0	22	27
North Inlet	NIN	3	6	880	57	6	48	3	0
Niwot Ridge	NWT	6	5	0.1	28	6	0	10	12
Northern Temperate Lakes	NTL	5-7	4-6	240	102	6	50	9	37
H. J. Andrews Experimental Forest	AND	3-9	4-17	15	33	21	9	3	0
Central Plains Experimental Range	CPR	4-9	4-6	2	13	0	0	12	1
Bonza Creek Experimental Forest	BNZ	7-14	12-14	0.1	4	3	0	1	0
Jornada	JRN	6-7	3-5	1.1	56	11	4	13	28
Coweeta Hydrologic Laboratory	CWT	3	6-17	96	18	12	6	0	0
Total					448	77	168	85	118

The Data Set

The data set, VARNAE (Variability in North American Ecosystems) was compiled in preparation for a workshop attended by representatives of each of the 12 sites (Kratz et al. 1991; Magnuson et al. 1991). The data set consists of a series of derived statistics for each of 448 parameters. For each parameter at least one LTER site had gathered data over several years (mean 5.9 years, range 3-17 years) at several locations at the site (mean = 5.8 locations, range 2-18 locations). Throughout this paper we use "location" to refer to one of several study areas within an individual LTER site, and "site" to refer to different LTER sites. Locations within an LTER site might, for example, refer to individual lakes, places along a transect, or different forest plots, whereas sites refer to Cedar Creek Natural History Area, Hubbard Brook Experimental Forest, etc. Data for each parameter were summarized to fill out a "year by location" matrix for each LTER site. Each entry in the matrix was the best estimate of the parameter for a given year at a given location, and may itself have been summarized from many individual measurements. For example, for mean summer chlorophyll a concentration at the North Temperate Lakes site, a five year by seven location (lake) matrix was completed, with each entry estimating the mean summer chlorophyll concentration in a particular lake at the North Temperate Lake site for a particular year. In this example, each of the 35 year-lake estimates in this matrix was derived from 14-50 individual measurements of chlorophyll.

This compilation resulted in 448 "year by location" matrices for the 12 LTER sites in total. For each of these matrices we used a two way analysis of variance framework to compute the variance associated with year and with location. The remaining variance is attributable to a combination of year-by-location interaction and error. Because our matrices had no replication in each of the cells, we could not separate error from interaction. Up to this point this is the same analytic framework used by Lewis (1978) and Matthews (1990) albeit at a different time and space scale. However, because we wanted to compare the variabilities of parameters measured in different units, before we computed the analysis of variance we relativized each matrix by dividing each element in the matrix by the grand mean of that matrix. The resulting "relative variances" are equivalent to the square of the coefficient of variation. The advantages of using relative measures of variation in ecological analyses have been reported by Kratz et al. (1987, 1991) and Rothschild and DiNardo (1987). None of the matrices had any missing data. The relative variances were computed using the following formulas:

$$V_L = (MS_L - MS_{YxL})/N_Y$$
$$V_Y = (MS_Y - MS_{YxL})/N_L$$
$$V_{YxL} = MS_{YxL}$$
$$V_T = V_L + V_Y + V_{YxL}$$

where, V refers to variance, MS refers to mean square, and L, Y, YxL, and T refer to location, year, interaction plus error, and total, respectively, from the two way analysis of variance; N_Y is the number of years in the matrix; and N_L is the number of locations in the matrix. Corresponding CV's are:

$$CV_L = \text{sqrt}(V_L)/\mu$$
$$CV_Y = \text{sqrt}(V_Y)/\mu$$
$$CV_{YxL} = \text{sqrt}(V_{YxL})/\mu$$
$$CV_T = \text{sqrt}(V_L + V_Y + V_{YxL})/\mu$$

where CV is coefficient of variance, μ is the grand mean of the matrix, and other terms are as defined above. V_Y is a measure of the variability that is due to the tendency for all locations to behave similarly in different years independent of the average for the locations. As an example, consider again chlorophyll concentrations at the Northern Temperate Lakes site. If in certain years (perhaps years of lower that average precipitation leading to lower than average nutrient loading) each lake had chlorophyll concentrations lower than the lake's average, and in years of plentiful rain each had values higher than the location's average, V_Y would be relatively large. On the other hand, if some lakes responded to drought years by having lower than normal values, while other lakes responded by having higher than average values, V_Y would be relatively low. Therefore, V_Y measures the degree to which a parameter is "coherent" in time (Magnuson et al. 1990). Similarly, V_L measures the degree to which a parameter is "coherent" in space, i.e. the degree to which difference in locations occurs independent of the year.

The interaction term is a measure of additional variability that is not associated with a fixed location effect nor a fixed year effect. This variance has been termed "ephemeral" because it is a measure of patchiness that differs on different dates (Platt and Fillion 1973, Lewis 1978). However, because the interaction term is not associated with fixed effects of location or years, it is also possible to interpret it as an additional amount of variability that is "incoherent" with respect to location or year. Therefore, it represents an additional amount of temporal variability, and symmetrically, an additional amount of spatial variance. We interpret the sum of the fixed year effect and interaction variabilities as total year variability, and similarly, the sum of the fixed location variability and interaction variability as the total location variability. These two terms are computed as follows:

$$V_{TY} = V_Y + V_{YxL}$$
$$V_{TL} = V_L + V_{YxL}$$

where V_{TY} is total year variability and V_{TL} is total location variability. In fewer than 5% of the parameters one of the computed variance terms was negative, and these negative values were set to zero.

For each of the 448 parameters we computed these six measures of relative variation: coherent year, coherent location, incoherent, total year, total location and total. Incoherent, total year, and total location are somewhat biased because they each contain the interaction term and we were unable to separate measurement error from the interaction term. Therefore, we overestimate each of these three terms. The effect of this bias is less important in our estimates of total year and total location variability because both terms contain the same bias and comparisons of the two terms will be valid, if not numerically exact.

We classified each of the 448 parameters into one of four categories: climatic, edaphic, plant related, or animal related. This grouping was necessary because, although each of the sites has similar long-term goals, the sites differ so much from each other that no single parameter was measured at all sites. The number of parameters available for this study in each of these categories at each LTER site is given in Table 1. An indication of the types of parameters grouped into the four categories for each site is given in Table 2.

Sensitivity to scale

Temporal Scale - Three of the LTER sites had data sets spanning at least 14 years. These longer data sets allowed us to test the sensitivity of our estimates of variability to the number of years in the data string. We tested the sensitivity by computing the average for each of the six variance estimates using all combinations of two consecutive years, three consecutive years, etc. For a 14 year data set, for example, there were 13 combinations of 2 consecutive years, 12 combinations of 3 consecutive years, 11 of 4..., 2 of 13. Finally, of course, there is just one combination of 14 consecutive years. We performed the sensitivity analysis for five parameters: two climatic (maximum and minimum streamflow from Coweeta), two edaphic (total N and Ca from H. J. Andrews), and one biological (basal area increments of trees from Bonanza Creek).

Spatial Scale - To understand the influence of spatial grain on variability we correlated the variability exhibited at each site with the size of the sampling unit. Determining the size of the sampling unit was problematic. The approach we took was to use the size of the location thought to be represented by the measurement. This size was determined using expert judgment by representatives from each site. For example, at the NTL site the samples are often taken at a central location in a lake. But because the lakes are generally well mixed horizontally the sample represents the entire lake. Therefore, the average location size for NTL was the mean area of the seven lakes. Measures of variability included in the analysis were coherent year, coherent location, incoherent, and total for each of climatic, edaphic, plant, and animal data, yielding a total of 16 measures of variance.

Aggregation Scale - We tested the relationship between observed variability and level of aggregation for edaphic, plant, and animal parameters. Edaphic parameters were grouped into three aggregation levels, and plant and animal parameters were grouped into four levels. Aggregation level was not a meaningful concept for the climatic data used in the study. For edaphic data, the finest level included parameters like ammonia or

Table 2. General description of the parameters by site and type. CPUE is catch per unit effort.

LTER Site	Climatic	Edaphic	Plant	Animal
Cedar Creek	-	-	Biomass of functional groups	CPUE of small mammals
Hubbard Brook	air temperature	major ions, nutrients, DOC, Al	-	-
Illinois River	water temperature and level	turbidity, pH, solids, conductance	-	CPUE of fish
Konza	soil moisture	-	% cover and diversity	relative abundance of grasshoppers and small mammals; diversity of small mammals
North Inlet	water temperature	nutrients, DOC, salinity, sediment characteristics	chlorophyll	-
Niwot Ridge	air temperature, precipitation, growing season length	-	biomass of species and groups	biomass and density of small mammals
Northern Temperate Lakes	water temperature	ions, nutrients, dissolved oxygen	chlorophyll	CPUE of fish, density of zooplankton
H.J. Andrews	streamflow, precipitation, air temperature, soil temperature	ions, nutrients in streams	litterfall	-
CPER	-	-	frequency, production	cattle weight gain
Bonanza Creek	air temperature, precipitation	-	basal increments of trees	-
Jornada	soil moisture, water input	soil nitrogen	cover and diversity of functional groups	density and diversity of small mammals and ant colonies
Coweeta	precipitation, streamflow	calcium, chloride	-	-

nitrate, the next level contained parameters such as total N. Finally, the coarsest level contained parameters such as total conductivity or total base cations. For the biological parameters the finest level contained species level information, the next contained guild level data such as C_3 grasses or zooplanktivorous fishes, the next coarsest included major groups such as all grasses or all fishes, and the coarsest level data included total animal or plant data such as total biomass of plants. To test for a relationship between variability and level of aggregation we computed the median of the total variability for each level and examined the relationship between variability and aggregation level for edaphic, plant, and animal data.

RESULTS

Sensitivity to Scale and Data Aggregation

Estimates of variability were relatively insensitive to the number of years in the data set when four or more years of data are included, at least to a maximum of 16 years (Figure 2). Estimates based on less than four years of data tended to be larger than those based on longer durations, for example, at Andrews for N and Ca. One possible exception to this general pattern in the data sets we analyzed was total nitrogen at Andrews which exhibited a slight decreasing trend for coherent location variance and a slight increasing trend for incoherent variance as more years were added. We conclude that 5 years of annual data for the 448 parameters we analyzed were adequate to estimate variability and that differences in the number of years available above 5 years did not bias the overall analyses.

We found little evidence for a relationship between spatial grain and variability. Of the sixteen possible correlations between variability and size of sampling unit, none had a correlation coefficient above 0.5 (in absolute value) and none were significant. However, 14 of the 16 correlation coefficients were negative, indicating that samples representing larger areas tended to be less variable spatially and temporally. We had difficulty assigning sizes to sampling units, and it is possible that we did not find a relationship between variability and size of sampling unit because of our definition of sampling unit. We concluded that differences in the size of sampling units among LTER sites should not proscribe further analyses, but that interpretation of differences in variability among sites may be biased.

There was a strong relationship between total variability and level of aggregation for biological parameters, but not for edaphic parameters (Figure 3). For edaphic parameters there was no statistical difference in variability of parameters at species, group, or major group levels of aggregation ($p > 0.1$, this and subsequent p values in this paragraph are based on Mann-Whitney tests). For plant parameters, species-level data were not more variable than guild-level data ($p = 0.745$), however, both species level data and guild level data were significantly more variable than group data ($p = 0.003$; $p = 0.004$, respectively). For animal parameters species-level data were more

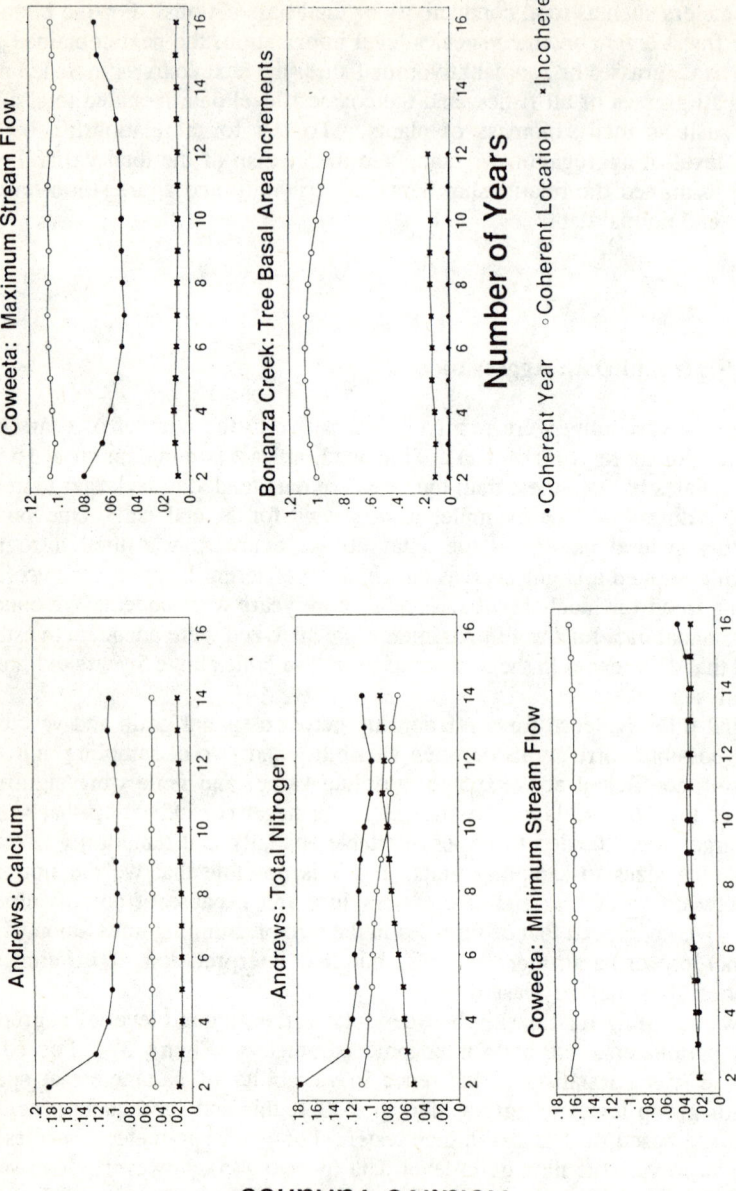

Figure 2. Sensitivity of coherent year, coherent location, and incoherent variance to number of consecutive years of data used for calculation.

variable than guild-level data (p < 0.0001) and group data (p < 0.0001); and guild-level data were more variable than group-level data (p = 0.03). We concluded that differences in level of biological aggregation could significantly bias our analyses and that comparisons should be made at the same level of biological aggregation.

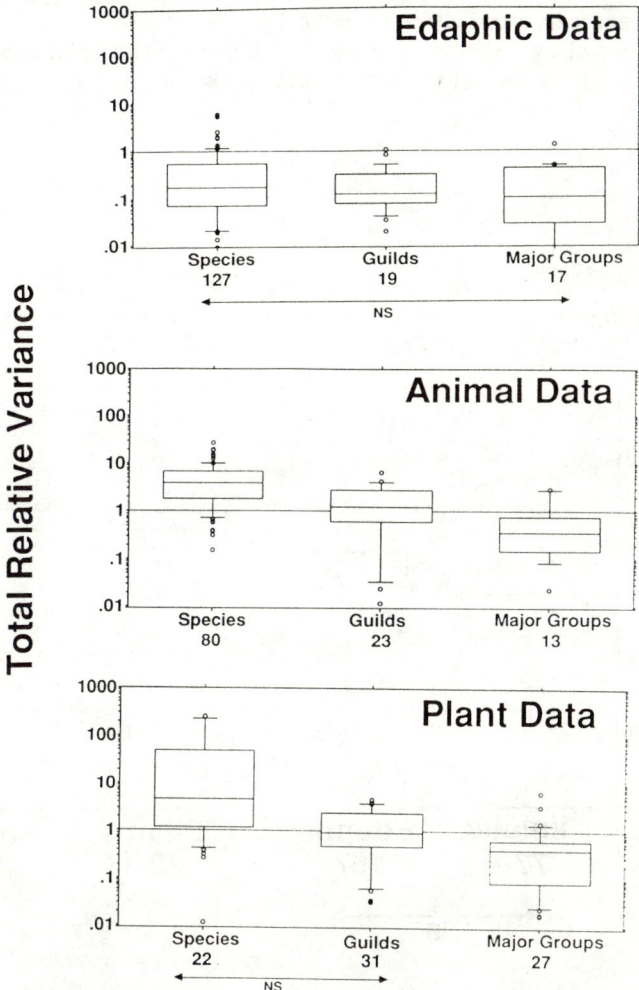

Figure 3. Relationship between total relative variance and aggregation level of data for edaphic, animal, and plant data. Numbers under group identifiers indicate sample sizes. Lines under plots for edaphic and plant data indicate groups that do not differ significantly at the P=0.05 level.

Overall Patterns of Variability

We compared the total relative variability of climatic, edaphic, plant, and animal data (Figure 4; Table 3). Because level of aggregation is an important determinant of variability in biological data, we used only species level data for plant and animal variables. Climatic data were all at the same level of aggregation and edaphic data showed no pattern with aggregation, so we used all climatic and edaphic data. We made pairwise comparisons for each of the four variable types.

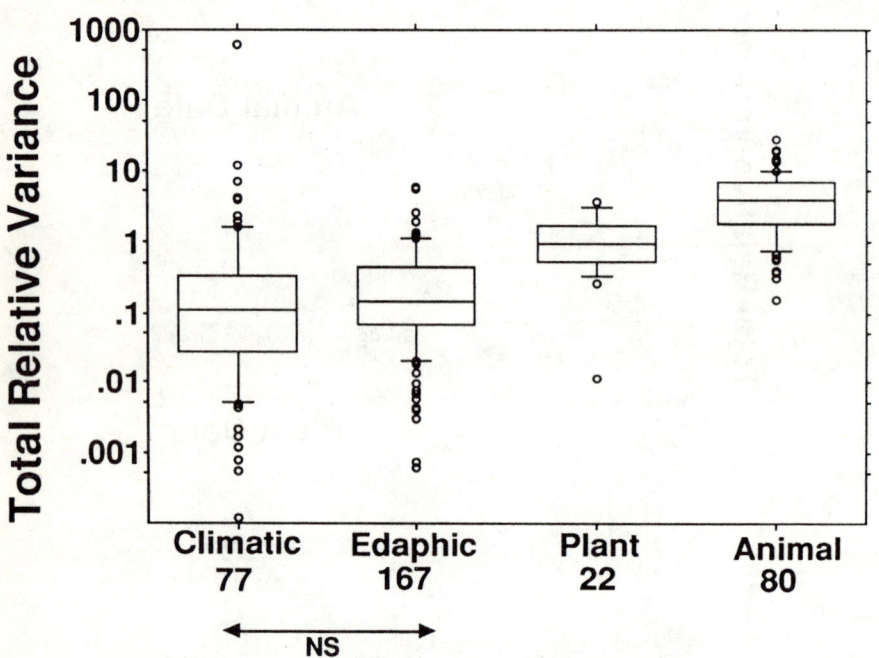

Figure 4. Total relative variance of parameters classified into climatic, edaphic, plant, and animal groupings. Numbers under parameter type indicate sample size. For plant and animal categories, only species-level data were used. Climatic and edaphic data did not differ significantly with respect to total relative variance.

Table 3. Comparison of variability of climatic, edaphic, plant, and animal data. Group with greater variability is indicated; numbers are p-values for Mann-Whitney test. Bold-faced indicates p-values less than 0.05.

Comparison	Coherent Location	Coherent Year	Incoherent	Total Location	Total Year	Total
Climatic vs Edaphic	**Edaphic 0.012**	Climatic 0.064	Edaphic 0.164	**Edaphic 0.025**	Climate 0.91	Edaphic 0.128
Climatic vs Plant	**Plant 0.0001**	Climatic 0.40	**Plant 0.0001**	**Plant 0.0001**	**Plant 0.001**	**Plant 0.0001**
Climatic vs Animal	**Animal 0.0001**	Animal 0.50	**Animal 0.0001**	**Animal 0.0001**	**Animal 0.0001**	**Animal 0.0001**
Edaphic vs Plant	**Plant 0.0001**	Plant 0.70	**Plant 0.0001**	**Plant 0.0001**	**Plant 0.0012**	**Plant 0.0001**
Edaphic vs Animal	**Animal 0.0001**	**Animal 0.012**	**Animal 0.0001**	**Animal 0.0001**	**Animal 0.0001**	**Animal 0.0001**
Plant vs Animal	Animal 0.056	Plant 0.27	**Animal 0.0001**	**Animal 0.0001**	**Animal 0.0001**	**Animal 0.0001**

For total variance, biological parameters were significantly more variable than nonbiological parameters. Climatic and edaphic parameters did not differ significantly in variability. However, both climatic and edaphic parameters were significantly less variable than either plant or animal data. Plant data were also significantly less variable than animal data. This general pattern also held for the other variability types, except for coherent year variability, where the only significant difference in variability was that animal parameters were more variable than edaphic parameters.

We also tested for differences among coherent year, coherent location, and incoherent variability (Figure 5). For these tests we stratified the data by climatic, edaphic, plant, and animal. The general pattern that emerged was that variability was ranked in the following order:

$$\text{coherent location} > \text{incoherent} > \text{coherent year}$$

This pattern held for edaphic and plant data. For animal data this pattern also held with the exception that there was no difference between coherent location and incoherent variability. Interestingly, for climatic data there was no significant difference among any of these three variability types. Total coherent variability (i.e. the sum of coherent year and coherent location variability) was larger than incoherent variability for climatic, edaphic and plant data, but there was no difference for animal data.

Desert vs Lakes

Deserts are exposed ecosystems with a highly variable, severe environment where variation in precipitation is critical. In contrast, a lake can be considered as a more constant environment, well buffered from thermal change by the mass and heat capacity of water and, depending on the organism, from biological invasions owing to isolation. To evaluate the prediction that deserts are more variable than lakes, we made a detailed comparison of the variabilities exhibited by the Jornada desert and the North Temperate Lakes sites. For these comparisons we stratified the data into the following, non-mutually exclusive groups: all data, abiotic, climatic, edaphic, and animal. Because of the relationship between aggregation level and variability in biotic data, we used species level data only, which meant that no comparisons for plant data could be made. For year-coherent variability, the desert site was significantly more variable than the lake site for all data considered together, climatic data and abiotic data (Table 4).

For location coherent variability, lakes were significantly more variable for abiotic data, but the two sites did not differ with respect to the other data groups. For incoherent variability, the lake site was more variable for animal data, whereas the desert was more variable for abiotic data. Finally, for total variance the desert was significantly more variable for the categories of all data, abiotic data, and climatic data.

Predator vs Prey

How variable in time are prey populations relative to those of their predators? One hypothesis, consistent with many ecological models, is that the species which is smaller-bodied, shorter-lived, and (or) has a faster turnover rate should respond most intensely to environmental variations and therefore be most variable. In contrast, larger-bodied, longer-lived, species with slower turnover rates should persist through short-term environmental fluctuations and therefore be less variable. We tested this hypothesis using annual variances of co-occurring predator-prey pairs.

Table 4. Comparison of variability between Jornada Desert and North Temperate Lake sites. Results of Mann-Whitney U test are shown. Site that is more variable is indicated. NS indicates p-value > 0.05. For significant differences p-value is given.

Variable	Location Coherent	Year Coherent	Incoherent	Total
All Data	NS	Desert 0.0001	Desert 0.0007	Desert 0.0018
Climatic	NS	Desert 0.0012	Desert 0.0009	Desert 0.0009
Edaphic	NS	NS	Desert 0.0075	NS
Animal	NS	NS	Lake 0.0208	NS
Abiotic	Lake 0.0058	Desert 0.0006	Desert 0.0001	Desert 0.0009

All species level data were examined to select predator-prey pairs. We required that (1) neither species be rare, (2) the prey be a major diet item for the predator, and (3) the predator be among the major causes of mortality for the prey. Fourteen pairs of predator and prey met these criteria. The pairs include 10 terrestrial plant-herbivore pairs, and 4 aquatic carnivore-carnivore pairs (Table 5).

Predator-prey pairs were evenly divided with regard to whether prey (7) or predator (7) was smaller and shorter-lived. With regard to which member had highest annual variance, the division was nearly even. In 6 pairs the prey was more variable; in 8 pairs the predator was more variable.

However, cross-classification of the predator-prey pairs showed that the smaller, shorter-lived species had higher annual variance regardless of whether it was predator or prey (Table 5). The null hypothesis that the probability of a species being shorter-lived is independent of its probability of having higher annual variance was tested using Fisher's exact test (Sokal and Rohlf 1981). The p value of 0.05 supports the inference that shorter-lived members of predator-prey pairs tend to have higher annual variance.

Table 5. Analysis of variability of predator vs prey as a function of life span. For each predator-prey pair, organisms with shorter lifespans are underlined and those that exhibited more variability are in bold.

Site	Predator	Prey
Konza	mouse	**<u>annual grass</u>**
Konza	**<u>grasshoppers</u>**	forbs
Konza	<u>insects</u>	grasses
Konza	**<u>mouse</u>**	*Poa*
Konza	**<u>mouse</u>**	C_3 grass
Northern Temperate Lakes	**northern pike**	<u>yellow perch</u>
Northern Temperate Lakes	muskellunge	**<u>yellow perch</u>**
Northern Temperate Lakes	yellow perch	**<u>*Leptodora*</u>**
Northern Temperate Lakes	**yellow perch**	<u>Chaoborus</u>
Jornada	mouse	**<u>annual C_3 forbs</u>**
Jornada	mouse	**<u>annual C_4</u>**
Jornada	<u>ants</u>	**perennial C_4 grass**
Cedar Creek	**<u>mouse</u>**	grasses
Cedar Creek	**<u>mouse</u>**	forbs

DISCUSSION

An important result of this work is the observation that, for biological data, level of aggregation (e.g., species, guild, major group) had greater effect on observed variability than did spatial or temporal extent of the data set. The level of aggregation at which data are collected is a neglected aspect of effects of scale in ecology (Frost et al. 1988; Rahel 1990). The level is often dictated by logistical rather than ecological reasons, even though the choice of aggregation level may be one of the most important determinants of a study's conclusions. For edaphic data, we observed no effect of aggregation on the relative variability. But for biological data the more aggregated the parameter, the lower the variability of the parameter relative to its mean. This difference in response to aggregation between biological and edaphic data could result from compensatory species interactions such as predation and competition. Increases or decreases in one species can be compensated for by the numerical response of another affected species. Such mechanisms, for example, result in density compensation in the abundance of species on islands (Wright 1980) and in lakes (Tonn 1985).

The effect of aggregation on observed variability has an important implication for detection of long-term trends or patterns. To detect trends and patterns it is necessary to monitor parameters that have two, potentially conflicting characteristics. Parameters must be sufficiently sensitive to environmental conditions to indicate changes that occur. But they also must not exhibit so much natural variability as to mask detection of changes in environmental conditions. Biological parameters which have a low degree of aggregation, such as species abundances, are sensitive indicators of environmental change (Schindler 1987), but exhibit so much variability that assessing the cause of change can be difficult. On the other extreme, parameters such as total plant biomass, having a high degree of aggregation, tend to be stable over time or space, and may not be sensitive of subtle environmental changes. Thus, understanding the relative variability and sensitivity of parameters as a function of aggregation level, become important in choosing optimal parameters for a monitoring program.

An important lesson for long-term ecological research is that spatial variability exceeds year-to-year variability. This robust result appears in the dominance of coherent location and incoherent variation over year coherent variation for all data groups except climatic data (Figures 5a and c) and by the dominance of coherent location variability over incoherent variability for edaphic and plant data (Figure 5b). Clearly, a single location within a landscape is insufficient to describe the full range of behaviours of systems within the landscape. One way to circumvent this problem is to assess the degree of variability among locations in a landscape for a limited period of time and then monitor a single location over the long term. This strategy would allow placement of the long-term measurement site within the context of the spatial variability exhibited by sites across the landscape. However, improvement of the long-term sampling is not likely to be this simple because incoherent variation is also large. Incoherent variation includes both error variability and variability owing to year by location interaction. Within our data framework these cannot be separated. If we assume that some percentage of the

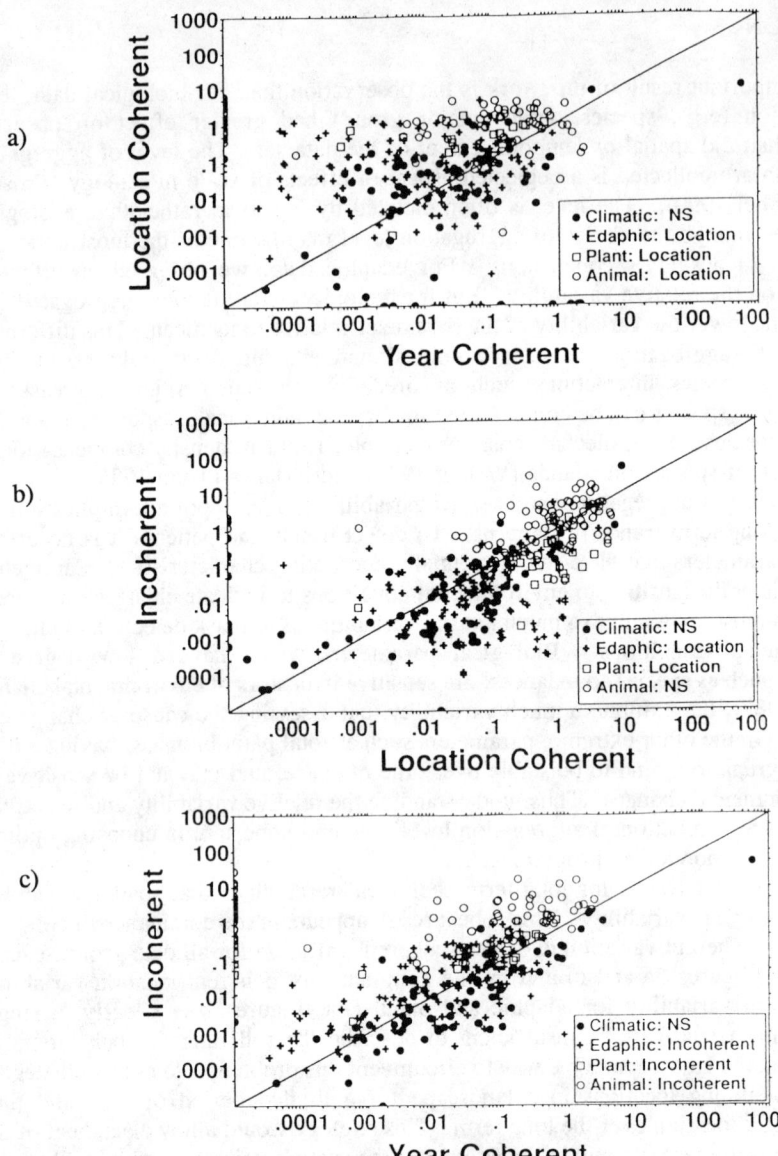

Figure 5. Relationships among year coherent, location coherent, and incoherent variability. Legend for each figure indicates for each variable type (climatic, edaphic, plant, or animal) which type of variance was significantly larger. (NS = no significant difference). All climatic and edaphic data were used, but only species-level biological data were used.

incoherent variation is from interaction, then it will be necessary to gather long-term data on a set of locations within a landscape to understand the dynamics of the spatial distribution of ecosystem properties among years.

An intriguing result of our study is that biotic data exhibit more variability than abiotic data both for animal and plant data in comparison with either climatic or edaphic data. The only exceptions to the twenty-four comparisons made in Table 3 are for the coherent year variation of climate versus plant or animal data, and the coherent year variation of edaphic data versus plant data. Variations in climatic and edaphic properties are magnified by the responses of organism to variation in physical-chemical factors among locations and owing to interaction between year and locations. An alternative explanation that measurement error is greater for biotic than abiotic data could be raised to explain the higher level of incoherent variation, but this would not be consistent with the greater location variation. Thus, at least some of these differences are generated from the behaviour of the ecological systems observed. The ability of organisms to exhibit exponential responses in growth and reproduction to incremental changes in physical and chemical factors is consistent with our result and provides one rationale. Growth and reproduction are often exponentially related to certain environmental variables. Responses to temperature provide a clear example for rate processes with ectothermic living systems (Regier et al. 1990). The vagility of animals and to some extent even plants to move among locations in a set of optimization responses also could contribute to the observation.

Within the abiotic data, the edaphic data are more variable among locations than are climatic data. Apparently, the common weather flowing across each landscape homogenizes some of the potential differences that can develop in microclimate. In contrast, the more stationary soil maintains edaphic differences in microhabitats associated with differences in morphology, hydrology, deposition or erosion, and elevation among locations.

Within the biological data, the animal data are more variable than plant data in terms of incoherent variability. Both explanations we present below rely on the greater vagility of animals compared with plants. The greater variability in animal data versus plant data could result from the greater mobility of animals that can respond quickly to spatial and temporal differences in the environment. This idea is supported by the fact that it is the incoherent variability rather than the coherent variability that is greater for animals than plants. Animals would have a greater possibility of altering their spatial distribution from year to year owing to their mobility than would the more slowly responding plants. Alternatively, we cannot eliminate the possibility that the greater incoherent variability for animals over plants results from greater sampling error with animals than with plants. The vagility of animals and their ability to hide from observers or to avoid and escape sampling devices could produce greater sampling variation than would be expected with plants that are essentially sessile by comparison. Both arguments are rational and we do not know which is more important in determining the greater incoherent variability of animal over plant data.

Comparisons Among Ecosystems

We had initially hoped to compare and contrast the properties of variability among each of the LTER sites included in this paper. We thought that some sites might be more dominated by location coherent variation, others by year coherent variation or by incoherent variation and that these again might differ in interesting ways for climatic, edaphic, plant and animal parameters. Only two sites had a rich enough array of biologic and abiotic parameters at the species level to make such a comparison with a relatively complete design. Fortunately, the two sites, deserts and lakes, differed dramatically enough in setting to make the comparison a challenge. The view that deserts and lakes are like the proverbial "apples and oranges" would discourage most conventional comparisons. Thus, the use of these dimensionless metrics of variation that are neither ecosystem nor parameter dependent were thought to have some promise. The concern that the differences in scale of spatial measurement were sufficient to bias any comparison of variation have been set aside for the one exploratory analysis on deserts versus lakes.

Desert vs Lakes

Deserts are exposed ecosystems with a highly variable, severe environment where year to year variation in precipitation is critical. In contrast, lakes are more constant environments, well buffered from thermal change by the mass and heat capacity of water, from certain chemical changes by bicarbonate buffering of pH, and from biological invasions owing to physical isolation. In terms of spatial connectivity locations along a desert catina appears more open to movement and connected than individual lakes in a lake district which are island-like and isolated from each other by land barriers and can take on their own unique behaviours. From such considerations we hypothesized that deserts would be more variable than lakes among years but that lakes would be more variable than deserts among locations. The locations were those along a catina at the Jornada Desert site and among lakes at the North Temperate Lakes site.

Our data generally supported the hypothesis that deserts are more variable than lakes among years and lakes more variable than deserts among locations (Table 4). The year coherent variability that determines the general conclusion is for the climatic data; thus, as expected, deserts are less buffered from year to year differences in climate than are lakes. The location coherent data that determines the general conclusion are for the abiotic data; thus a suite of isolated lakes differ more in physical-chemical properties than do locations along a catina.

With the exception of animal data, incoherent variability also was greater for deserts than lakes. Thus the interaction between year and location for physical-chemical factors appears to be greater in deserts while the same interaction for animals appears greater for lakes. The alternative explanation that physical-chemical variables are measured with more error in deserts than lakes and animals with more error in lakes could be possible; lakes are mixed by wind and buffered from short term change; thus the measurements could be more integrative for lakes than deserts. We see no inherent reason why sampling error for animals should be greater for lakes than for deserts. One explanation for a large incoherent variation for animals in lakes is that year to year variation in fish recruitment (reproduction) is notoriously high (Wootton 1990) and is not coherent among lakes (Magnuson et al. 1990). For example, yellow perch, *Perca flavescens*, at the North Temperate Lakes Site exhibit strong year classes every few years, but populations are not synchronous in the timing of the strong year classes in the different lakes.

This analysis points out that hypothesized differences between radically different ecosystem types can be tested by comparison of variability metrics. The analysis was consistent and to some extent explainable even though we were concerned by absence of certain types of variables, by poor representation of some variable types, and by differences in the spatial scale of measurements. Our suggestions for further work include using variables that can be measured at the same spatial scales in a wide variety of ecosystems, designing a measurement system that would allow separation of interaction from sampling error included in the incoherent variation and designing studies with a higher overlap in parameter types.

For example, Kratz et al. (1991) compared the variability of limnological parameters across a gradient of lakes ranging from high to low in landscape position. By comparing the same parameters measured at the same spatial and temporal scales they were able to show that lakes higher in the landscape of the Northern Highland Lake District of northern Wisconsin were more variable than those lower in the landscape (Figure 6). By analyzing the variability patterns of over 60 parameters they were also able to infer mechanisms that might lead to such a gradient in variability.

Predator Prey

Despite the small sample size, the analysis of predator-prey variability relationships suggests a significant pattern. There is no evidence that the relative variability of predator-prey pairs can be explained by trophic position. Rather, the smaller, shorter-lived species tends to have higher interannual variance, regardless of whether it is predator or prey.

The "shorter-lived, higher-variance" rule merits further investigation using a larger data set that includes a wider range of trophic levels from both terrestrial and aquatic systems. Several intriguing questions remain unanswered. How general is the rule? In view of the fact that estimates of population variance increase with the duration of the data set (Pimm and Redfearn 1988), will the rule hold at even longer time scales than those examined here? If the rule is general, then what causes exceptions? Are exceptions to the rule indicative of strong regulation of one population by the other? Do exceptions occur at regular levels of the trophic hierarchy, indicating alternate control by competition and predation (Hairston et al. 1960; Oksanen et al. 1981; Persson et al. 1988)?

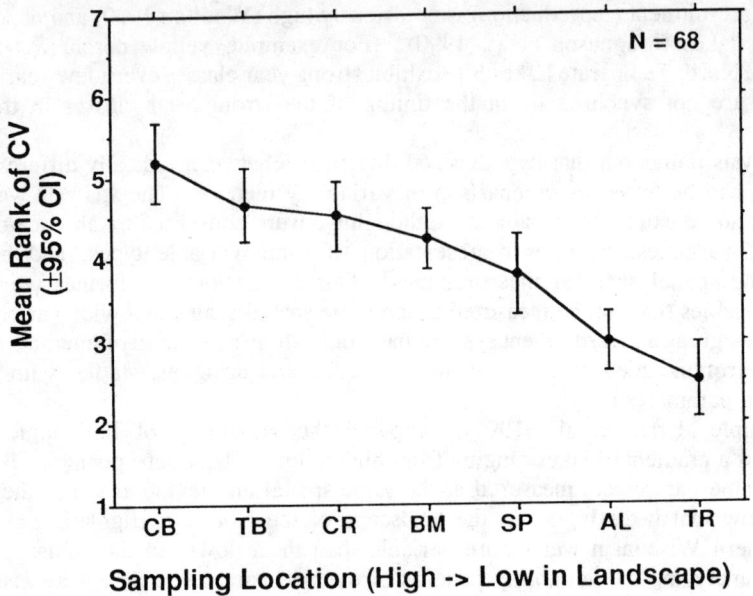

Figure 6. Average ranks of coefficients of variation (CV) as a function of landscape position for seven lakes in the Northern Higland Lake District of northern Wisconsin (adapted from Kratz et al. 1991).

IMPLICATIONS

Comparative studies in ecology vary in scale from how a single organism lives under a variety of different environments to how abstract, system-level properties vary in different system types. This study is an example of the latter scale. We found that there were few precedents to act as guideposts in conducting comparative ecology at such a general scale, yet we believe that powerful understanding can derive from comparison at the most general levels. Here we list what we consider to be some of the broader implications of our study:

1) The wide range of ecological systems on earth should not act as a deterrent to cross-system comparisons. However, the greater the variety of systems

considered, the more general the metric of comparison must be. In our study, for example, there was no single parameter, even for climate, that was measured identically at all 12 sites, so it was not possible to compare the behaviour of any single parameter across the sites. However, by grouping parameters into classes, we were able to make comparisons at a more general level. Comparative ecology requires standardized parameters, but these parameters may be at more general levels than the original data. The emphasis should be on comparable measurements in a broad sense, not simply comparable methods.

2) Even though common sets of parameters are not necessarily a prerequisite for useful cross-system comparisons, it is important to try to make measurements at similar temporal, spatial, and aggregation scales. We were able to show a strong dependence of variability on the level of aggregation of the data. However, we were limited in our study by the lack of control over spatial scale of measurement. One way to enhance control over spatial scale for some parameters would be to use remotely sensed data which are taken at clearly defined spatial resolutions and extents. Clearly, the limitations that scale dependency puts on comparative studies is not a function of the diverse array of systems, but of how they are studied.

3) The richest comparisons are made when sites to be compared have multi-level hierarchical data. In our study, data on all four major groups of parameters (climatic, edaphic, plant, and animal) were available to us only for two sites, the Jornada desert and Northern Temperate Lakes. Even then, species level plant data could not be compared between lakes and deserts because the lake site had not enumerated phytoplankton at the species level.

4) Long-term research must deal explicitly with spatial heterogeneity of ecosystems. Location coherent variation and incoherent variation is, in general, large relative to temporally coherent variation. Long-term, landscape-level studies must fully incorporate this strong spatial variation and the interaction between spatial and temporal variation in order to understand ecological processes operating at these larger spatial and temporal scales.

5) Finally, we found variability to be an interesting and informative system property that could be used to make meaningful comparisons across a wide range of ecosystems. It is likely that there are other relatively simple, but general, properties of systems that also could be used for comparisons. Some of broadest generalizations in ecology will come from including in our comparisons the full range of systems available, rather than just lakes or just deserts, or just animals or just plants, for examples.

ACKNOWLEDGMENTS

This paper evolved from discussions at the LTER intersite workshop on Variability in North American Ecosystems held at the Trout Lake Station and we acknowledge the use of the station's facilities. We thank the LTER coordinating committee for partially supporting the workshop through NSF grant #BSR8996172 A02. We thank all those who helped collect and collate the original data on which this paper is based. There are too many to list independently. Partial support for the project was also provided by the Northern Temperate Lakes LTER through NSF grant # BSR 8514330, and 9011660.

REFERENCES

Allen T.F.H., Starr T.B. (1982) Hierarchy: Perspectives from Ecological Complexity. University of Chicago Press, Chicago 310 pp.

Frost T. M., Carpenter S. R., Kratz T. K. (1992). Choosing ecological indicators: effects of taxonomic aggregation on sensitivity to stress and natural variability. In McKenzie D. H., Hyatt D. E., McDonald V. J. (eds.) Ecological Indicators (volume 1), Elsevier Applied Science Publishers, Essex, England pp:215-227

Frost T. M., DeAngelis D. L., Bartell S. M, Hall D. J, Hurlbert S. H. (1988) Scale in the design and interpretation of aquatic community research. In Carpenter S. R. (ed.) Complex Interactions in Lake Communities. Springer-Verlag, New York, pp:229-258

Hairston N.G.,Smith F.E., Slobodkin L.B. (1960) Community structure, population control, and competition. Am. Nat. 94: 421-425

Kratz T. K., Benson B. J., Blood E., Cunningham G. L., Dahlgren R.A. (1991) The influence of landscape position on temporal variability in four North American ecosystems. American Naturalist 138:355-378

Kratz T.K., Frost T.M., Magnuson J.J. (1987) Inferences from spatial and temporal variability in ecosystems: analyses of long-term zooplankton data from a set of lakes. American Naturalist 129:830-846

Lewis W. M. Jr. (1978) Comparison of temporal and spatial variation in the zooplankton of a lake by means of variance components. Ecology 59:666-671

Magnuson J. J., Benson B. J., Kratz T. K. (1990) Temporal coherence in the limnology of a suite of lakes in Wisconsin, U.S.A. Freshwater Biology 23:145-159

Magnuson J.J., Kratz T.K., Frost T.M., Benson B.J., Nero R., Bowser C.J. (1991) Expanding the temporal and spatial scales of ecological research and comparison of divergent ecosystems: roles for LTER in the United States. In Risser P. G. (ed.) Long-term Ecological Research. Wiley & Sons pp:45-70

Matthews W. J. (1990) Spatial and temporal variation in fishes of riffle habitats: a comparison of analytical approaches for the Roanoke River. American Midland Naturalist 124:31-45

O'Neill R.V., DeAngelis D.L., Waide J.B., Allen. T.F.H. (1986) A Hierarchical Concept of Ecosystems. Princeton Univ. Press, Princeton, NJ

Oksanen L., Fretwell S.D., Arruda J., Niemela P.(1981) Exploitation ecosystems in gradients of primary productivity. Am. Nat. 118: 424-444

Persson L., Andersson G., Hamrin S.F, Johansson L. (1988) Predator regulation and primary production along the productivity gradient of temperate lake ecosystems. In Carpenter S.R. (ed.) Complex Interactions in Lake Communities. Springer-Verlag, NY. pp:45-65

Pimm S.L., Redfearn A. (1988) The variability of population densities. Nature 334: 613-614

Platt T., Filion C. (1973) Spatial variability of the productivity:biomass ratio for phytoplankton in a small marine basin. Limnology and Oceanography 18:743-749

Rahel F. J. (1990) The hierarchial nature of community persistence: a problem of scale. The American Naturalist. 136 (3): 328-344

Regier H. A., Holmes J. A. (1990) Influence of temperature changes on aquatic ecosystems: an interpretation of empirical data. Trans. Amer. Fisheries Soc. 119:374-389

Rothschild B. J., DiNardo G. T. (1987) Comparison of recruitment variability and life history data among marine and anadromous fishes. American Fisheries Society Symposium 1:531-546

Schindler D. W. (1987) Detecting ecosystem responses to anthropogenic stress. Canadian Journal of Fisheries and Aquatic Sciences, 44 (Supp. 1):6-25

Sokal R.R.,Rohlf F.J. (1981) Biometry. Second ed. Freeman, San Francisco.

Tonn W. E. (1985) Density compensation in the Umbra-Perca fish assemblages of northern Wisconsin lakes. Ecology: 66(2):415-429

Turner M. G. (1989) Landscape ecology: the effect of pattern on process. Annual Review of Ecology and Systematics 20:171-197

Van Cleve K., Martin S. (1991) Long-term ecological research in the United States. LTER Publication No. 11., Long-Term Ecological Research Network Office, University of Washington, Seattle, Washington

Wiens J. A. (1989) Spatial scaling in ecology. Functional Ecology 1989 3:385-397

Wootton R. J. (1990) Ecology of teleost fishes. Chapman and Hall, New York, xii + 404 pp.

Wood E. F., Sivapalan M., Beven K. (1990) Similarity and scale in catchment storm response. Reviews of Geophysics 28:1-18

Wright S. J. 1980. Density compensation in island avifaunas. Oecologia (Berlin) 45:385-389

20. ASSESSMENT AND MONITORING OF LARGE MARINE ECOSYSTEMS

Kenneth Sherman and Donna A. Busch
National Marine Fisheries Service, Narragansett Laboratory
Narragansett, Rhode Island 02882-1199
USA

INTRODUCTION

A significant milestone in marine resource development was achieved in July 1992 with the adoption by a majority of coastal countries of follow-on actions to the United Nations' Conference on Environment and Development (UNCED). The UNCED declarations on the ocean explicitly recommended that nations of the globe: (1) prevent, reduce, and control degradation of the marine environment so as to maintain and improve its life-support and productive capacities; (2) develop and increase the potential of marine living resources to meet human nutritional needs, as well as social, economic, and development goals; and (3) promote the integrated management and sustainable development of coastal areas and the marine environment. UNCED also recognized the general importance of capacity building, as well as the important linkage between monitoring and the achievement of marine resource development goals.

Achievement of UNCED recommendations will require the implementation of a new paradigm in ocean monitoring and management that can overcome traditional geopolitical and interdisciplinary sectorization. Such an approach should be based on principles of ecology and sustainable development. The large marine ecosystem (LME) (Sherman and Alexander, 1986) concept provides the framework for achievement of UNCED commitments. LMEs are areas which are being subjected to increasing stress from growing exploitation of fish and other renewable resources, coastal zone damage, river basin runoff, dumping of urban wastes, and fallout from aerosol contaminants. The LMEs are regions of ocean space encompassing near-coastal areas from river basins and estuaries on out to the seaward boundary of continental shelves and the seaward margins of coastal current systems. They are relatively large regions on the order of 200,000 km^2 or larger, characterized by distinct bathymetry, hydrography, productivity, and trophically dependent populations. The theory, measurement, and modelling relevant to monitoring the changing states of LMEs are imbedded in reports on multistable ecosystems, and on the pattern formation and spatial diffusion within ecosystems.

Because LMEs usually subsume the coastal waters of more than one state, coordination between those states in monitoring and resource management is desirable. At present, however, no single international organization is authorized to work with coastal states to monitor the changing ecological state of LMEs and to reconcile the needs of individual nations where mitigation actions are necessary to reverse the deleterious impacts of stress on productivity and biomass yields. The need for a regional approach for implementation of marine research, monitoring, and stress mitigation has been recognized. Of course, coastal states maintain ultimate responsibility for their territorial sea and exclusive economic zone resources. Recognition of the LME concept engenders no legal commitment by coastal states to share their resources, although it is in their enlightened self interest to do so.

It is within the nearshore coastal domains of the LMEs that the human-induced stress on ecosystems requires mitigating actions to ensure the continued productivity and economic viability of marine resources. Although the management of LMEs is an evolving scientific and geopolitical process, sufficient progress has been made to allow for useful comparisons among the primary, secondary, and tertiary driving forces influencing large-scale changes in the biomass yields and long-term sustainability of LMEs. Results from a series of LME studies are presented in this report. They depict a linkage between the efforts of Intergovernmental Oceanographic Commission (IOC) to initiate a Global Ocean Observing System (GOOS) that includes modules for monitoring changes in ocean health and living resources. The report is intended to encourage dialogue and debate on strategies for linking scientific and societal interests. It is aimed at assisting in the short-term and long-term development and sustainability of coastal ocean ecosystems in the post-UNCED decade of the 1990s, bearing in mind the need to meet the objectives of Agenda 21 aimed at reducing the degradation of marine ecosystems and promoting their integrated management and sustainable development.

UTILITY OF COASTAL ECOSYSTEM ASSESSMENT AND MONITORING

Human intervention and climate change are sources of increasing variability in the natural productivity of coastal marine ecosystems. Within the near shore areas and extending seaward around the margins of the global land masses, coastal ecosystems are being subjected to increased stress from toxic effluents, habitat degradation, excessive nutrient loadings, fallout from aerosol contaminants, and episodic losses of living marine resources from pollution effects, and overexploitation. The growing awareness that the quality of the global coastal ecosystems are being adversely impacted by multiple driving forces has accelerated efforts to assess, monitor, and mitigate coastal stressors from an ecosystem perspective. The IOC of the United Nations Educational, Scientific, and Cultural Organization (UNESCO) is encouraging coastal nations to establish national programs for assessing and monitoring coastal ecosystems so as to enhance the ability of national and regional management organizations to develop and implement effective remedial programs for improving the quality of degraded ecosystems (IOC 1992a).

For purposes of this report, marine ecosystem monitoring is defined as a component of a management system that includes: (1) regulatory, (2) institutional, and (3) decision-making aspects relating to marine ecosystems, and therefore, would include a range of activities needed to provide management information about ecosystem conditions, contaminants, and resources at risk. Based on successful experiences in North America, Europe, and elsewhere, the core component of a comprehensive ecosystem monitoring system that consists of conceptual and numerical modelling capability, laboratory and field research, time-series measurements, data analysis, synthesis and interpretation, and a capacity for initiating the effort with preliminary or scoping studies is most likely to be successful. The principal characteristic of a comprehensive ecosystem monitoring program is the integration and coordination of the component parts of the effort into a total ecosystems approach designed to produce scientific information in support of coastal resources management.

We provide information and guidelines that focus on an integrated regional and global approach to coastal marine ecosystem assessment and monitoring. The strategy is consistent with the conclusion that monitoring efforts at the regional scale need to be strengthened to improve understanding of broader-scale trends in marine ecosystem quality. In addressing sampling design alternatives for a regional strategy for coastal ecosystem monitoring, we have reviewed options, with respect to: (1) an independent, fixed-station monitoring effort, or (2) an integrated network of monitoring stations based on a statistical design on a regional scale. A regional network that links both strategies and includes areas of special attention is preferred. Among the options to be employed is the use of stratified sampling for long-term trend assessments of ecosystem parameters using standard and intercalibrated protocols in areas with different levels of ecosystem stress augmented by high-intensity sampling in areas at high-risk from environmental degradation.

We have drawn from several recent reports in the preparation of this document, including the United Nations' report on the status of the global marine environment (GESAMP, 1990), the IOC's report on the GOOS presented to the UNCED in 1992 (IOC 1992a), the reports of several international commissions, including the Helsinki Commission (HELCOM), the Oslo-Paris Commission (OSPARCOM), the North Sea Task Force (NSTF 1991), and the report of the International Council for the Exploration of the Sea (ICES) Working Group on Environmental Assessments and Monitoring Strategies (WGEAMS 1992).

COASTAL ECOSYSTEM COMPONENT OF THE GLOBAL OCEAN OBSERVING SYSTEM

The IOC, based on guidance from member nations, is encouraging the development of a comprehensive GOOS to provide information needed for oceanic and atmospheric forecasting for ocean management by coastal nations and for the needs of global environmental change research, and related education, training, and technical assistance programs to ensure that all countries can participate and benefit from the effort (UNESCO 1992). The coastal ecosystem component of the GOOS is being planned to provide a basis for the assessment, monitoring, and mitigation of the present ecological conditions in the coastal areas of nations already experiencing significant economic losses of resource sustainability from degraded water quality, contaminated fisheries resources, and loss of important habitat (e.g. coral reefs, mangrove lagoons, estuaries, and embayments). Timely and effective responses to these changes in ecosystem conditions will depend in part on the quality and availability of information on the rate and magnitude of these changes at the regional ecosystem level.

The strategic design of a coastal ecosystem component for the GOOS should include pertinent existing research and monitoring investigations that measure-up to program criteria that will constitute a "core" of the program. Included in the "core" will be selected components of existing national networks of laboratories and vessels of marine resource ministries that are already engaged in coastal ecosystems monitoring, assessment, mitigation, and information transfer operations relating to marine environmental quality, living marine resources, and ecosystem health.

Mitigating actions to reduce stress on marine ecosystems are required to ensure the long-term sustainability of marine resources. The principles adopted by coastal states under the terms of the United Nations Convention for the Law of the Sea (UNCLOS) have been interpreted as supportive of the management of living marine resources and coastal habitats from an ecosystems perspective (Belsky 1986, 1989). However, at present no single international institution has been empowered to monitor the changing ecological states of marine ecosystems and to reconcile the needs of individual nations with those of the community of nations in taking appropriate mitigation actions (Myers 1990). In this regard, the need for a regional approach to implement research, monitoring, and stress mitigation in support of marine resources development and sustainability at less than the global level has been recognized from a strategic perspective (Taylor and Groom 1989; Malone 1991).

From the ecological perspective, the concept that critical processes controlling the structure and function of biological communities can best be addressed on a regional basis (Ricklefs 1987) has been applied to ocean space in the utilization of marine ecosystems as distinct global units for marine research, monitoring, and management. The concept of monitoring and managing renewable resources from a regional ecosystem perspective has been the topic of a series of symposia and workshops initiated in 1984 and continuing through 1992, wherein the geographic extent of each region is defined on the basis of ecological criteria (Table 1). As the regional units under consideration are large, the term Large Marine Ecosystem (LME) is used to characterize them. Several occupy semi-enclosed seas, such as the Black Sea, the Mediterranean Sea, and the Caribbean Sea. Some of these can be divided into domains, or subsystems - for example the Adriatic Sea, a subsystem of the Mediterranean Sea LME. In other LMEs geographic limits are defined by the scope of continental margins. Among these are the U.S. Northeast Continental Shelf, the East Greenland Sea, the Northwestern Australian Shelf. The seaward limit of the LMEs extends beyond the physical outer limits of the shelves, themselves, to include all or a portion of the continental slopes. Care was taken to limit the seaward boundaries to the areas affected by ocean currents, rather than relying simply on the 200-mile Exclusive Economic Zone (EEZ) or fisheries zone limits. Among the ocean current LMEs are the Humboldt Current, Canary Current, and Kuroshio Current.

It is the coastal ecosystems adjacent to the land masses that are being stressed from habitat degradation, pollution, and overexploitation of marine resources. Nearly 95% of the usable annual global biomass yield of fish and other living marine resources is produced in 49 LMEs within, and adjacent to, the boundaries of the EEZs of coastal nations located around the margins of the ocean basins, where levels of primary production are persistently higher than for the open-ocean pelagic areas of the globe (Figure 1).

Pollution at the continental margins of marine ecosystems can impact on natural productivity cycles, including eutrophication from high nitrogen and phosphorus effluent from estuaries. The presence of toxins in poorly treated sewage discharge, and loss of wetland nursery areas to coastal development are also ecosystem-level problems that need to be addressed (GESAMP 1990). Overfishing has caused biomass flips among the dominant pelagic components of fish communities resulting in multimillion metric ton losses in potential biomass yield (Fogarty et al. 1991). The biomass flip, wherein a dominant species rapidly drops to a low level to be succeeded by another species, can generate cascading effects among other important components of the ecosystem, including marine birds (Powers and Brown 1987), marine mammals, and zooplankton (Overholtz and Nicolas 1979; Payne et al. 1990). Recent studies implicate climate and natural environmental changes as prime driving forces of variability in fish population levels (Kawasaki et al. 1991; Bakun 1993; Alheit and Bernal 1993). The growing awareness that biomass yields are being influenced by multiple driving forces in marine ecosystems around the globe has accelerated efforts to broaden research strategies to encompass food chain dynamics and the effects of environmental perturbations and pollution on living marine resources from an ecosystem perspective.

MONITORING STRATEGY

During the twenty-fifth session of the IOC Executive Council meeting in Paris, an overview of the LME Concept was presented by the National Oceanic and Atmospheric Administration (NOAA) as a contribution to the discussion of candidate strategies for assessing and monitoring the changing states, or "health," of coastal ecosystems (IOC 1992b). This document expands on the earlier presentation. It is focused on the LMEs around the margins of the landmasses where the pressures of overexploitation, pollution, and habitat degradation from a growing global population are stressing marine resources. Strategies are outlined in this report for monitoring the changing-states of marine ecosystems using methods that are designed to provide ecosystem-level measurements contributing to the coastal ecosystem component of the GOOS. Included in the "core" monitoring strategy is the use of the Continuous Plankton Recorder (CPR) for plankton and water quality assessment, bottom trawling for measuring changes in the fish community, and environmental/pollution assessments. These strategies can be used to measure the changing ecological states of LMEs as described in an earlier IOC report (UNESCO 1992).

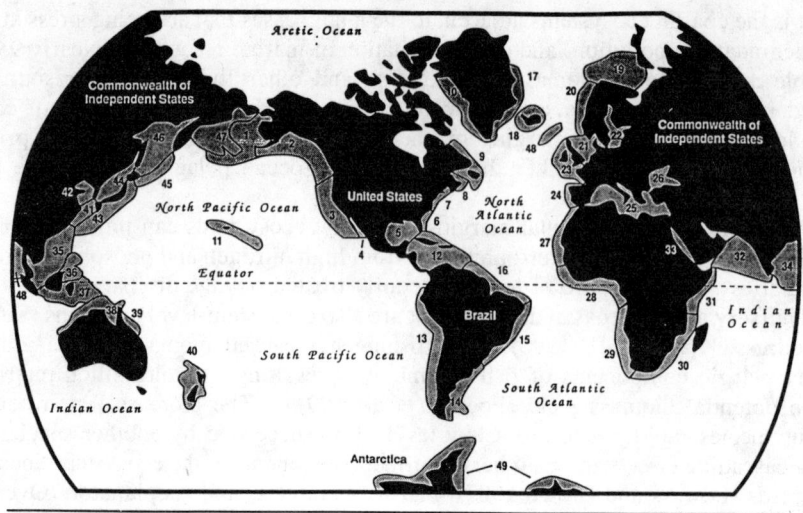

WORLD MAP OF LARGE MARINE ECOSYSTEMS

1. Eastern Bering Sea
2. Gulf of Alaska
3. California Current
4. Gulf of California
5. Gulf of Mexico
6. Southeast U.S. Continental Shelf
7. Northeast U.S. Continental Shelf
8. Scotian Shelf
9. Newfoundland Shelf
10. West Greenland Shelf
11. Insular Pacific--Hawaiian
12. Caribbean Sea
13. Humboldt Current
14. Patagonian Shelf
15. Brazil Current
16. Northeast Brazil Shelf
17. East Greenland Shelf
18. Iceland Shelf
19. Barents Sea
20. Norwegian Shelf
21. North Sea
22. Baltic Sea
23. Celtic-Biscay Shelf
24. Iberian Coastal
25. Mediterranean Sea
26. Black Sea
27. Canary Current
28. Guinea Current
29. Benguela Current
30. Agulhas Current
31. Somali Coastal Current
32. Arabian Sea
33. Red Sea
34. Bay of Bengal
35. South China Sea
36. Sulu-Celebes Seas
37. Indonesian Seas
38. Northern Australian Shelf
39. Great Barrier Reef
40. New Zealand Shelf
41. East China Sea
42. Yellow Sea
43. Kuroshio Current
44. Sea of Japan
45. Oyashio Current
46. Sea of Okhotsk
47. West Bering Sea
48. Faroe Plateau
49. Antarctic

Figure 1. Boundaries of 49 large marine ecosystems.

The scientific "hallmark" of the monitoring strategy recommended for the coastal ecosystem component of GOOS is a holistic approach. The areas to be monitored include river drainage basins, estuaries, bays and coastal waters out to the seaward boundary of the ecosystem. The design of the sampling strategy will be developed for the distinctive characteristics of each LME. The strategy is to assess and monitor the changing states of LMEs at all relevant scales of interest contained within their boundaries. The program strategy should employ a hierarchical ecosystem sampling design within each LME, placing emphasis on nearshore areas under greatest stress from contaminants, eutrophication and degraded living marine resources, and relatively less emphasis on areas of low risk. This monitoring system will provide the scientific basis for remedial action at the appropriate scale in relation to environmental stress. The challenge for determining the cause and effect of natural and human induced processes of ecosystem level variability is significant. For example, within the 258,000 km^2 Northeast U.S. Continental Shelf Ecosystem, the impact of pollution from 54 million people generating about 7 million gallons of wastewater per day, and contaminants from atmospheric deposition, and 478,000 km^2 of watersheds, will need to be quantified to determine the effects of these stressors on the billion dollar/year living resource industries so as to provide a firm basis for exercising appropriate mitigation actions. At finer scales of resolution within the LME, the impacts will need to be quantified as relevant; for example, determining the water quality parameters of a specific bay or harbor, or quantifying effluent discharges from a particular watershed of interest to local officials.

Consideration should be given to the use of standard and intercalibrated protocols in areas with levels of effort focused on areas of the LME under the greatest environmental and biological stress including the watersheds, bays and estuaries, and coastal water of LMEs. Long-term historical time series data on living marine resources (some up to forty years), coupled with measured or inferred long-term pollutant loading histories, have proven useful for relating the results of intensive monitoring to the quantification of "cause and effect" mechanisms based on first principles of aquatic toxicological and ecological theory on the changing states of health and environmental stress on LMEs (Sherman et al. 1993). Maximum use of historical long-term time series data, using cutting-edge analytical techniques, supplemented with intensive monitoring, is an important component in the implementation of coastal ecosystems monitoring. Numerical models of environmental impacts need to be developed and applied in a manner that will link information on natural characteristics of coastal ecosystems and the impacts of human-induced changes to the decision-making process of resource managers. The models will be used to assist managers and scientists in designing efficient monitoring strategies. Statistical models should be utilized to evaluate the most efficient use of sampling and analytical resources.

Temporal and spatial scales influencing biological production and changing ecological states in marine ecosystems have been the topic of a number of theoretical and empirical studies. The selection of scale in any study is related to the processes under investigation. An excellent treatment of this topic can be found in Steele (1988). He indicates that in relation to general ecology of the sea, the best known work in marine population dynamics includes studies by Schaefer (1954), and Beverton and Holt (1957), following the earlier pioneering approach of Lindemann (1942). A heuristic projection was produced by Steele (1988) to illustrate scales of importance in monitoring pelagic components of the ecosystem including phytoplankton, zooplankton, fish, frontal processes, and short-term but large-area episodic effects (Figure 2). However, as noted by Steele (1988), this array of models is unsuitable for consideration of temporal or spatial variability in the ocean. The LME approach defines a spatial domain based on ecological principles and, thereby, provides a basis for focused temporal and spatial scientific research and monitoring efforts in support of management aimed at the long-term productivity and sustainability of marine habitats and resources. The theory and modelling relevant to measuring the changing states of LMEs are imbedded in contemporary studies of: (1) multistable ecosystems (Holling 1973, 1986; Pimm 1984; Beddington 1986), and (2) pattern formation and spatial diffusion in ecosystems (Levin 1978, 1990, 1993).

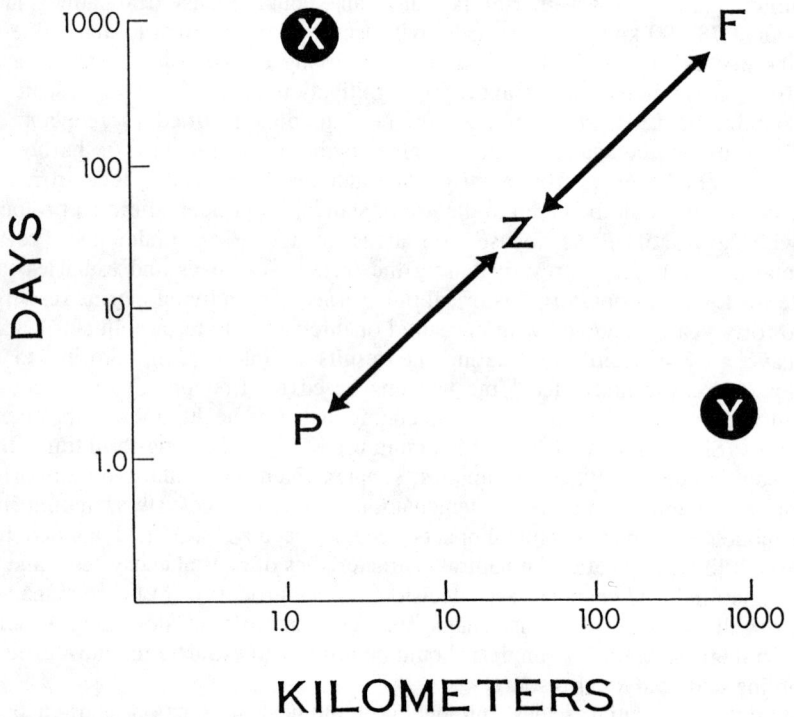

Figure 2. A simple set of scale relations for the food web P (phytoplankton), Z (zooplankton) and F (pelagic fish). Two physical processes are indicated by X, predictable fronts with small cross-front dimensions and (Y) unpredictable weather events occurring on relatively large scales (Steele 1988).

PERTURBATIONS AND DRIVING FORCES IN LMEs

Among the marine resources at risk from global climate change and human intervention of natural productivity cycles are the fish components of coastal LMEs. Increasing attention has been focused over the past few years on synthesizing available biological and environmental information influencing the natural productivity of the fishery biomass within LMEs in an effort to identify the primary, secondary, and where important, the tertiary driving forces causing major shifts in species composition (Table 1).

For nearly 75 years since the turn of the century, biological oceanographers did not achieve any great success in predicting fish yield based on food chain studies. As a result, through the mid 1970s, the predictions of the levels of biomass yields for different regions of the world ocean were open to disagreement (Ryther 1969; Alverson et al. 1970; Lasker 1988). It is clear that "experts" have been off the mark in earlier estimates of global yield of fisheries biomass. Projections given in *The Global 2000 Report* (U. S. Council on Environmental Quality 1980) indicated that the world annual yield was expected to rise little, if at all, by the year 2000 from the 60 million metric tons (mmt) reached in the 1970s. In contrast, estimates given in *The Resourceful Earth* (Wise 1984) argued for an annual yield of 100-120 mmt by the year 2000. The trend is upward; the 1989 level of marine global fishery yields reached 86.5 mmt (FAO 1992). The lack of a clear definition of actual and/or potential global yield is not unexpected, given the limited efforts presently underway to improve the global information base on living marine resource yields. A milestone in fishery science was achieved in 1975 with the convening of a symposium by the International Council for the Exploration of the Sea that focused on changes in the fish stocks of the North Sea and their causes. The symposium, which dealt with the North Sea as an ecosystem, following the lead of Steele (1974), Cushing (1975), Andersen and Ursin (1977), and others, was prompted by a rather dramatic shift in the finfish community of the North Sea from a balance between pelagic and demersal finfish prior to 1960 to demersal domination from the mid-1960s through the mid-1970s. Although no consensus on cause and effect was reached by the participants, it was suggested by the convener (Hempel 1978) that the previous studies of seven-and-a-half decades may have been too narrowly focused, and that future studies should take into consideration fish stocks, their competitors, predators and prey, and interactions of the fish stocks with their environments, the fisheries, habitat change, and pollution from an ecosystems perspective.

The LMEs that together produce approximately 95% of the annual global fisheries biomass yield are listed in Table 2. Although the United Nations Food and Agriculture Organization (FAO) world fishery statistics have shown an upward trend in annual biomass yields for the past three decades, it is largely the clupeoids that are increasing in abundance (FAO 1992). A large number of stocks have been and continue to be fished at levels above long-term sustainability. The variations in abundance levels among the species constituting the annual global biomass yields are indicative of changing regional ecosystem states caused by natural environmental perturbations, overexploitation, and pollution. Although the spatial dimensions of LMEs preclude a strictly controlled experimental approach to their study, they are amenable to the comparative method of science as described by Bakun (1993). Since 1984, 30 case studies investigating the major causes of large-scale perturbations in biomass yields of LMEs have been completed (Table 1). The principal driving forces for biomass changes vary among ecosystems.

Changes in the ocean climate of the northern North Atlantic during the late 1960s and early 1970s have been considered by some marine scientists as the dominant cause of change in the food chain structure and biomass yields of at least three northern North Atlantic LMEs, including the Norwegian Sea, Barents Sea, and West Greenland Sea ecosystems where large-scale declines in the population levels of important fish stocks (e.g. capelin, cod) have been observed.

In the West Greenland Sea Ecosystem, cod stocks were displaced southward since 1980, attended by a decrease in their average size and abundance. Biomass yields declined from about 300,000 metric tons (mt) per year in the mid-1960s to less than 15,000 mt in 1985. Both changes appear to have been due to short-term cooling that influenced stability of water masses, dynamics of the plankton community, and adversely affected the growth and survival of early developmental stages of cod resulting in a reduction in recruitment. Observations since the 1920s have shown that catches of cod are correlated with temperature - increasing during warm periods and declining during cool periods. The effects of fishing mortality on the decline of the cod are secondary to the major influence of climatic conditions over the North Atlantic (Hovgaard and Buch 1990; Blindheim and Skjoldal 1993).

Changes in the temperature structure of the Norwegian Sea Ecosystem also appear to be the major driving force controlling the recruitment of the commercially important cod stocks. Strong or medium production of cod biomass is related to warmer temperatures. The conditions for growth and survival of early developmental stages of cod are enhanced during warmer years when the larval cod are maintained for longer periods within coastal nursery grounds, where their most important prey organism, the copepod, *Calanus finmarchicus*, swarms in high densities under conditions of well-defined thermocline structure and consequently optimal feeding conditions of abundant phytoplankton.

Primarily, changes in hydrographic conditions, and secondarily, excessive fishing mortality, appear to have caused changes in the biomass yields of the Barents Sea Ecosystem. The average annual biomass yield (fish, crustaceans, molluscs, algae) of the ecosystem in the 1970s was about two million metric tons. However, by the 1980s annual yields declined to approximately 350,000 mt. Reduced flow of warm Atlantic water into the Barents Sea Ecosystem, coupled with excessive levels of fishing effort led to: (1) the collapse of the major fisheries of the region (cod, capelin, haddock, herring, redfish, shrimp); (2) subsequent disruption in the structure of the food chain; and (3) an increase in the abundance levels of the shrimp-like euphausiids representing a significant amount of biomass that is underutilized in relation to the potential sustained yield of this ecosystem. Given the depressed state of the fish stocks, any restoration management would need to consider significant reduction in fishing effort by fisherman of Norway and

Table 1. List of 30 Large Marine Ecosystems and sub-systems for which syntheses relating to principal, secondary, or tertiary driving forces controlling variability in biomass yields have been completed by February 1993.

Large Marine Ecosystem	Volume No.*	Authors
U.S. Northeast Continental Shelf	1	M. Sissenwine
	4	P. Falkowski
U.S. Southeast Continental Shelf	4	J. Yoder
Gulf of Mexico	2	W. J. Richards and M. F. McGowan
	4	B. E. Brown et al.
California Current	1	A. MacCall
	4	M. Mullin
	5	D. Bottom
Eastern Bering Shelf	1	L. Incze and J. D. Schumacher
West Greenland Shelf	3	H. Hovgaard and E. Buch
	5	J. Blindheim and H. R. Skjoldal
Norwegian Sea	3	B. Ellertsen et al.
	5	J. Blindheim and H. R. Skjoldal
Barents Sea	2	H. R. Skjoldal and F. Rey
	5	J. Blindheim and H. R. Skjoldal
	4	V. Borisov
North Sea	1	N. Daan
Baltic Sea	1	G. Kullenberg
Iberian Coastal	2	T. Wyatt and G. Perez-Gandaras
Mediterranean-Adriatic Sea	5	G. Bombace
Mediterranean-Black Sea	5	J. F. Caddy
Canary Current	5	C. Bas
Gulf of Guinea	5	D. Binet and E. Marchal
Benguela Current	2	R.J.M. Crawford et al.
	5	A. Bakun
Patagonian Shelf	5	A. Bakun
Caribbean Sea	3	W. J. Richards and J. A. Bohnsack
South China Sea-Gulf of Thailand	2	T. Piyakarnchana
Yellow Sea	2&5	Q. Tang
Sea of Okhotsk	5	V. V. Kusnetsov
Humboldt Current	5	J. Alheit and P. Bernal
Indonesia Seas-Banda Sea	3	J. J. Zijlstra and M. A. Baars
Bay of Bengal	5	S. N. Dwivedi
Antarctic Marine	1&5	R. T. Scully et al.
Weddell Sea	3	G. Hempel
Kuroshio Current	2	M. Terazaki
Oyashio Current	2	T. Minoda
Great Barrier Reef	2	R. H. Bradbury and C. N. Mundy
	5	G. Kelleher
South China Sea	5	D. Pauly and V. Christensen

* AAAS Selected Symposia (**Vol. 1**: 1986; **Vol. 2**: 1989; **Vol. 3**: 1990; **Vol. 4**: 1991; **Vol. 5**: 1993). See Reference list for further details on Volumes.

the former USSR, the coastal nations that share the resources of the Barents Sea Ecosystem (Skjoldal and Rey 1989; Borisov 1991). For further discussion of the dynamics and interrelationships among the Barents Sea, Norwegian Sea, and West Greenland Sea LMEs, see Blindheim and Skjoldal (1993).

In the North Sea Ecosystem, important species "flipped" from a dominant to a subordinate position over the decade of the 1960s. The finfish stocks of the North Sea Ecosystem have been subjected to intensive fishing mortality, resulting in a decrease in pelagic herring and mackerel yields from 5 mmt to 1.7 mmt, whereas small, fast-growing and commercially less desireable sand lance (also called sand eel), Norway pout, and sprat yields increased by 1.5 mmt along with an approximate 36% increase in gadoid yields. The causes for the biomass flips are not well understood. Several arguments suggest that the "flip" is caused by changing oceanographic conditions as the principal driving force. Other explanations support overexploitation as the major cause. However, none of the arguments can be considered more than speculative at this time, pending rigorous analysis of more recent information (Hempel 1978; Postma and Zijlstra 1988).

Further to the south of the North Atlantic are the Iberian Shelf and Benguela Current ecosystems, where the abundance of important fishery resources also seems to be related to climate-induced changes in the physical dynamics within each system. The alternation in abundance levels of horse mackerel and sardine within the Iberian Coastal Ecosystem is attributed to changes in natural environmental perturbation of its thermal structure rather than to any density dependent interaction among the two species (Wyatt and Perez-Gandaras 1989).

Similarly, the environment has been a principal driving force for large-scale shifts in abundance among the fish species in the Benguela Current Ecosystem off the southwest coast of Africa (Shelton et al. 1985). The long-term fluctuations in the abundance levels of pilchard, horse-mackerel, hakes, and anchovy are attributed to changes in the oceanographic regime (Crawford et al. 1989). The Benguela LME is bounded by warm water regimes at both the equatorward and polarward extremities. Cold, nutrient-rich water upwells intensely in the central section and less intensely and seasonally in the northern and southern areas. In general, warmer environmental conditions favor the epipelagic species, and cooler conditions favor the demersal species. The effects of the fisheries on changes in species abundance are secondary. Changes in abundance of pilchard stocks have led to detectable effects in the abundance level of dependent predator species, particularly marine bird populations, and led to flips in dominant species such as anchovy replacing pilchard (Crawford et al. 1989).

In the Pacific, the greatest increases in biomass yields have been reported at the area of confluence between the Oyashio and Kuroshio Current ecosystems off Japan (Minoda 1989; Terazaki 1989) and in the Humboldt Current Ecosystem off Chile. In the Oyashio and Kuroshio Current ecosystems the yield of Japanese sardines increased from less than one-half million metric tons (mmt) in 1975 to just over 4 mmt in 1984. The yield of the Chilean sardine in the Humboldt Current Ecosystem also increased from about 500,000 mt in 1974 to 4.3 mmt in 1986. The increased yields have been attributed to density-independent processes involving an increase in lower food chain productivity,

Table 2. Contributions by country, and large marine ecosystem (LME) representing 95 percent of the annual global catch in 1990.

Country	Percentage of[1] world marine nominal catch yield	LMEs producing annual biomass	Cumulative percentages
Japan	12.25	Oyashio Current, Kuroshio Current; Sea of Okhotsk, Sea of Japan, Yellow Sea, East China Sea, W. Bering Sea, E. Bering Sea, and Scotia Sea	
USSR	11.37	Sea of Okhotsk, Barents Sea, Norwegian Shelf, W. Bering Sea, E. Bering Sea, and Scotia Sea	
China	8.28	W. Bering Sea, Yellow Sea, E. China Sea, and S. China Sea	
Peru	8.27	Humboldt Current	
USA	6.76	Northeast US Shelf, Southeast US Shelf, Gulf of Mexico, California Current, Gulf of Alaska, and E. Bering Sea	
Chile	5.98	Humboldt Current	52.91
Korea Republic	3.28F*	Yellow Sea, Sea of Japan, E. China Sea, and Kuroshio Current	
Thailand	2.96F	South China Sea, and Indonesian Seas	
India	2.78	Bay of Bengal and Arabian Sea	
Indonesia	2.76	Indonesian Seas	
Norway	2.11	Norwegian Shelf and Barents Sea	
Korea D. P. Rep.	1.98F	Sea of Japan and Yellow Sea	
Philippines	1.96	S. China Sea, Sulu-Celebes Sea	
Canada	1.90	Scotian Shelf, Northeast U.S. Shelf, Newfoundland Shelf	

Table 2 continued

Country	Percentage of[1] world marine nominal catch	LMEs producing annual biomass yield	Cumulative percentages
Iceland	1.82	Icelandic Shelf	
Denmark	2.07	Baltic Sea and North Sea	76.25
Spain	1.73	Iberian Coastal Current and Canary Current	
Mexico	1.46	Gulf of California, Gulf of Mexico, and California Current	
France	1.03F	North Sea, Biscay-Celtic Shelf, Mediterranean Sea	80.47
Viet Nam	0.74	South China Sea	
Myanmar	0.72	Bay of Bengal, Andaman Sea	
Brazil	0.71F	Patagonian Shelf and Brazil Current	
Malaysia	0.71F	Gulf of Thailand, Andaman Sea, Indonesian Seas, and S. China Sea	
UK-Scotland	0.70	North Sea	
New Zealand	0.68	New Zealand Shelf Ecosystem	
Morocco	0.68	Canary Current	
Argentina	0.66	Patagonian Shelf	
Italy	0.57	Mediterranean Sea	
Netherlands	0.52	North Sea	
Poland	0.52	Baltic Sea	
Ecuador	0.47	Humboldt Current	
Pakistan	0.44	Bay of Bengal	

Table 2 continued

Country	Percentage of[1] world marine nominal catch	LMEs producing annual biomass yield	Cumulative percentages
Turkey	0.41	Black Sea, Mediterranean Sea	
Germany (F.R. and N.L.)	0.41	Baltic Sea and Scotia Sea	
Ghana	0.40	Gulf of Guinea	
Portugal	0.39	Iberian Shelf and Canary Current	90.20
Venezuela	0.38	Caribbean Sea	
Namibia	0.35	Benguela Current	
Faeroe Islands	0.34	Faeroe Plateau	
Senegal	0.34	Gulf of Guinea and Canary Current	
Sweden	0.31	Baltic Sea	
Bangladesh	0.31	Bay of Bengal	
Ireland	0.28	Biscay-Celtic Shelf	
Hong Kong	0.28	S. China Sea	
Nigeria	0.26	Gulf of Guinea	
Australia	0.25	N. Australian Shelf and Great Barrier Reef	
Iran, I.R.	0.24F	Arabian Sea	
UK Eng., Wales	0.21	North Sea	
Cuba	0.20	Caribbean Sea	
Panama	0.19	California Current and Caribbean Sea	
Greenland	0.17	East Greenland Shelf, West Greenland Shelf	

Table 2 continued

Country	Percentage of[1] world marine nominal catch	LMEs producing annual biomass yield	Cumulative percentages
Sri Lanka	0.19	Bay of Bengal	
Greece	0.16	Mediterranean Sea	
Oman	0.15	Arabian Sea	
Angola	0.12	Guinea Current, Angola Basin	
United Arab Em.	0.11	Arabian Sea	95.01

[1]Percentages based on fish catch statistics from FAO 1990 Yearbook, vol. 70 FAO, 1992.
*F = Percentage calculated using FAO estimate from available sources of information.

made possible by coastward shifts in the boundary areas of the Oyashio and Kuroshio systems and water mass shifts in the Humboldt Current Ecosystem. The effects of fishing on the sardines in both areas were of secondary importance compared to the enhanced productivity of the phytoplankton and zooplankton components of the ecosystems that provided an improved environment for growth and recruitment. Studies are underway to determine the extent of the teleconnection between the Pacific-wide El Niño events of the past decade and: (1) the multimillion metric-ton increases in yields of sardines occurring nearly simultaneously in the northern and southern hemispheres, and (2) the dramatic decline in the biomass yields of anchovy in the northern areas of the Humboldt Current Ecosystem in the early 1970s from about 12 mmt in 1970 to less than 2 mmt by 1976 (Canon 1986; Figure 3).

Although less dramatic, the long-term shifts in abundance levels of both sardines and anchovies within the California Current Ecosystem are considered the result primarily of natural environmental change and secondarily of intensive fishing, rather than of any density-dependent competition between the two species (MacCall 1986; Holmgren-Urba and Baumgartner 1993).

To the north and west of Australia lies the relatively pristine Banda Sea Ecosystem, where no large-scale fisheries are presently conducted. The ecosystem is under the influence of monsoon-induced seasonal periods of large-scale upwelling and downwelling. Biological feedback to these environmental signals is reflected in the changes in phytoplankton, mesozooplankton, micronekton, and fish. During upwelling events, productivity of the ecosystem is enhanced by a factor of 2 to 3. The biomass of pelagic fish resources is also higher during the upwelling period. The fish biomass of the ecosystem is estimated at between 600,000 mt to 900,000 mt in the peak upwelling season (August), and 150,000 mt to 250,000 mt in the downwelling period (February). The estimated sustained annual biomass yield of the ecosystem is approximately 30,000 mt of pelagic fish (Zijlstra and Baars, 1990).

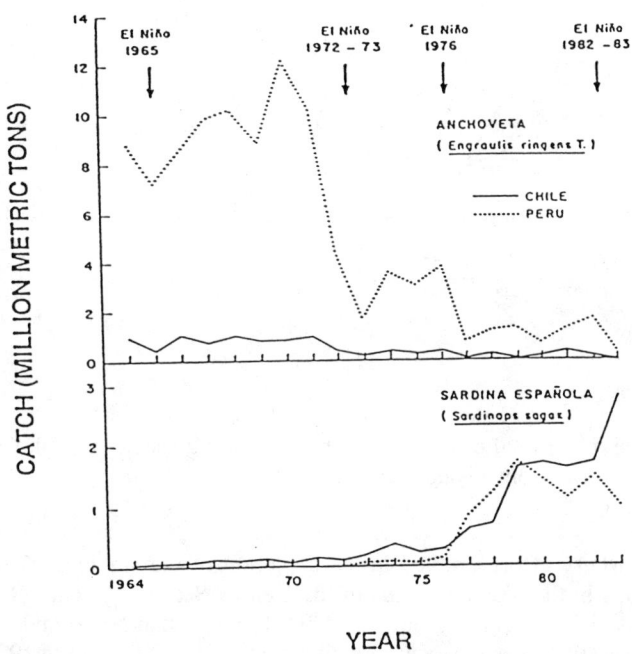

Figure 3. Fluctuations in the catch of anchovies and sardines from the waters of the Humboldt Current Ecosystem off the coasts of Chile and Peru (from Canon 1986).

Changes in biomass yields of two other Pacific rim LMEs have been the result of overexploitation. The introduction of highly efficient modern trawlers to the Gulf of Thailand Ecosystem in an effort to increase fishing efficiencies, led to excessive fishing mortality and a marked reduction in annual yields of biomass of fish for human consumption between 1977 and 1982 (Piyakarnchana 1989; Figure 4).

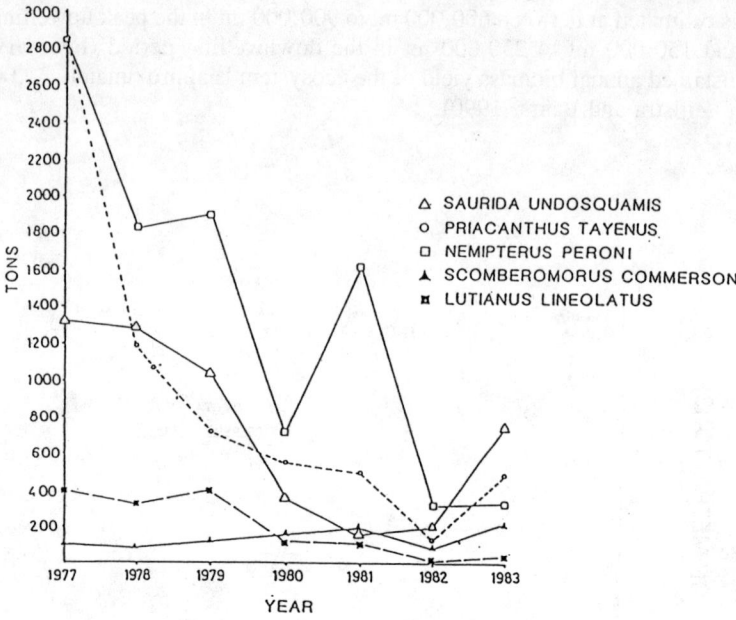

Figure 4. Decline in the total catch of carnivorous feeding species of fish from the Gulf of Thailand Ecosystem (from Piyakarnchana 1989).

Intensive fishery effort resulted in the depletion of the demersal fish stocks and dramatic reductions in the biomass yields of the Yellow Sea Ecosystem. Between 1958 and 1968 fisheries yields declined from 180,000 mt to less than 10,000 mt. The fishery then shifted to harvesting pelagic stocks reaching a level of 200,000 mt in 1972, followed by a reduction to less than 20,000 mt in 1981. The fisheries of the Yellow Sea in 1982 shifted principally to anchovy and sardine with a total annual yield of all species 40% lower than the 1958 level. The demersal fishery remains in a depleted state (Tang 1989; Figure 5).

The importance of a natural predator driving an ecosystem is evidenced in the large-scale changes in the community structure of the Great Barrier Reef Ecosystem that extends over 230,000 km^2 of the Queensland continental shelf. The predation by the crown-of-thorns starfish in the 1960s and 1970s, resulted in a shift in the biomass of corals, community structure of the benthos, and a decoupling of energy transfer to several fish stocks (Bradbury and Mundy 1989).

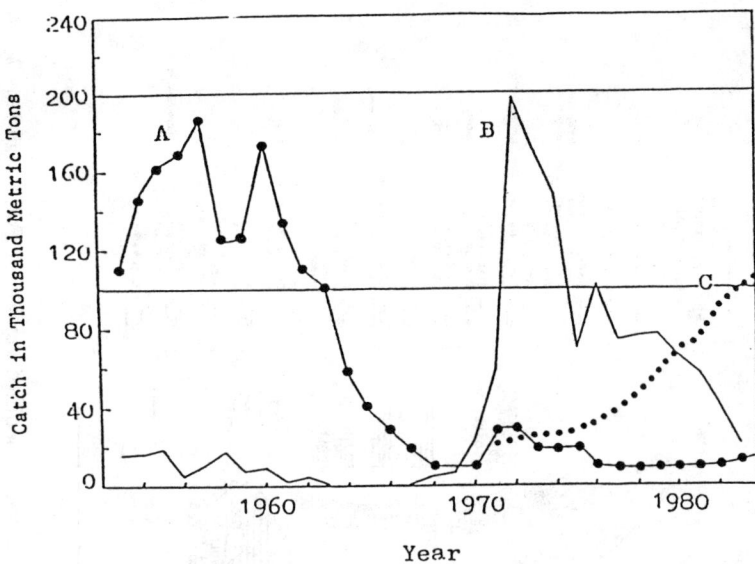

Figure 5. Annual catch of dominant species: (A) small yellow croaker and hairtail; (B) Pacific herring and Japanese mackerel; (C) *Setipinna taty*, anchovy and scaled sardine of the Yellow Sea Ecosystem, 1953 through 1984 (from Tang 1989).

In the enclosed and semi-enclosed marine ecosystems, the effects of pollution in the form of coastal eutrophication attibuted to high levels of nitrate and phosphate inputs from population centers have resulted in unusual phytoplankton blooms, oxygen depletion, biotoxin generated mortalities, and changes in ecosystem trophodynamics (Smayda 1991). Among the impacted ecosystems are the Black Sea (Zaitsev 1992; Caddy 1993), Baltic Sea (Kullenberg 1986), and Adriatic Sea (Bombace 1993).

MANAGEMENT CONSIDERATIONS

Empirical and theoretical aspects of yield models for large marine ecosystems have been reviewed by several ecologists. According to Beddington (1986), Daan (1986), Levin (1990), and Mangel (1991), published dynamic models of marine ecosystems offer

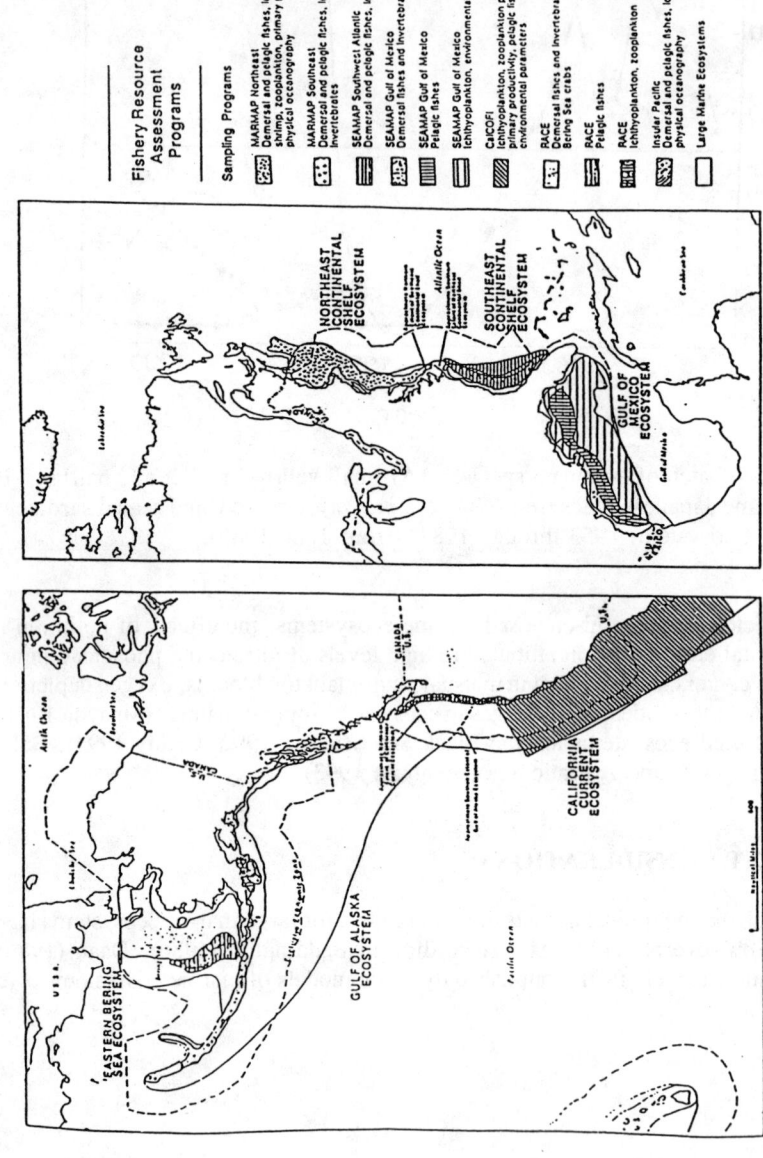

Figure 6. Large marine ecosystems of the United States. [This figure is a modified version of Folio Map No. 7, "A National Atlas: Health and Use of Coastal Waters, United States of America." U.S. Department of Commerce, National Oceanic and Atmospheric Administration, National Ocean Service, Office of Oceanography and Marine Assessment, Washington DC 1988]

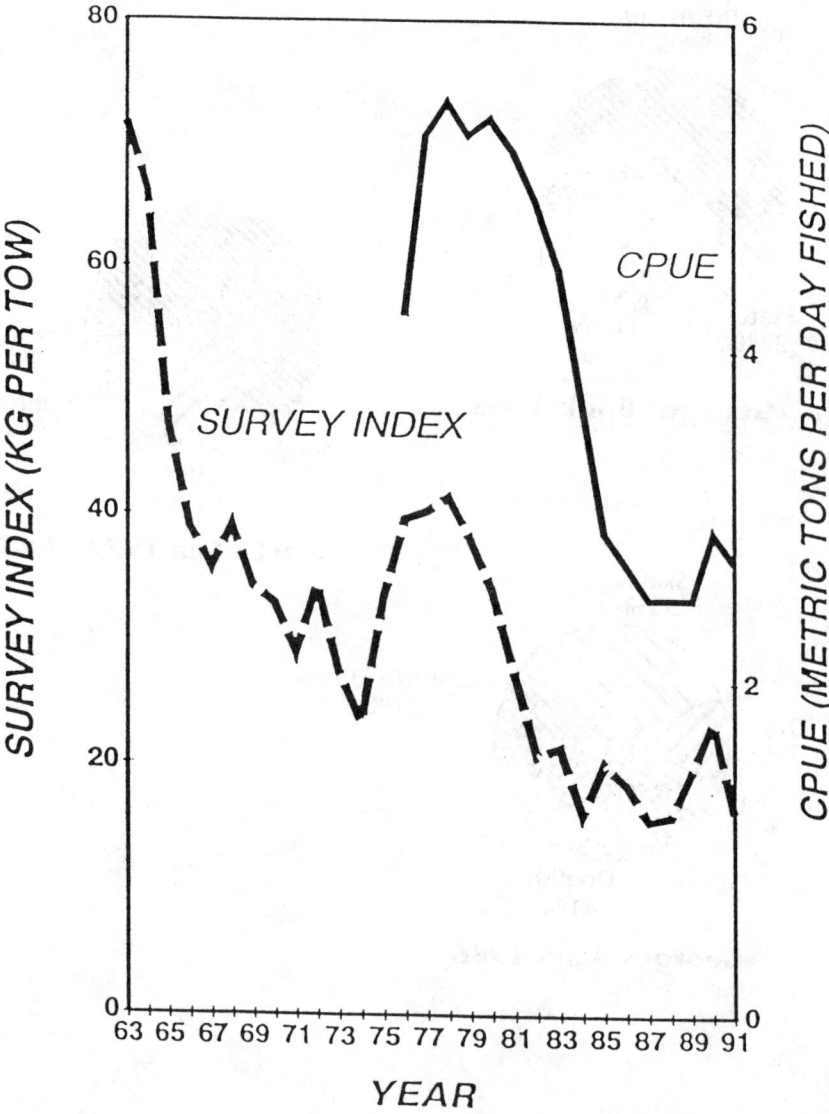

Figure 7. Abundance indices for principal U.S. Northeast Shelf groundfish resources from 1963 to 1991 (from Anthony 1993).

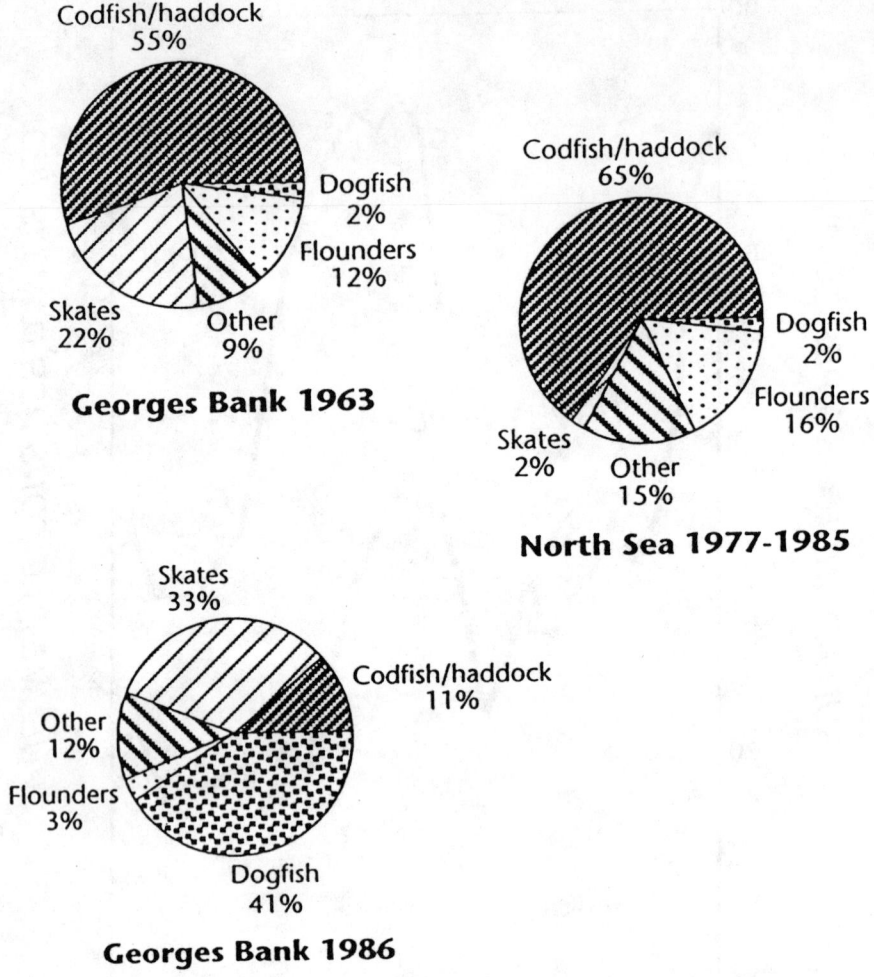

Figure 8. Species shifts and abundance of small elasmobranchs (dogfish and skates) on Georges Bank within the Northeast Shelf Ecosystem of the United States compared with the North Sea Ecosystem (from Anthony 1988).

little guidance on the detailed behavior of communities. However, these authors concur on the need for covering the common ground between observation and theory by implementing monitoring efforts on the large spatial and long temporal scales (decadal) of "key" components of the LMEs. The sequence for improving the understanding of the possible mechanisms underlying observed patterns in LMEs is described by Levin (1990) as examination of: (1) statistical analyses of observed distributional patterns of physical and biological variables; (2) construction of competing models of variability and patchiness based on statistical analyses and natural scales of variability of critical processes; (3) evaluation of competing models through experimental and theoretical studies of component systems; and (4) integration of validated component models to provide predictive models for population dynamics and redistribution. The approach suggested by Levin (1990) is consistent with the recent observation by Mangel (1991) that empirical support for the currently used models of LMEs is relatively weak, and that a new generation of models is needed that serves to enhance the linkage between theory and empirical results.

There is a growing awareness among marine scientists, geographers, economists, government representatives, and lawyers, of the utility of a more holistic ecosystem approach to resource management (Byrne 1986; Christy 1986; Alexander 1989; Belsky 1989; Crawford et al. 1989; Morgan 1989; Prescott 1989). On a global scale the loss of sustained biomass yields from LMEs from mismanagement and overexploitation has not been fully investigated but is likely very large (Gulland 1984).

Effective management strategies for LMEs will be contingent on the identification of the major driving forces causing large-scale changes in biomass yields. Management of species responding to strong environmental signals will be enhanced by improving the understanding of the physical factors forcing biological changes; whereas in other LMEs, where the prime driving force is predation, options can be explored for implementing adaptive management strategies. Remedial actions are required to ensure that the "pollution" of the coastal zone of LMEs is reduced and does not become a principal driving force in any LME. For at least one LME, the Antarctic, a management regime has evolved based on an ecosystem perspective in the adoption and implementation of the Convention for the Conservation of Antarctic Marine Living Resources (Sherman and Ryan 1988). Efforts are also underway to implement ecosystem management within the LMEs of the U.S. EEZ, for example, in the northern California Current Ecosystem (Bottom et al. 1989). Concerns remain regarding the socioeconomic and political difficulties in management across national boundaries as in the case of the Sea of Japan Ecosystem where the fishery resources are shared by five countries (Morgan 1988), or the North Sea Ecosystem, or the Caribbean Sea Ecosystem where 38 nations share the resources.

A systems approach to the management of LMEs is depicted in Table 3. The LMEs represent the link between local events (e.g. fishing, pollution, environment) occurring on the daily-to-seasonal temporal scale and their effects on living marine resources and the more ubiquitous global effects of climate changes on the multidecadal timescale. The regional and temporal focus of season to decade is consistent with the evolved spawning and feeding migrations of the fishes. These migrations are seasonal and occur over hundreds to thousands of kilometers within the unique physical and biological characteristics of the regional LME to which they have adapted. As the fisheries represent most of the usable biomass yield of the LMEs and fish populations consist of several age classes, it follows that measures of variability in growth, recruitment, and mortality should be conducted over multiyear timescales. Consideration of the naturally occurring environmental events and the human-induced perturbations affecting demography of the populations within the ecosystem is necessary. Based on scientific inferences of the principal causes of variability in abundance and with due consideration to socioeconomic needs, management options from an ecosystems perspective can be considered for implementation. The final element in the system, with regard to the concept of resource maintenance and sustained yield, is the feedback loop that allows for evaluation of the effects of management actions at the fisheries level (single species, multispecies) and the ecosystem level.

It will be necessary to conduct supportive research on the processes controlling sustained productivity of LMEs. Within several of the LMEs, including the Northeast Shelf, Gulf of Mexico, California Current, and Eastern Bering Sea, important hypotheses concerned with the growing impacts of pollution, overexploitation, and environmental changes on sustained biomass yields are under investigation (Table 4). By comparing the results of research among the different systems, it should be possible to accelerate an understanding of how the systems respond and recover from stress. The comparisons should allow for narrowing the context of unresolved problems and capitalizing on research efforts underway in the various ecosystems. Recent reports describing the effects of biological and physical perturbations on the fisheries biomass yields of 30 large marine ecosystems (Table 1) address questions similar to those posed a few years ago by Beddington (1984):

> *There are a number of scientific questions which are central to the rational management of marine communities, but all revolve around the question of sustainability. What levels of mortality imposed by a fishery will permit a sustainable yield? Are there levels below which a fish population will not recover? Can judicious manipulation of the catch composition of the fishery alter the potential of the community to produce yields of a particular type, e.g., high value species? Can a community be depleted to a level where its potential for producing a harvestable resource is reduced? With the exception of the first question, these questions and others like them are rarely explicitly addressed in the scientific bodies of the various fisheries' organizations. Instead, such bodies concentrate on the estimation of stock abundance and the calculation of allowable catch levels, although often implicit in the advice given by these bodies to management are a set of beliefs about the answers to such questions.*

Given the increasing number of responsibilities of government agencies for: (1) managing fisheries, (2) mitigating pollution, (3) reducing environmental stress, and (4) restoring lost habitat, it is not surprising that interest is growing to pursue resource management problems from an ecosystem perspective.

ECOSYSTEM ASSESSMENT AND MONITORING

In the USA, greater emphasis has been focused over the past decade within the National Marine Fisheries Service of the NOAA, on approaching fisheries research from a regional ecosystem perspective in LMEs within and adjacent to the EEZ of the United States--The Northeast Continental Shelf, the Southeast Continental Shelf, the Gulf of Mexico, the California Current, the Gulf of Alaska, the Eastern Bering Sea, and the Insular Pacific including the Hawaiian Islands. These ecosystems, in 1991, yielded 4.3 million metric tons of fisheries biomass valued at approximately $16.5 billion to the economy of the United States.

The sampling programs providing the biomass assessments within the U. S. EEZ have been described in Folio Map 7 produced by the Office of Oceanography and Marine Assessment of NOAA's National Ocean Service. The map depicts the seven ecosystems under investigation (Figure 6). Sampling programs supporting biomass estimates in LMEs within and adjacent to the EEZ of the United States are designed to: (1) provide detailed statistical analyses of fish and invertebrate populations constituting the principal yield species of biomass, (2) estimate future trends in biomass yields, and (3) monitor changes in the principal populations. The information obtained by these programs provides managers with a more complete understanding of the dynamics of marine ecosystems and how these dynamics affect harvestable stocks. Additionally, by tracking components of the ecosystems, these programs can detect changes, natural or human-induced, and warn of events with possible economic repercussions. Although sampling schemes and efforts vary among programs (depending on habitats, species present, and specific regional concerns), they generally involve systematic collection and analysis of catch-statistics; the use of NOAA vessels for fisheries-independent bottom and midwater trawl surveys for adults and juveniles; ichthyoplankton surveys for larvae and eggs; measurements of zooplankton standing stock, primary productivity, nutrient concentrations, and important physical parameters (e.g., water temperature, salinity, density, current velocity and direction, air temperature, cloud cover, light conditions); and, in some habitats, measurements of contaminants and their effects. At the shoreward margin of the LMEs, monitoring efforts include the use of mussels and other biological indicator species to measure pollution effects as part of NOAA's Status and Trends Program. The pilot Environmental Monitoring Assessment Program (EMAP) of the Environmental Protection Agency that focused on the estuarine and nearshore monitoring of contaminants in the water column, substrate, and selected groups of organisms, will be extended to more open waters of LMEs in cooperation with NOAA during 1993 and 1994.

Table 3. Key spatial and temporal scales and principal elements of a systems approach to the research and management of large marine ecosystems (from Sherman 1991).

1. Spatial-Temporal Scales

	Spatial	Temporal	Unit
1.1	**Global** (World Ocean)	Millennia-Decadal	Pelagic Biogeographic
1.2	**Regional** (Exclusive Economic Zones)	Decadal-Seasonal	Large Marine Ecosystems
1.3	**Local**	Seasonal-Daily	Subsystems

2. Research Elements

   2.1 Spawning Strategies
   2.2 Feeding Strategies
   2.3 Productivity, Trophodynamics
   2.4 Stock Fluctuations/Recruitment/Mortality
   2.5 Natural Variability
        (Hydrography, Currents, Water Masses, Weather)
   2.6 Human Perturbations
        (Fishing, Waste Disposal, Petrogenic Hydrocarbon Impacts,
        Aerosol Contaminants, Eutrophication Effects)

3. Management Elements--Options and Advice--International, National, Local

   3.1 Bioenvironmental and Socioeconomic Models
   3.2 Management to Optimize Fisheries Yields

4. Feedback Loop

   4.1 Evaluation of Ecosystem Status
   4.2 Evaluation of Fisheries Status
   4.3 Evaluation of Management Practices

Table 4. Selected Hypotheses Concerning Variability in Biomass Yields of Large Marine Ecosystems. Note that references can be found in Table 1 (from Sherman 1991).

Ecosystem	Predominant Variables	Hypothesis
Oyashio Current Kuroshio Current California Current Humboldt Current Benguela Current Coastal	Density-independent natural environmental perturbations	Clupeoid Population Increases: Predominant variables influencing changes in biomass of clupeoids are major increases in water-Iberian column productivity resulting from shifts in the direction and flow velocities of the currents and changes in upwelling within the ecosystem.
Yellow Sea U. S. Northeast Continental Shelf Gulf of Thailand	Density-dependent predation	Declines in Fish Stocks: Precipitous decline in biomass of fish stocks is the result of excessive fishing mortality, reducing the probability of reproductive success. Losses in biomass are attributed to excesses of human predation expressed as overfishing.
Great Barrier Reef	Density-dependent predation	Change in Ecosystem Structure: The extreme predation pressure of crown-of-thorns starfish has disrupted normal food chain linkage between benthic primary production and the fish component of the reef ecosystem.
East Greenland Sea Barents Sea Norwegian Sea	Density-independent natural environmental perturbations	Shifts in Abundance of Fish Stock Biomass: Major shifts in the levels of fish stock biomass within the ecosystems are attributed to large-scale environmental changes in water movements and temperature structure.

Table 4 continued

Ecosystem	Predominant Variables	Hypothesis
Baltic Sea	Density-independent pollution	Changes in Ecosystem Productivity Levels: The apparent increases in productivity levels are attributed to the effects of nitrate enrichment resulting from elevated levels of agricultural contaminant inputs from the bordering land masses.
Antarctic Marine	Density-dependent perturbations	Status of Krill Stocks: Annual natural production cycle of krill is in balance with food requirements of dependent predator populations. Surplus production is available to support economically significant yields, but sustainable level of fishing effort is unknown.
	Density-independent natural environmental perturbations	Shifts in Abundance in Krill Biomass: Major shifts in abundance levels of krill biomass within the ecosystem are attributed to large-scale changes in water movements and productivity.

A monitoring strategy for measuring the changing states of LMEs was recommended by a panel of international experts that met at Cornell University in July 1991 (Table 5) (Sherman and Laughlin 1992). The two monitoring methods recommended are: (1) regular trawling using a stratified random sampling design, and (2) plankton surveys. The large-scale changes in the fisheries of the North Sea and the Northeast Continental Shelf of the United States have been successfully detected using trawling techniques for several decades (Azarovitz and Grosslein 1987). The surveys have been conducted by relatively large research vessels. However, standardized sampling procedures, when deployed from small calibrated trawlers, can provide important information on fish stocks. The fish catch provides biological samples for stomach analyses, age and growth, fecundity, and size comparisons (ICES 1991), and data for clarifying and quantifying multispecies trophic relationships. Samples of trawl-caught fish can also be used to monitor gross pathological conditions that may be associated with coastal pollution. The need for both biological and environmental monitoring in the North Sea Ecosystem has been emphasized following the Symposium on Long-Term Changes in the Fish Stocks of the North Sea Ecosystem (Hempel 1978). In this regard, physical measurements can be made from small trawlers or ships-of-opportunity, using readily available and relatively inexpensive systems for measuring temperature and salinity of the water column. Standard logs for weather observations, important in detecting global change, are an important component of the data-collecting effort. The monitoring of changes in fish stocks is ongoing in LMEs across the North Atlantic basin, including the Northeast U. S. Shelf, the Canadian Scotian Shelf, Newfoundland Shelf; and on the Greenland Shelf, Icelandic Shelf, Norwegian Shelf, Barents Sea Shelf, and the North Sea. The plankton of LMEs can be measured at a relatively low cost by deploying CPR systems from commercial vessels of opportunity (Glover, 1967). The advanced plankton recorders can be fitted with sensors for temperature, salinity, chlorophyll, nitrate/nitrite, light, bioluminescence, zooplankton, and ichthyoplankton (Aiken, 1981; Aiken and Bellan, 1990; Williams and Aiken 1990; Kolber and Falkowski 1992; UNESCO 1992), providing the means to monitor changes in phytoplankton, zooplankton, primary productivity, species composition and dominance, and long-term changes in the physical and nutrient characteristics of the LME, as well as longer term changes relating to the biofeedback of the plankton to the stress of climate change (Colebrook 1986; Dickson et al. 1988; Jossi and Smith 1990; Sherman et al. 1990b). Plankton monitoring using the CPR system is at present expanding in the North Atlantic.

A critical feature of the LME monitoring strategy is the development of a consistent long-term data base for understanding interannual changes and multi-year trends in biomass yields for each of the LMEs. For example, during the late 1960s and early 1970s, when there was intense foreign fishing within the Northeast U. S. Continental Shelf Ecosystem, marked alterations in fish abundances were recorded. Significant shifts among species abundances were observed. The finfish biomass of commercially important species (e.g., cod, haddock, flounders, herring, and mackerel) declined by approximately 50% (Figure 7). This was followed by increases in the biomass of sand lance (Fogarty et al. 1991) and elasmobranchs (dogfish and skates) (Figure 8) and led to the conclusion that the overall carrying capacity of the ecosystem for finfish did not change. The excessive fishing effort on highly valued species allowed for low-valued species to increase in abundance. Analyses of catch-per-unit-effort and fishery-independent bottom trawling survey data were critical sources of information used to implicate overfishing as the cause of the shifts in relative abundance among the species of the fish community within the shelf ecosystem. It is important to note, however, that the lower-end of the food chain in the offshore waters of the ecosystem remained unchanged, largely as described by Bigelow (1926) and Riley et al. (1949), suggesting that ecosystem productivity remained high during a period of species dominance shifts among the fish community caused by human interventions through fishing (Sherman et al. 1983). The natural "resilience" of the ecosystem in relation to recovery from stress can be documented in the recovery of mackerel to former (pre-1960) levels of abundance and the apparent recovery of herring to 1960's level of abundance on Georges Bank (Murawski 1991; Smith and Morse 1990). The gyre systems of the Gulf of Maine and Georges Bank subsystems and the nutrient enrichment of the estuaries in the southern half of the Northeast Shelf Ecosystem contribute to the maintenance on the shelf of relatively high levels of phytoplankton and zooplankton prey fields for planktivores including fish larvae, menhaden, herring, mackerel, sand lance, butterfish, and marine birds and mammals.

Table 5. Core marine ecosystem monitoring program. The Core Program is based on transects sampled by UOR or instrumented CPR, supplemented by satellite oceanography and systematic trawl and acoustic surveys (from Sherman and Laughlin 1992).

Candidate parameters for the Core Program include:

*Chlorophyll Fluorescence	*Salinity structure	*Temperature structure
*+Primary Production	*Nutrients	
*Diatom/Flagellate Ratio	NO_2	*Stratification index
Zooplankton composition and Biomass	NO_3	*transparency
*Copepod Diversity	Pollution index (e.g., hydrocarbons, sewage)	*+PAR
Fisheries Survey		Rainfall or Runoff, Wind strength and direction

Assessment
    Changes in Abundance and Distribution
Biology
    Length
    Age and Growth
    Predator-Prey
    Pathology
Acoustics for Pelagics
Nets for Demersals
Physical Measurements
    Temperature
    Salinity
Chemical Measurements
    Water samples (nutrients, productivity, pollutants)

---

*Measurements derived from instrumented CPR/UOR sensors.
*+Based on inclusion of fast repetition rate (FRR) fluorometer (Kobler and Falkowski, 1992).

CHANGING ECOSYSTEM STATES AND "HEALTH" INDICES

The topic of change and persistence in marine communities and the need for multispecies and ecosystem perspectives in fishery management relate to the reports of changing states of marine ecosystems (Sugihara et al. 1984). Collapses of the Pacific sardine in the California Current Ecosystem, the pilchard in the Benguela Current Ecosystem, and the anchovy in the Humboldt Current Ecosystem, are but a few examples of cascading effects on other ecosystem components including marine birds (MacCall 1986; Croxall 1987; Burger 1988; Crawford et al. 1989). Ecosystem "health" is a concept of wide interest for which a single precise scientific definition is problematical. Ecosystem health is used herein, to describe the resilience, stability, and productivity of the ecosystem in relation to the changing states of ecosystems. In present practice, assessing the health of LMEs relies on a series of indicators and indices (Costanza 1992; Rapport 1992; Norton and Ulanowicz 1992; Karr 1992). The overriding objective is to monitor changes in health from an ecosystem perspective as a measure of the overall performance of a complex system (Costanza 1992). The health paradigm is based on the multiple-state comparisons of ecosystem resilience and stability (Pimm 1984; Holling 1986; Costanza 1992) and is an evolving concept. Definitions of several variables important to the changing states and health of marine ecosystems are given in Table 6. Following the definition of Costanza (1992), to be healthy and sustainable, an ecosystem must maintain its metabolic activity level, its internal structure and organization, and must be resistent to external stress over time and space frames relative to the ecosystem. These concepts were discussed at a workshop convened by NOAA/NMFS at the Northeast Fisheries Science Center's Narragansett Laboratory in April 1992. Among the indices discussed by the participants were five that are being considered as experimental measures of changing ecosystem states and health--(1) diversity; (2) stability; (3) yields; (4) production; and (5) resilience (Sherman 1993).

The data from which to derive the experimental indices are obtained from time-series monitoring of key ecosystem parameters. A prototype effort to validate the utility of the indices is under development by NOAA at the Northeast Fisheries Science Center. The ecosystem sampling strategy is focused on parameters relating to the resources at risk from overexploitation, species protected by legislative authority (marine mammals), and other key biological and physical components at the lower end of the food chain (plankton, nutrients, hydrography) (Sherman and Laughlin 1992). The parameters of interest depicted in Figure 9 include zooplankton composition, zooplankton biomass, water column structure, photosynthetically active radiation (PAR), transparency, chlorophyll-a, NO_2, NO_3, primary production, pollution, marine mammal biomass, marine mammal omposition, runoff, wind stress, seabird community structure, seabird counts, finfish composition, finfish biomass, domoic acid, saxitoxin, and paralytic shellfish poisoning (PSP). The experimental parameters selected incorporate the behavior of individuals, the resultant responses of populations and communities, as well as their interactions with the physical and chemical environment. The selected parameters, if measured in all LMEs, will permit comparison of relative changing states and health status among ecosystems. The interrelations between the datasets and the selected parameters are indicated by the arrows leading from column 1 to column 2 in the figure. The measured ecosystem components are depicted in relation to ecosystem structure in a diagrammatic conceptualization of patterns and activities within the LME at different levels of complexity (Figure 10).

Table 6. Definitions of some important variables for use in the indexing of changing ecosystem states (health) (adapted and expanded from Costanza 1992).

Variable	Definition	Units
Stability		
Homeostasis	Maintenance of a steady state in living organisms by the use of feedback control processes.	
Stable	A system is stable if, and only if, the variables all return to the initial equilibrium following their being perturbed from it. A system is locally stable if this return applies to small perturbations, and globally stable if it applies to all possible perturbations.	Binary
Sustainable	A system that can maintain its structure and function indefinitely. All non-successional (i.e., climax) ecosystems are sustainable, but they may not be stable (see resilience below). Sustainability is a policy goal for economic systems.	Binary
Resilience	1. How fast the variables return towards their equilibrium following a perturbation. Not defined for unstable systems (Pimm 1984). 2. The ability of a system to maintain its structure and Patterns of behavior in the face of disturbance (Holling 1986).	Time
Resistance	The degree to which a variable is changed, following a perturbation.	Nondimensional and continuous
Variability	The variance of population densities over time, or allied measures such as the standard deviation or coefficient of variation (sd/mean).	

Table 6 continued

Variable Units	Definition	Units
Complexity		
Species richness	The number of species in a system.	Integer
Connectance	The number of actual interspecific interactions divided by the possible interspecific interactions.	Dimensionless
Interaction strength	The mean magnitude of interspecific interaction: the size of the effect of one species' density on the growth rate of another species.	
Evenness	The variance of the species abundance distribution.	
Diversity indices	Measures that combine evenness and richness with a particular weighting for each. One important member of this family is the information theoretic index, H.	Bits
Ascendency	An information theoretic measure that combines the average mutual information (a measure of connectedness) and the total throughput of the system as a scaling factor (see Ulanowicz, 1992).	
Other Variables		
Perturbation	A change to a system's inputs or environment beyond the normal range of variation.	Varies
Stress	A perturbation with a negative effect on a system.	
Subsidy	A perturbation with a positive effect on a system.	

Initial efforts to examine changing ecosystem states and relative health within a single ecosystem are underway for four subareas of the U.S. Northeast Continental Shelf Ecosystem--Gulf of Maine, Georges Bank, Southern New England, Mid-Atlantic Bight. Initial studies of the structure, function, and productivity of the system have been reported (Sherman et al., 1988). It appears that the principal driving force in relation to sustainable ecosystem yield is fishing mortality expressed as predation on the fish stocks of the system, and that long-term sustainability of high economic yield species will be dependent on the application of adaptive management strategies (Sissenwine and Cohen 1991; Murawski 1991).

Several alternative management strategies for the fish stocks of the U.S. Northeast Continental Shelf Ecosystem are under consideration by the New England Fisheries

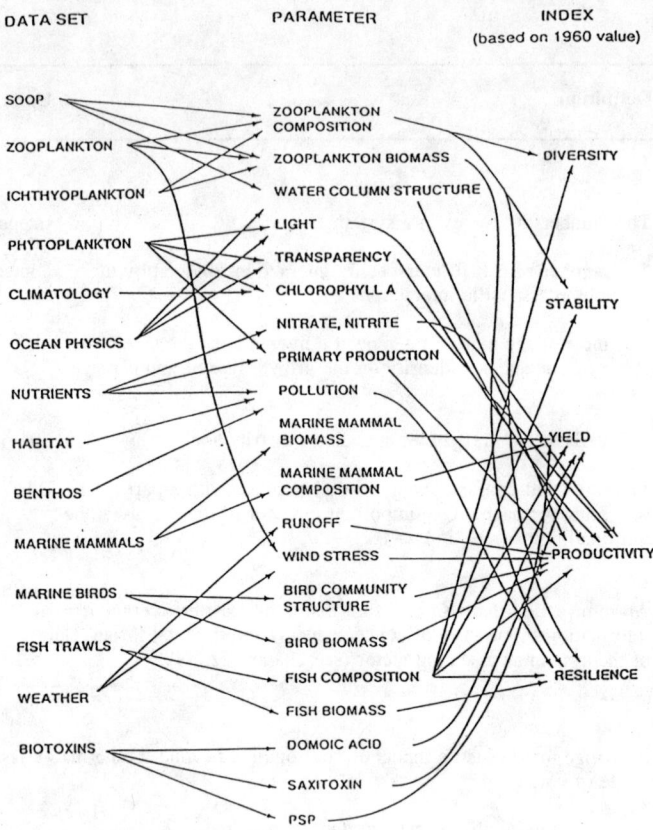

Figure 9. A schematic representation of the data bases and experimental parameters for indexing the changing states of large marine ecosystems. The data base represents time-series measurements of key ecosystem components from the U.S. Northeast Continental Shelf Ecosystem. Indices will be based on changes compared with the ecosystem state in 1960.

Management Council and the Atlantic States Marine Fisheries Commission. In addition to fisheries management issues and significant biomass flips among dominant species, the Northeast Continental Shelf Ecosystem is also under stress from the increasing frequency of unusual plankton blooms, and eutrophication within the nearshore coastal zone resulting from high levels of phosphate and nitrate discharges into drainage basins. Whether the increases in the frequency and extent of nearshore plankton blooms are responsible for the rise in incidence of biotoxin related shellfish closures and marine mammal mortalities, remains an important open question that is the subject of considerable concern to state and federal management agencies (Sherman et al. 1992a; Smayda 1991).

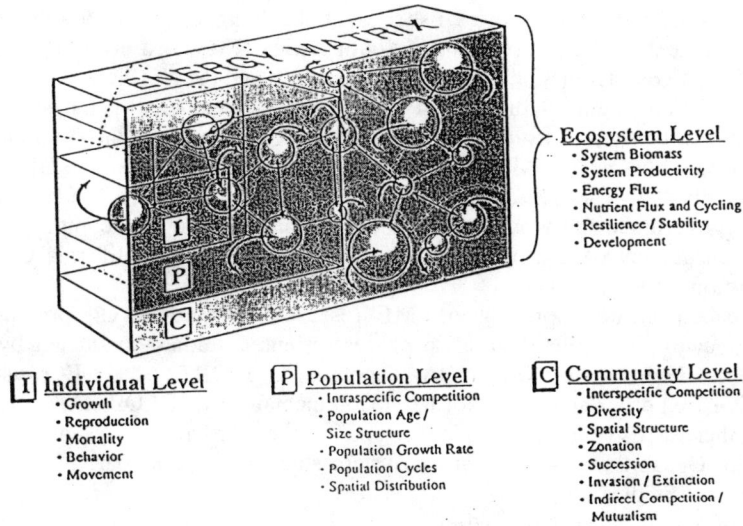

Figure 10. Diagrammatic conceptualization of patterns and activities at different levels of complexity. Each sphere represents an individual abiotic or biotic entity. Abiotic is defined as nonliving matter. Broad, double-headed arrows indicate feedback between entities and the energy matrix for the system. The thin arrows represent direct interactions between individual entities. Much of ecology is devoted to studying interactions between biotic and abiotic entities with a focus on the effects of such interactions on individuals (I), populations (P), or communities (C) of organisms. Ecosystem ecology studies these interactions from the viewpoint of their effect on both the biotic and abiotic entities and within the context of the system. The boundaries of the system must be established to conduct quantitative studies of flux. Figure 1 depicts the boundaries of LMEs, located around the margins of the ocean basins, where the influence of overexploitation, pollution, and habitat degradation and climate change are affecting the structure and function of the ecosystems (from Likens 1992).

PRESENT AND FUTURE EFFORTS

The topics of change and persistence in marine communities, and the need for multispecies and ecosystem perspectives in fisheries management were reviewed at the Dahlem Conference on Exploitation of Marine Communities in 1984 (May, 1984). The designation and management of LMEs is, at present, an evolving scientific and geopolitical process (Morgan 1988; Alexander 1989). Sufficient progress has been made to allow for useful comparisons among different processes influencing large-scale changes in the biomass yields of LMEs (Bax and Laevastu 1990).

Among the ecosystems being managed from a more holistic perspective are: the Yellow Sea Ecosystem, where the principal effort is underway by the Peoples Republic of China (Tang 1989); the multispecies fisheries of the Benguela Current Ecosystem under the management of the government of South Africa (Crawford et al. 1989); the Great Barrier Reef Ecosystem (Bradbury and Mundy 1989) and the Northwest Australian Continental Shelf Ecosystem (Sainsbury 1988) under management by the state and federal governments of Australia; the Antarctic marine ecosystem under the Commission for the Conservation of Antarctic Marine Living Resources (CCAMLR) and its 21-nation membership (Scully et al. 1986; Sherman and Ryan 1988). Within the EEZ of the United States, the state governments of Washington and Oregon have developed a comprehensive plan for the management of marine resources within the Northern California Current Ecosystem (Bottom et al. 1989).

The broad-spectrum approach to LME research and monitoring provides a conceptual framework for collaboration in process-oriented studies conducted by the National Science Foundation (NSF)-NOAA sponsored GLOBal ocean ECosystems dynamics (GLOBEC) program on the Northeast Continental Shelf (GLOBEC, 1991) and proposed for other LMEs (e.g., California Current, Antarctic marine ecosystem) and the proposed Indian Ocean-Somalia Current Ecosystem study planned as part of the Joint Global Ocean Flux Studies (JGOFS).

With a minimum of expense and effort, ongoing FAO fisheries programs can be strengthened by refocusing them around the natural boundaries of regional LMEs. The United Nations Environment Program (UNEP) Regional Seas programs can be enhanced by taking a more holistic ecosystems approach to pollution issues as part of an overall effort to improve the health of the oceans. The LME research and monitoring strategies are compatible with the proposed GOOS, and will, in fact, strengthen GOOS by adding an ecosystem module to the existing physical and meteorological modules (IOC, 1992a). The LME approach will complement the GLOBEC studies and provide useful data inputs to JGOFS. The concept has been discussed at meetings of ICES, International Council for the Scientific Exploration of the Mediterranean Sea (ICSEM), FAO, IOC, International Union for Conservation of Nature and Natural Resources (IUCN), and United Nations Environment Program (UNEP) with generally favorable responses for developing the concept more fully and implementing it more widely within the United Nations framework of ongoing programs. The concept is wholly compatible with the FAO interest in studying "catchment basins" and quantifying their impact on enclosed and semi-enclosed seas (e.g. LMEs). The observations presently underway under the IOC/OSLR (Ocean Studies Related to Living Resources) program now operating within the California Current, Humboldt Current, and Iberian Coastal ecosystems provide an important framework for expanded LME studies of these systems in relation to not only fisheries issues, but also problems of pollution and coastal zone management.

Future effort directed at an improved definition of ecosystem health that will consider both natural environmental perturbation as well as the effects of human intervention on the changing states of ecosystems will focus on:

(1) The development of ecosystem change and health indices and indicators for LMEs.
(2) The development of component models of LMEs incorporating measurements of changing states and health indicators rather than single, large models that generally have limited prediction capability.
(3) The development and evaluation of models using health indicators that are directly applicable to management decisions. They should be simple in construction, allow for interaction with resource managers, and provide sufficient flexibility for testing hypotheses for a range of scenarios.

Efforts are underway to link scientific and societal needs to support long-term, broad-area coastal ocean assessment and monitoring studies. If the proposition for time-series monitoring of changing ecosystem states is to be realized in this period of shrinking budgets, it would be in the best interests of science and society to be tightly linked in the endeavor. The basis for the linkage can be found in a series of recent developments revolving around: (1) recent interest in global climate change; (2) legal precedent for international cooperation implicit in the Law of the Sea; (3) a growing interest in marine ecosystems as regional units for marine research, monitoring, and management; (4) and the effort of the IOC to encourage the implementation of a GOOS. Global climate change has become a factor in the sustainability of biomass production in LMEs. The rather large-scale fluctuations in marine biomass yields of LMEs over the past several decades when considered in the light of the growing concerns over coastal pollution and habitat loss, are serving to accelerate movement toward the development and implementation of a coastal component for GOOS. The core monitoring strategy for large marine ecosystems proposed for the coastal GOOS is designed to provide biological, physical, and chemical data pertinent to the development of indices to monitor changing states of LMEs. The indices are the basis for improving the dialogue between scientists and resource managers, for implementing mitigation strategies where appropriate, and for reinforcing the need for the long-term (multidecadal) ecosystem-wide monitoring programs.

The more holistic, ecosystem-wide approach to broad-area coastal and ocean marine resource assessment through monitoring studies is a means for fostering international cooperation and support in regions where the resources of the ecosystem are shared by several countries. The 49 large marine ecosystems that have been identified are located around the margins of the ocean basins and extend over the coastlines of several countries. They are in regions of the world ocean most affected by overexploitation, pollution, and habitat degradation, and collectively represent target areas for mitigation effort. The Global Environment Facility (GEF) of the World Bank, in collaboration with NOAA, IOC, UNEP, FAO, Natural Environment Research Council (NERC), the Sir Alister Hardy Foundation for Ocean Science, and scientists from national marine resource agencies of several countries (e.g., Belgium, Cameroon, China, Denmark, Estonia, Germany, Ivory Coast, Japan, Kenya, Korea, The Netherlands, Nigeria, Norway, Philippines, Poland, Thailand) are prepared to support LME assessment, mitigation, and coastal monitoring activities of the kind proposed recently in a commentary by Duarte et al. (1992). These efforts will include a comparative approach to long-term monitoring of the environment which allows for examination of data sets from various areas, and is supported by "robust international management and funding systems." It appears that resource managers and scientists are being responsive in reversing the cancellation of monitoring programs described by Duarte et al. (1992), and that monitoring initiatives related to the long-term sustainability of marine resources are underway in Europe and elsewhere (Sherman et al. 1992b).

REFERENCES

AAAS Selected Symposia (1986-1993) **Vol. 1**, Variability and Management of Large Marine Ecosystems. AAAS Selected Symposium 99, Westview Press Inc., Boulder, CO, 1986 319 pp.; **Vol. 2**, Biomass Yields and Geography of Large Marine Ecosystems. AAAS Selected Symposium 111, Westview Press Inc., Boulder, CO, 1989 493 pp.; **Vol. 3**, Large Marine Ecosystems: Patterns, Processes, and Yields. AAAS Symposium, AAAS Press, Washington DC, 1990 242 pp.; **Vol. 4,** Food Chains, Yields, Models, and Management of Large Marine Ecosystems. AAAS Symposium, Westview Press Inc., Boulder, CO. 1991 320 pp.; **Vol. 5**, Large Marine Ecosystems: Stress, Mitigation, and Sustainability. AAAS Press Washington DC, 1993 376 pp.

Aiken J. (1981) The Undulating Oceanographic Recorder Mark 2. J. Plankton Res. 3:551-560

Aiken J., Bellan I. (1990) Optical oceanography: An assessment of a towed method. In Herring P.J., Campbell A.K., Whitfield M., Maddock L. (eds.) Light and Life in the Sea. Cambridge Univ. Press, London pp:39-57

Alexander L.M. (1989) Large marine ecosystems as global management units. In . Sherman K. Alexander L.M. (eds.) Biomass Yields and Geography of Large Marine Ecosystems. American Association for the Advancement of Science (AAAS) Selected Symposium 111. Westview Press Inc Boulder CO pp:339-344

Alheit J., Bernal P. (1993) Effects of physical and biological changes on the biomass yield of the Humboldt Current ecosystem. In Sherman K., Alexander L.M., Gold B.D. (eds.) Large Marine Ecosystems: Stress, Mitigation, and Sustainability. AAAS Press, Washington DC pp:53-68

Alverson D.L., Longhurst A.R., Gulland J.A. (1970) How much food from the sea? Science 168:503-505

Andersen K.P., Ursin E. (1977) A multispecies extension to the Beverton and Holt Theory of Fishing with accounts of phosphorus circulation and primary production. Medd. Dan. Fish-Havunders NS 7:319-435

Anthony V.C. (1988) The New England fisheries in the 21st Century. Nat. Mar. Fish. Serv., Northeast Fish. Sci. Ctr., Woods Hole MA, mimeo

Anthony V.C. (1993) The state of groundfish resources off the northeastern United States. Fisheries 18(3):12-17

Azarovitz T.R., Grosslein M.D. (1987) Fishes and squids. In Backus R.H. (ed.) Georges Bank. MIT Press, Cambridge, MA pp:315-346

Bakun A. (1993) The California Current, Benguela Current, and Southwestern Atlantic Shelf ecosystems: A comparative approach to identifying factors regulating biomass yields. In Sherman K., Alexander L.M., Gold B.D. (eds.) Large Marine Ecosystems: Stress, Mitigation, and Sustainability. AAAS Press, Washington DC pp: 199-221

Bax N.J., Laevastu T. (1990) Biomass potential of large marine ecosystems: a systems approach. In Sherman K., Alexander L.M., Gold B.D. (eds.) Large Marine Ecosystems: Patterns, Processes and Yields. AAAS Press, Washington DC pp:188-205

Beddington J.R. (1984) The response of multispecies systems to perturbations. In May R.M.(ed.) Exploitation of Marine Communities. Springer-Verlag, Berlin pp:209-255

Beddington J.R. (1986) Shifts in resource populations in large marine ecosystems. In Sherman K., Alexander L.M. (eds.) Variability and Management of Large Marine Ecosystems. AAAS Selected Symposium 99. Westview Press Inc., Boulder, CO pp:9-18

Belsky M.H. (1986) Legal constraints and options for total ecosystem management of large marine ecosystems. In Sherman K., Alexander L.M. (eds.) Variability and Management of Large Marine Ecosystems. AAAS Selected Symposium 99. Westview Press Inc., Boulder, CO pp:241-261

Belsky M.H. (1989) The ecosystem model mandate for a comprehensive United States ocean policy and Law of the Sea. San Diego L. Rev. 26(3):417-495

Beverton R.J.H., Holt S.J. (1957) On the dynamics of exploited fish populations. Fish Invest. Minist. Agric. Fish Food (GB) Ser II 19:1-533

Bigelow H.B. (1926) Plankton of the offshore waters of the Gulf of Maine. Bull. Bureau of Fish XL Part 2. Gov Printing Office Washington DC

Blindheim J., Skjoldal H.R. (1993) Effects of climate changes on the biomass yield of the Barents Sea, Norwegian Sea, and West Greenland Large Marine Ecosystems. In Sherman K., Alexander L.M., Gold B.D. (eds.) Large Marine Ecosystems: Stress, Mitigation, and Sustainability. AAAS Press Washington DC pp:185-198

Bombace G. (1993) Ecological and Fishing Features of the Adriatic Sea. In Sherman K., Alexander L.M., Gold B.D. (eds.) Large Marine Ecosystems: Stress,

Mitigation, and Sustainability. AAAS Press, Washington DC pp:119-136

Borisov V. (1991) The state of the main commercial species of fish in the changeable Barents Sea ecosystem. In Sherman K., Alexander L.M., Gold B.D. (eds.) Food Chains, Yields, Models, and Management of Large Marine Ecosystems. Westview Press Inc., Boulder, CO pp:193-203

Bottom D.L., Jones K.K., Rodgers J.D., Brown R.F. (1989) Management of living resources: A research plan for the Washington and Oregon continental margin. Nat. Coastal Resources Res. Development Inst. Newport, OR, NCRI-T-89-004

Bradbury R.H., Mundy C.N. (1989) Large-scale shifts in biomass of the Great Barrier Reef ecosystem. In Sherman K., Alexander L.M. (eds.) Biomass Yields and Geography of Large Marine Ecosystems. AAAS Selected Symposium 111. Westview Press Inc., Boulder, CO pp:143-167

Burger J. (1988) Interactions of marine birds with other marine vertebrates in marine environments. In Burger J. (ed.) Seabirds and Other Marine Vertebrates. Columbia Univ. Press, New York, NY pp:3-28

Byrne, J. (1986) Large marine ecosystems and the future of ocean studies. In Sherman K., Alexander L.M. (eds.) Variability and Management of Large Marine Ecosystems. AAAS Selected Symposium 99. Westview Press Inc Boulder CO

Caddy J. (1993) A contrast between recent fishery trends and evidence for nutrient enrichment in two large marine ecosystems: The Mediterranean and the Black Seas. In Sherman K., Alexander L.M., Gold B.D. (eds.) Large Marine Ecosystems: Stress, Mitigation, and Sustainability. AAAS Press Washington DC pp:137-147

Canon J.R. (1986) Variabilidad ambiental en relacion con la pesqueria neritica pelagica de la zona Norte de Chile. In Arana P. (ed.) La Pesca en Chile. Escuela de Ciencias del Mar, Facultad de Recursos Naturales, Universidad Catolica de Valparaiso, Chile pp:195-205

Christy F.T. Jr. (1986) Can large marine ecosystems be managed for optimum yields? In Sherman K., Alexander L.M. (eds.) Variability and Management of Large Marine Ecosystems. AAAS Selected Symposium 99. Westview Press Inc., Boulder, CO pp: 263-267

Colebrook J.M. (1986) Environmental influences on long-term variability in marine plankton. Hydrobiologia 142:309-325

Costanza R. (1992) Toward an operational definition of ecosystem health. In Costanza R., Norton B.G., Haskell B.D. (eds.) Ecosystem Health: New Goals for Environmental Management. Island Press, Washington DC pp:239-256

Crawford R.J.M., Shannon L.V., Shelton P.A. (1989) Characteristics and management of the Benguela as a large marine ecosystem. In Sherman K., Alexander L.M. (eds.) Biomass Yields and Geography of Large Marine Ecosystems. AAAS Selected Symposium 111. Westview Press Inc Boulder CO pp:169-219

Croxall J.P. (ed.) (1987) Seabirds: Feeding Ecology and Role in Marine Ecosystems. Cambridge Univ. Press, London

Cushing D.H. (1975) Marine Ecology and Fisheries. Cambridge Univ. Press, London

Daan N. (1986) Results of recent time-series observations for monitoring trends in large marine ecosystems with a focus on the North Sea. In Sherman K., Alexander L.M. (eds.) Variability and Management of Large Marine Ecosystems. AAAS Selected Symposium 99. Westview Press Inc., Boulder, CO pp: 145-174.

Dickson R.R., Kelly P.M., Colebrook J.M., Wooster W.S., Cushing D.H. (1988) North winds and production in the eastern North Atlantic. J. Plankton Res. 10:151-169

Duarte C.M., Cebrian J., Marba N. (1992) Uncertainty of detecting sea change. Nature 356:190

FAO [Food and Agriculture Organization of the UN] (1992) FAO yearbook of fishery statistics. Vol. 70 (for 1990) FAO, Rome

Fogarty M., Cohen E.B., Michaels W.L., Morse W.W. (1991) Predation and the regulation of sand lance populations: An exploratory analysis. ICES Mar. Sci. Symp. 193:120-124

GESAMP [Group of Experts on the Scientific Aspects of Marine Pollution] (1990) The state of the marine environment. UNEP Regional Seas Reports and Studies No 115, Nairobi

Global Ocean Ecosystems Dynamics [GLOBEC] (1991) Report Number 1. Initial science plan. February 1991. Produced by Joint Oceanographic Institutions Inc., Washington DC

Glover R.S. (1967) The continuous plankton recorder survey of the North Atlantic. Symp. Zool. Soc. Lon. 19:189-210

Gulland J.A. (1984) Epilogue. In May R.M. (ed.) Exploitation of Marine Communities. Springer-Verlag, Berlin pp:335-337

Hempel G. (ed.) (1978) Symposium on North Sea fish stocks - Recent changes and their causes. Rapp. P-v Reun. Cons. int. Explor. Mer. 172

Holling C.S. (1973) Resilience and stability of ecological systems. Institute of Resource Ecology, Univ. British Columbia, Vancouver, Canada

Holling C.S. (1986) The resilience of terrestrial ecosystems local surprise and global change. In Clark W.C., Munn R.E. (eds.) Sustainable Development of the Biosphere. Cambridge Univ. Press, London pp:292-317

Holmgren-Urba D., Baumgartner T.R. (1993) A 250-year history of pelagic fish abundances from the anaerobic sediments of the central Gulf of California. Calif. Coop. Oceanic Fish Inv. Rep 34:60-68

Hovgaard H., Buch E. (1990) Fluctuation in the cod biomass of the West Greenland Sea ecosystem in relation to climate. In Sherman K., Alexander L.M., Gold B.D. (eds.) Large Marine Ecosystems: Patterns, Processes and Yields. AAAS Press, Washington DC pp:36-43

ICES [International Council for the Exploration of the Sea] (1991) Report of the Multispecies Working Group. ICES CM 1991/Assess:7

IOC [International Oceanographic Commission of UNESCO] (1992a) GOOS. Global Ocean Observing System, An initiative of the Intergovernmental Oceanographic Commission (of UNESCO). IOC, UNESCO, Paris, France

IOC (1992b) The use of large marine ecosystem concept in the Global Ocean Observing System (GOOS). Twenty-fifth Session of the IOC Executive Council, Paris, 10-18 March 1992. IOC/EC-XXV/Inf.7

Jossi J.W., Smith D.E. (1990) Continuous plankton records: Massachusetts to Cape Sable, NS, and New York to the Gulf Stream, 1989. NAFO Ser. Doc. 90/66:1-11

Karr J. (1992) Ecological Integrity: Strategies for protecting earth's life support system. In Costanza R., Norton B.G., Haskell B.D. (eds.) Ecosystem Health: New Goals for Environmental Management. Island Press, Washington DC pp. 223-238

Kawasaki T., Tanaka S., Toba Y., Taniguchi A. (eds.) (1991) Long-term variability of pelagic fish populations and their environment. Proceedings of the International Symposium. Sendai Japan 14-18 November 1989. Pergamon Press, Tokyo, Japan

Kolber Z.S., Falkowski P.G. (1992) Fast repetition rate (FRR) fluorometer for making in situ measurements of primary productivity. Oceans 92: Mastering the oceans through technology, 26-29 October 1992, Newport, RI. Proceedings volume 2. IEEE 0-7803-0838-7/92 pp: 637-641

Kullenberg G. (1986) Long-term changes in the Baltic Ecosystem. In Sherman K., Alexander L.M. (eds.) Variability and Management of Large Marine Ecosystems. AAAS Selected Symposium 99. Westview Press Inc., Boulder, CO pp:19-32

Lasker R. (1988) Food chains and fisheries: An assessment after 20 years. In Rothschild B.J. (ed.) Toward a Theory on Biological-Physical Interactions in the World Ocean. pp 173-182. NATO ASI Series. Series C: Mathematical and Physical Sciences, Vol. 239, Kluwer Academic Publishers, Dordrecht, The Netherlands

Levin S.A. (1978) Pattern formation in ecological communities. In Steele J.A. (ed.) Spatial pattern in plankton communities. Plenum Press, New York pp: 433-470

Levin S.A. (1990) Physical and biological scales, and modelling of predator-prey interactions in large marine ecosystems. In Sherman K., Alexander L.M., Gold B.D. (eds.) Large Marine Ecosystems: Patterns, Processes, and Yields. AAAS Press, Washington DC pp:179-187

Levin S. (1993) Approaches to forecasting biomass yields in large marine ecosystems. In Sherman K., Alexander L.M., Gold B.D. (eds.) Large Marine Ecosystems: Stress, Mitigation, and Sustainability. AAAS Press, Washington DC pp:36-39.

Likens G.E. (1992) The ecosystem approach: Its use and abuse. In Kinne O. (ed.) Excellence in Ecology. Vol 3. Ecology Institute W2124 Oldendorf/Luhe, Germany

Lindemann R.L. (1942) The trophic dynamic aspect of ecology. Ecology 23:399-418

MacCall A.D. (1986) Changes in the biomass of the California Current system. In Sherman K., Alexander L.M. (eds.) Variability and Management of Large Marine Ecosystems. AAAS Selected Symposium 99. Westview Press Inc., Boulder, CO pp:33-54

Malone T.C. (1991) River flow, phytoplankton production and oxygen depletion in Chesapeake Bay. In Tyson R.V., Pearson T.H. (eds.) Modern and Ancient Continental Shelf Anoxia. Geological Society Spec. Publ. No. 58

Mangel M. (1991) Empirical and theoretical aspects of fisheries yield models for large marine ecosystems. In Sherman K., Alexander L.M., Gold B.D. (eds.) Food Chains, Yields, Models, and Management of Large Marine Ecosystems. Westview Press Inc., Boulder, CO pp:243-261

May R.M. (ed.) (1984) Exploitation of Marine Communities. Springer-Verlag, Berlin

Minoda T. (1989) Oceanographic and biomass changes in the Oyashio Current ecosystem. In Sherman K., Alexander L.M. (eds.) Biomass Yields and Geography of Large Marine Ecosystems. AAAS Selected Symposium 111. Westview Press Inc Boulder CO pp: 67-93

Morgan J.R. (1988) Large marine ecosystems: an emerging concept of regional management. Environment 29(10):4-9 and 26-34

Morgan J.R. (1989) Large marine ecosystems in the Pacific Ocean. In Sherman .K, Alexander L.M. (eds.) Biomass Yields and Geography of Large Marine Ecosystems. AAAS Selected Symposium 111. Westview Press Inc., Boulder, CO pp: 377-394.

Murawski S.A. (1991) Can we manage our multispecies fisheries? Fisheries 16(5):5-13

Myers N. (1990) Working towards one world. Book review. Nature 344(6266):499-500

NSTF [North Sea Task Force] (1991) Scientific Activities in the Framework of the North Sea Task Force. North Sea Environment Report No 4. North Sea Task Force, Oslo and Paris Commissions, International Council for the Exploration of the Sea London

Norton B.G., Ulanowicz R.E. (1992) Scale and biodiversity policy: A hierarchical approach. AMBIO 21(3):244-249

Overholtz W.J., Nicolas J.R. (1979) Apparent feeding by the fin whale *Balaenoptera physalus*, and humpback whale, *Megoptera novaeangliae*, on the American sand lance, *Ammodytes americanus*, in the Northwest Atlantic. Fish. Bull. US 77:285-287

Payne P.M., Wiley D.N., Young S.B., Pittman S., Clapham P.J., Jossi J.W. (1990) Recent fluctuations in the abundance of baleen whales in the southern Gulf of Maine in relation to changes in selected prey. Fish. Bull. US 88:687-696

Pimm S.L. (1984) The complexity and stability of ecosystems. Nature 307:321-326

Piyakarnchana T. (1989) Yield dynamics as an index of biomass shifts in the Gulf of Thailand ecosystems. In Sherman K., Alexander L.M. (eds.) Biomass Yields and Geography of Large Marine Ecosystems. AAAS Selected Symposium 111. Westview Press Inc., Boulder, CO pp: 95-142

Postma H., Zijlstra J.J. (eds.) (1988) Ecosystems of the world 27 - Continental shelves. Elsevier Amsterdam, The Netherlands.

Powers K.D., Brown R.G.B. (1987) Seabirds. In Backus R.H.. (ed.) Georges Bank. MIT Press Cambridge MA pp:359-371

Prescott J.R.V. (1989) The political division of large marine ecosystems in the Atlantic Ocean and some associated seas. In Sherman K., Alexander L.M. (eds.) Biomass Yields and Geography of Large Marine Ecosystems. AAAS Selected Symposium

111. Westview Press Inc., Boulder, CO pp: 395-442
Rapport D.J. (1992) Defining the Practice of Clinical Ecology. In Costanza R., Norton B.G., Haskell B.D. (eds.) Ecosystem Health: New Goals for Environmental Management. Island Press, Washington DC pp:144-156
Ricklefs R.E. (1987) Community diversity: Relative roles of local and regional processes. Science 235(4785):167-171
Riley G.A., Stommel H., Bumpus D.F. (1949) Quantitative ecology of the plankton of the western North Atlantic. Bull. Bingham Oceanogr. Coll. XII (3)
Ryther J.H. (1969) Relationship of photosynthesis to fish production in the sea. Science 166:72-76
Sainsbury K.J. (1988) The ecological basis of multispecies fisheries, and management of a dermersal fishery in tropical Australia. In Gulland J.A. (ed.) Fish Population Dynamics, 2nd ed. John Wiley & Sons, New York pp:349-382
Schaefer M.B. (1954) Some aspects of the dynamics of populations important to the management of the commercial marine fisheries. Bull. Inter-Am. Trop. Tuna Comm. 1:27-56
Scully R.T., Brown W.Y., Manheim B.S. (1986) The Convention for the Conservation of Antarctic Marine Living Resources: A model for large marine ecosystem management. In Sherman K., Alexander L.M. (eds.) Variability and Management of Large Marine Ecosystems. AAAS Selected Symposium 99. Westview Press Inc., Boulder, CO pp:281-286.
Shelton P.A., Boyd A.J., Armstrong M.J. (1985) The influence of large-scale environmental processes on neritic fish populations in the Benguela Current system. Calif. Coop. Oceanic Fish. Invest. Rep. 26:72-92
Sherman K. (1991) The large marine ecosystem concept: A research and management strategy for living marine resources. Ecol. Applications 1(4):349-360
Sherman K. (1993) Emerging theoretical basis for monitoring changing states (health) of large marine ecosystems. US Dep. Commer., NOAA Tech. Mem. NMFS-F/NEC-100
Sherman K., Alexander L.M. (eds.) (1986) Variability and Management of Large Marine Ecosystems. AAAS Selective Symposium 99. Westview Press Inc., Boulder, CO
Sherman K., Laughlin T. (eds.) (1992) Large marine ecosystems monitoring workshop report. US Dep. Commer., NOAA Tech. Mem. NMFS-F/NEC-93
Sherman K., Ryan A.F. (1988) Antarctic marine living resources. Oceanus 31(2):59-63
Sherman K., Alexander L.M., Gold B.D. (eds) (1990a) Large Marine Ecosystems: Patterns, Processes and Yields. AAAS Press, Washington DC
Sherman K., Alexander L.M., Gold B.D. (eds.) (1993) Large Marine Ecosystems: Stress, Mitigation, and Sustainability. AAAS Publishers, Washington DC
Sherman K., Cohen E.B., Langton R.W. (1990b) The northeast continental shelf: An ecosystem at risk. In Konrad V., Ballard S., Erb R., Morin A. (eds.) Gulf of Maine: Sustaining our Common Heritage. Proceedings of an International Conference held at Portland, Maine, December 10-12, 1989. Published by Maine

State Planning Office and the Canadian-American Center of the University of Maine pp: 120-167

Sherman K., Green J.R., Goulet J.R., Ejsymont L. (1983) Coherence in zooplankton of a large Northwest Atlantic Ecosystem. Fish. Bull., US 81:855-62

Sherman K., Grosslein M., Mountain D., Busch D., O'Reilly J., Theroux R. (1988) The continental shelf ecosystem off the northeast coast of the United States. In Postma H., Zilstra J.J. (eds.) Ecosystems of the World 27: Continental Shelves. Elsevier Amsterdam, The Netherlands pp:279-337.

Sherman K., Jaworski N., Smayda T. (1992a) The Northeast Shelf Ecosystem: Stress, Mitigation, and Sustainability, 12-15 August 1991 Symposium Summary. US Dep. Commer., NOAA Tech. Mem. NMFS-F/NEC-94

Sherman K., Skjoldal H., Williams R. (1992b) Global ocean monitoring. Nature 359:769

Sissenwine M.P., Cohen E.B. (1991) Resource productivity and fisheries management of the northeast shelf ecosystem. In Sherman K., Alexander L., Gold B.D. (eds.) Food Chains, Yields, Models, and Management of Large Marine Ecosystems. Westview Press Inc., Boulder, CO pp:107-123

Skjoldal H.R., Rey F. (1989) Pelagic production and variability of the Barents Sea ecosystem. In Sherman K., Alexander L.M. (eds.) Biomass Yields and Geography of Large Marine Ecosystems. Westview Press Inc., Boulder, CO pp:241-286

Smayda T. (1991) Global epidemic of noxious phytoplankton blooms and food chain consequences in large ecosystems. In Sherman K., Alexander L.M., Gold B.D. (eds.) Food Chains, Yields, Models, and Management of Large Marine Ecosystems. Westview Press Inc., Boulder, CO pp: 275-307

Smith W.G., Morse W.W. (1990) Larval distribution patterns: evidence for the collapse/recolonization of Atlantic herring on Georges Bank. ICES CM 1990/H:17

Steele J.H. (1974) The Structure of Marine Ecosystems. Harvard Univ. Press, Cambridge, MA

Steele J.H. (1988) Scale selection for biodynamic theories. In Toward a Theory on Biological-Physical Interactions in the World Ocean. Rothschild B.J. (ed.) NATO ASI Series C: Mathematical and Physical Sciences Vol 239. Kluwer Academic Publishers, Dordrecht, The Netherlands pp:513-526

Sugihara G., Garcia S., Gulland J.A., Lawton J.H., Maske H., Paine R.T., Platt T., Rachor E., Rothschild B.J., Ursin E.A., Zeitzschel B.F.K. (1984) Ecosystem dynamics: group report. In May R.M. (ed.) Exploitation of Marine Communities. Springer-Verlag, Berlin pp:130-153

Tang Q. (1989) Changes in the biomass of the Yellow Sea ecosystem. In Sherman K., Alexander L.M. (eds.) Biomass Yields and Geography of Large Marine Ecosystems. AAAS Selected Symposium 111. Westview Press Inc., Boulder, CO pp: 7-35

Taylor P., Groom A.J.R. (eds.) (1989) Global Issues in the United Nation's

Framework. Macmillan, London
Terazaki M. (1989) Recent large-scale changes in the biomass of the Kuroshio Current ecosystem. In Sherman K., Alexander L.M. (eds.) Biomass Yields and Geography of Large Marine Ecosystems. AAAS Selected Symposium 111. Westview Press Inc., Boulder, CO pp: 37-65
UNESCO [United Nations Educational, Scientific and Cultural Organization] (1992) Monitoring the Health of the Oceans: Defining the Role of the Continuous Plankton Recorder in Global Ecosystems Studies. The Intergovernmental Oceanographic Commission and The Sir Alister Hardy Foundation for Ocean Science. IOC/INF-869, SC-92/WS-8
U.S. Council on Environmental Quality and the Department of State, Gerald O. Barney (director) (1980) The global 2000 Report to the President: Entering the Twenty-First Century. Vols I-III. US Government Printing Office, Washington DC
Williams R., Aiken J. (1990) Optical measurements from underwater towed vehicles deployed from ships-of-opportunity in the North Sea. In Nielsen H.O. (ed.) Environment and Pollution Measurement, Sensor and Systems. Proc. SPIE 1269 pp:186-194
Wise J.P. (1984) The future of food from the sea. In Simon J.L., Kahn H. (eds.) The Resourceful Earth. Basil Blackwell Inc., New York pp:113-127
WGEAMS [Working Group on Environmental Assessments and Monitoring Strategies] (1992) Report of the Working Group on Environmental Assessments and Monitoring Strategies. ICES CM 1992/Poll:9, Sess V ICES Copenhagen, Denmark
Wyatt T., Perez-Gandaras G. (1989) Biomass changes in the Iberian ecosystem. In Sherman K., Alexander L.M. (eds.) Biomass Yields and Geography of Large Marine Ecosystems. AAAS Selected Symposium 111. Westview Press Inc., Boulder, CO pp: 221-239
Zaitsev Y.P. (1992) Recent changes in the trophic structure of the Black Sea. Fish. Oceanogr. 1(2):180-189
Zijlstra J.J., Baars M.A. (1990) Productivity and fisheries potential of the Banda Sea Ecosystem. In Sherman K., Alexander L.M., Gold B.D. (eds.) Large Marine Ecosystems: Patterns, Processes, and Yields, AAAS Press, Washington DC pp:54-65

21. REMOTE SENSING AND ECOSYSTEM HEALTH: AN EVALUATION OF TIME-SERIES AVHRR NDVI DATA

David A. Mouat
Biological Sciences Center
Desert Research Institute, University of Nevada System
Reno, Nevada
USA

INTRODUCTION

The concept of ecosystem health is one which could proceed from a positive or negative standpoint; most easily, in fact, from the latter - that is, from the absence of disease. Costanza et al. (1992) however, prefer a definition that proceeds from a more positive direction, a definition that states more positively the characteristics of healthy ecosystems. Their definition involves a set of ecologically-based criteria of sustainability within a generally hierarchical approach and defines ecosystem health as a complex of natural systems. Within this context, Costanza et al. (1992) use a broad medical model to develop the ecosystem health paradigm. Hargrove (1992), on the concept of ecosystem health writes "The notion of ecosystem health has developed on the model of human health as an aesthetic perspective. Everybody can tell if a person 'looks' healthy, and usually that is what is meant to 'be' healthy. Nevertheless, doctors can frequently determine that a person who looks healthy really is not. Seen in this way, the ability to perceive health in a person depends largely on technical knowledge". The theme of this section is to describe the use of a technology, remote sensing, that, when used with ancillary evidence, assists in the determination of health in ecosystems.

Remote sensing measurements integrated with other measurements of the environment can result in a strategy for estimating and monitoring ecosystem health. Remote sensing techniques provide the opportunity to make measurements of objects and sets of objects. The measurements rely on the physical interactions of sensor and object (or set of objects). The sensor detects energy (radiance) reflected and emitted by the objects across the electromagnetic spectrum in discrete bands or wavelength intervals. This collected radiation is unique to the object (or set of objects) and may be used to characterize the objects; that is, to identify or make some statement about the condition of the object. The ability to differentiate environmental phenomena depends in part on the phenomena (that is, the characteristics of what is being sensed) and on the characteristics of the sensors: the spatial, spectral, radiometric (ability to detect levels of brightness), and the sensor system's temporal resolution. The latter, a function of the orbit and other characteristics of the host satellite, can allow discrimination of phenomena as a function of how characteristics of the phenomena change with time.

This paper discusses the use of time series remote sensing data acquired by satellites of ecosystem structure and function which, when integrated over time, might serve as a measure of ecosystem health, a potential stress test which can show how the dynamic aspects of ecosystem structure and function can serve to illustrate condition of health. Indices such as LAI (leaf area index), NPP (net primary productivity), NDVI (Normalized Difference Vegetation Index), and other indices taken over time can be considered to effectively monitor stress provided that 1) the nature of the stress is clearly understood; 2) the indices are acquired over a sufficiently long time interval to be able to adequately characterize ecosystem response; 3) suitable land use histories, vegetation data, climate data, and other, ancillary data are available for an equally sufficient time period to allow for adequate index/ground data comparison; 4) sufficient ground data are acquired to adequately verify the relationships which are being stated. In our time series analyses, then, we are looking for elements of sustainability, diversity, nutrient retention, and to a lesser extent, productivity, which, when examined over time provide an indication of health. When these elements of health remain within ecosystems following perturbations at "acceptable" or "nominal" levels, then we can consider them to be indicative of health. On the other hand, if these indicators of health, as measured by long term indices, are not resilient in the face of perturbations, we might be able to conclude that the ecosystem response did not adequately respond to the stress test.

SOME RELATIONSHIPS AMONG ECOSYSTEM STRUCTURAL AND FUNCTIONAL VARIABLES RELEVANT TO REMOTE SENSING

A number of researchers have shown very strong relationships between ecosystem structural (such as biomass and Leaf Area Index or LAI) and functional (such as gas exchange, and Net Primary Productivity or NPP) variables (e.g. Gholz 1982; Waring et al. 1978). Gholz, in reporting research on a diverse array of vegetation types ("zones") in the states of Oregon and Washington, observed a strong correlation ($R^2 = 0.96$), between LAI and NPP. Sellers (1985, 1987) described a significant relationship between LAI and absorbed photosynthetically active radiation (APAR), approaching, asymptotically, an LAI of 6 where nearly all incident short wave radiation is absorbed by the canopy. Sellers showed that:

$$APAR = f[LAI, ISR, Canopy\ geometry]$$

where ISR = incident short wave radiation.

Many researchers have utilized combinations of spectral radiance observations acquired over vegetated surfaces by satellite or aircraft-mounted sensors as indicators of density, health, or biomass of the vegetation (Sellers et al. 1992). These empirical applications of remote sensing take advantage of the large difference between the light scattering properties of green leaves in the visible and near infrared wavelength intervals (Sellers et al. 1992). A strategy using remote sensing then, is that if LAI or another ecosystem variable can be measured with a known degree of certainty from the synoptic perspective of an aircraft or satellite sensor, then an effective method for analyzing ecosystems from a spatial perspective will be possible. Furthermore, if these measurements can be made over time, then a dynamic evaluation of ecosystems might be possible.

Consequently, given a canopy of known structure and light scattering and absorbing properties, any one measure of the canopy can be used interchangeably with the others with some algebraic manipulation of the formulae (Running 1990). The Normalized Difference Vegetation Index, NDVI, is illustrative of a remote sensing measurement having a very close relationship with APAR (e.g. Myneni et al. 1992). Differences in canopy biophysical rates (e.g. photosynthesis and conductance), as indicators of ecosystem health, can be discriminated through the use of the NDVI acquired over time.

USE OF AVHRR NDVI FOR ECOSYSTEM STUDIES

The *Advanced Very High Resolution Radiometer* (AVHRR) is a five-channel sensor on the NOAA series of satellites in sun synchronous orbit (NOAA 1986). The scanner records measurements of the earth's surface with 1.1 km resolution at nadir in the following spectral bands: red (Channel 1, 0.58-0.68 μm), near infrared (Channel 2, 0.725-1.10 μm), the middle infrared (Channel 3, 3.55-3.93 μm), and two thermal infrared bands (Channel 4, 10.3-11.3 μm and Channel 5, 11.5-12.5 μm). While the spatial resolution of the AVHRR might be considered coarse for monitoring ecosystem dynamics on a regional or global scale, it is much more attractive on account of its twice daily temporal resolution (i.e. data are acquired twice daily for the entire earth's surface). The 1.1 km pixel size (referred to as high resolution picture transmission, for data collected directly by ground receiving stations, and as local area coverage, for data gathered using on-board satellite tape recorders) of the AVHRR results potentially in the integration of diverse land cover types averaging diverse spectral responses. As the AVHRR pixel size degrades away from nadir or as a result of pixel aggregation (for example, 4 or 16 km pixels, referred to as global area coverage) the likelihood for integration of diverse cover types increases. Yet, the importance of examining ecosystems from a regional, continental, and global context often requires the use of these coarse data sets. Malingreau and Belward (1992) state that the choice of scales is dependent on the property of the ecological process of interest. It is hypothesized here that the spatial scale afforded by the AVHRR is directly applicable to the examination of ecosystem health, especially when ecosystem components are integrated and ecosystem processes over large areas are considered.

Tucker recognized the importance of using a vegetation index developed from AVHRR data early on. With it, he and others have analyzed vegetation dynamics on continental and global scales (e.g. Tucker et al. 1985). Perhaps the most widely used (Graetz 1990) vegetation index is the Normalized Difference Vegetation Index (NDVI) expressed as:

$$NDVI = (NIR-RED)/(NIR+RED)$$

where NIR is a near infrared channel (e.g. AVHRR Channel 2) and RED is a red channel (e.g. AVHRR Channel 1). Transformations such as the NDVI have advantages over the original data as they minimize soil and litter background effects, reduce data dimensionality, provide a degree of standardization for comparison, and enhance the vegetation signal. Furthermore, the normalization provided by the NDVI partially compensates for changing illumination conditions and surface terrain effects (Reed et al. 1994).

The use of spectral indices for measuring LAI was first attempted for crops and grasslands, correlating spectral reflectances against direct measurement of LAI. Running (1990) and others have found strong relationships between NIR/RED ratios from Landsat Thematic Mapper data and LAI as well as between AVHRR NDVI and LAI in coniferous forests (Figure 1). While NDVI has enjoyed widespread use in ecosystem studies, a number of problems need to be addressed when considering its use. Malingreau and Belward (1992) discuss problems of changes in scale from the nadir outward; Huete and Tucker (1991) discuss the importance of soil condition (including color) variations; Spanner et al. (1990) describe problems associated with quantization truncation (from 10 bits to 8), changes in solar zenith angle, atmospheric effects, and sensor scan angle; to list just a few of the problems which can be encountered. Nevertheless, if the difficulties are clearly kept in mind, a wide diversity of ecosystem studies can be made with the data. The effects can be somewhat mitigated, however, through the use of maximum value compositing (e.g. Spanner et al. 1990). In maximum value compositing, the highest AVHRR NDVI value of a site for a period of time (one week to one month) of registered imagery is calculated. The highest value of each pixel from the registered imagery is retained for analysis. The maximum value composite selects the least hazy pixel for the compositing period.

USE OF TIME-SERIES AVHRR NDVI IN THE ANALYSIS OF ECOSYSTEM HEALTH

Information on environmental issues is a function of both the spatial and temporal domains. By conducting research with data sets on the order of those afforded by the AVHRR, the spatio-temporal structure of the measurements, and in many cases of the ground features themselves, can be better approached (Malingreau and Belward 1992). It is proposed that one measure of ecosystem health can be the temporal response pattern of an ecosystem function, for example NPP, when evaluated against a closely related and causative factor such as climate. As such, examination of time-series measurements of ecosystem structure and function is suggested as a technique for the evaluation of ecosystem health.

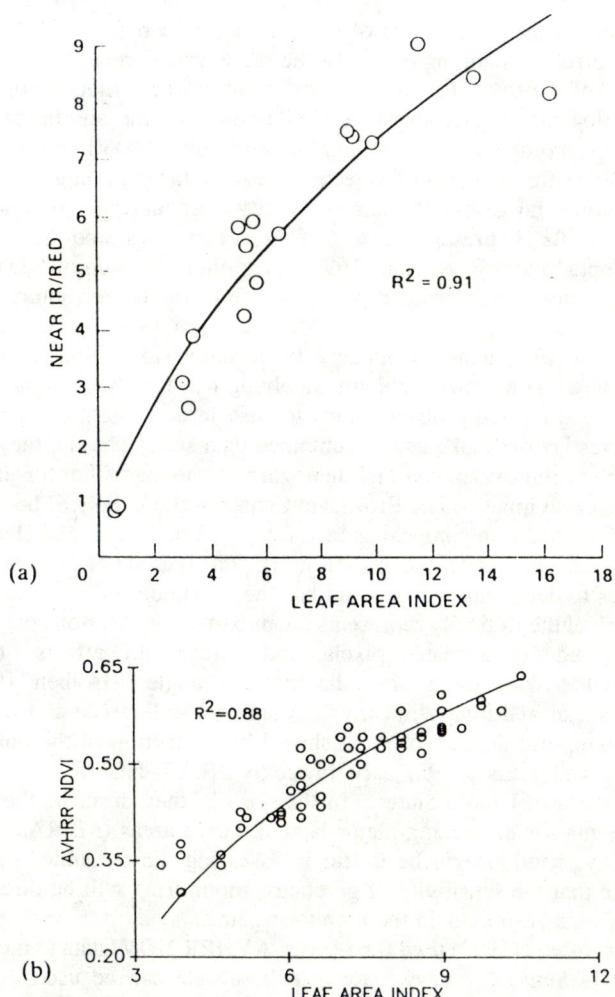

Figure 1. Relationships between leaf area index (LAI) and Landsat TM (NIR/RED) and AVHRR NDVI (adapted from Running 1990).

Time-series analyses of ecosystem processes (e.g. Wharton et al. 1990) gives an indication of the ecosystem's phenological response or pattern. The phenological pattern of an ecosystem might be considered as an indicator of its health. If, for example, land use histories within an ecosystem diverge such that erosion rates and vegetation degradation accelerate in one vis-a-vis the other, it might be suggested that one was less healthy than the other. It would be expected that the phenology, that is, the "timing of recurring biological events, the causes of their timing with regard to biotic and abiotic forces, and the interrelation among phases of the same or different species" (Lieth 1974; Reed et al. 1994), of the two sites would be different. An alternative possibility would be that the phenological pattern might be similar but that the amplitude of measured responses along a phenological gradient might be different. NDVI values acquired from the AVHRR satellite offer a means of objectively evaluating phenological characteristics of land cover regions and assessing their variability over large geographic areas (Reed et al. 1994). Table 1 presents a list of NDVI metrics and their phenological interpretation (adapted from Reed et al. 1994). The interpretation of NDVI time-series data sets discussed later in this chapter follows much of the nomenclature developed by Reed.

While the AVHRR acquires imagery twice daily (four times a day when two satellites are orbiting), numerous problems involving off-nadir viewing and, especially, cloud cover, preclude the daily observations for use in ecosystem comparison studies. Consequently, investigators make use of combined data sets. That is, they select a time interval involving several overpasses and then work out an algorithm for the selection of an optimal time interval image. The Eros Data Center of the U.S.G.S. began developing biweekly vegetation condition composites in 1989 (e.g. Eidenshink and Haas 1992) from AVHRR NDVI. For each weekly compositing period, the NDVI is examined for each of the daily passes to determine which pixel has the maximum value. Using the highest NDVI value per pixel theoretically represents the maximum vegetation condition, reduces the number of cloud contaminated pixels, and reduces the effects of atmospheric conditions associated with large off-nadir viewing angles (Holben 1986). Other researchers use shorter and longer intervals. Again, as Malingreau and Belward (1992) pointed out, the temporal domain or scale should be a function of the questions asked.

Eidenshink and Haas (1992) used biweekly NDVI data to examine vegetation dynamics in the western United States. In their work, they describe the variability of NDVI measurements for examining major land resource areas (MLRAs), and consider that this variability, itself, might be useful in assessing the dynamics of the systems. They further state that the sensitivity of greenness monitoring will be directly related to the uniformity of land resources in the monitoring unit.

Tappan and others (1992) used time-series AVHRR NDVI data to monitor seasonal range conditions in Senegal. They state that these data can be used as indicators of grazing conditions and drought and further state that identification of drought anomaly assessment is best accomplished through the comparison of production potential within homogeneous resource units.

Table 1. NDVI metrics and their interpretation (adapted from Reed et al. 1994).

NDVI Metrics	Phenological Interpretation
Composite period of onset of greenness	Time when measurable photosynthetic activity begins
Value of onset of greenness	Level of photosynthetic activity when growing season begins
Composite period of end of greenness	Time when measurable photosynthetic activity ceases
Value of end of greenness	Level of photosynthetic activity when growing season ends
Duration of greenness	Length of measurable photosynthetic activity
Composite period of maximum NDVI	Time of maximum measurable level of photosynthetic activity
Value of maximum NDVI	Maximum measurable value of photosynthetic activity
Range of NDVI values	Range of measurable level of photosynthetic activity
Time-integrated NDVI	Gross primary production
Modality	Periodicity of photosynthetic activity
Rate of greenup	Acceleration of photosynthesis
Rate of senescence	Deceleration of photosynthesis

Figure 2, adapted from their work, illustrates changes in the NDVI of a formation type, Sahelian shrub steppe, in a site located in northern Senegal. While the figure is highly illustrative of the phenological development of this site, it is more illustrative of the extreme variability of this site over the period, 1984-1989, from which the data came.

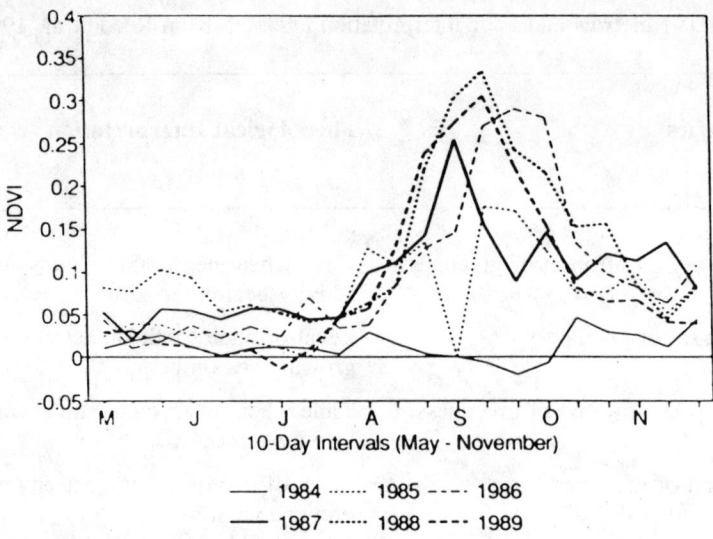

Figure 2. NDVI time-series (1984-1989) profiles of Sahelian shrub steppe from northern Senegal (adapted from Tappan et al. 1992).

Malo and Nicholson (1990) investigated the relationship between the NDVI of six formation types and rainfall over a four year period in west Africa. They found that both the temporal and spatial patterns of monthly NDVI closely replicate those of rainfall. They report a high linear correlation between NDVI and rainfall, although NDVI lags behind rainfall by approximately two months. They further report that, for most of their study area, NDVI is a sensitive indicator of interannual fluctuations, with the strongest response occurring as long as annual rainfall exceeds about 150mm (Malo and Nicholson, 1990). The ratio of NDVI to rainfall provides a rough quantitative measure of the efficiency of water use. In their study, the highest efficiencies were found in plant formations of the driest environments.

Peters and others (1993) compared existing drought indices (i.e. the Palmer Drought Severity Index, PDSI) to an AVHRR-based vegetation index to identify measures of plant/moisture relationships within native semiarid ecosystems in New Mexico. They found that the grassland plant community type was more spectrally responsive to moisture stress than were shrublands or forests. Burgan and Hartford (1993) have illustrated how NDVI responses of different vegetation formations (forest, grassland, and shrubland) trend differently. They showed that differences in timing and extent of greenness within a vegetation type can be observed at specific sites across different years.

Water use efficiencies and other ecosystem processes provide an indicator to ecosystem health (for additional discussion on the use of indicators for monitoring and assessing ecosystems see Mouat et al. 1992). That is, healthy ecosystems (within similar environmental constraints) have higher water use efficiencies, nutrient availability, higher C fixation rates, and other processes than less healthy ecosystems. Schlesinger and others (1990) in reporting on research conducted at the Jornada Experimental Range in southern New Mexico, found that when net, long-term desertification (or degradation) of productive grasslands occurs, a relatively uniform distribution of water, nitrogen, and other soil resources, is replaced by an increase in their spatial and temporal heterogeneity. This heterogeneity leads to invasion of the grasslands by shrubs. The net effect of land degradation in arid and semiarid (including subhumid) regions, resulting from reallocation of resources, changes in species composition, decreases in plant cover and biomass (typically), and changes in ecosystem process efficiencies is that those less healthy systems behave as if they are occurring in more xeric environments. Those ecosystems, then, when examined by time-series AVHRR NDVI data, should behave differently than will healthier systems. If two or more sites having similar origins but diverging in terms of their land use histories are examined through the use of time-series data, their response patterns with respect to perturbations occurring within the time-series will differ. It is hypothesized that these different responses might be used as indicators of health.

Three sets of time-series AVHRR NDVI data within the United States have been analyzed by a hypertemporal image processing algorithm (Elvidge and Jansen 1991) for the purpose of establishing temporal response of NDVI within healthy and less healthy ecosystems. Sites have been chosen from the perspective of common origin, that is, they (two or more sites) are selected such that they originated from similar conditions (climatic, edaphic, physiographic) but diverged as a result of land use practices including grazing, chaining, herbicide application, fencing, etc. Sites having common origins were compared with respect to the manner in which their NDVI response pattern varied with time. An attempt was then made to determine if differences in pattern could be related to differences in health. Examples have been chosen from California, Utah, and New Mexico.

Figure 3 depicts NDVI responses of two sites in central California; Figure 4 depicts NDVI responses of two sites in the Colorado Plateau within the state of Utah; and Figure 5 depicts NDVI responses of two sites in the Jornada Experimental Range region of southern New Mexico. Each set of AVHRR NDVI responses makes use of 44 months of Global Vegetation Index data by the Global Ecosystem Database produced in 1991 by the U.S. Environmental Protection Agency Environmental Research Laboratory and the NOAA National Geophysical Data Center and are composited from 4 km Global Area Coverage NDVI data with aggregation to 10 minute pixels. They are in 8 bit form and cover the period May 1985 to December 1988.

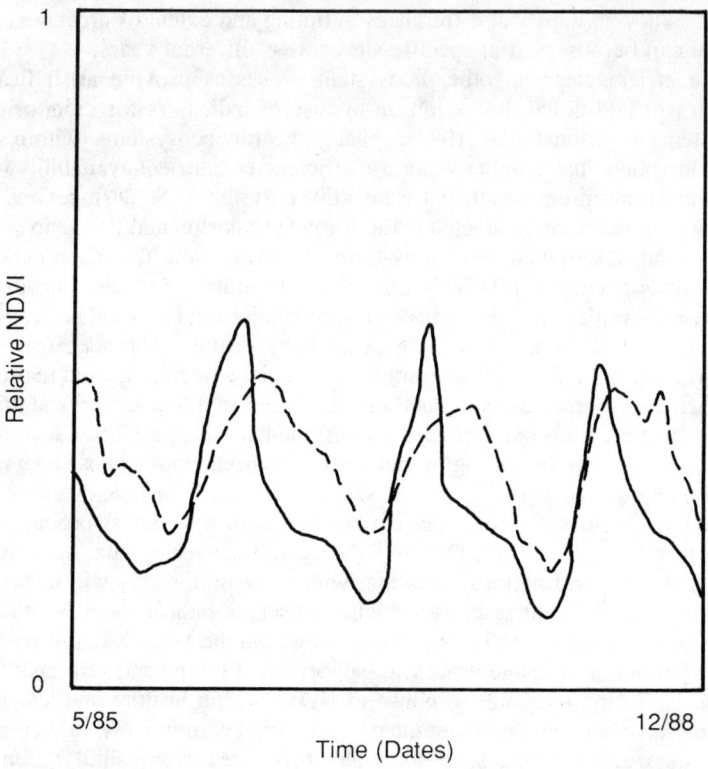

Figure 3. NDVI time-series (1985-1988) curves for two grassland sites in central California (the annual grassland is indicated by the solid line while the perennial grassland is indicated by the dashed line).

Analysis of the data sets depicted in Figures 3 to 5 does not reveal differences in the phenological dynamics of different formation types, but rather differences in the temporal responses of vegetation originating within the same formation type in similar ecological sites and differing as a result of changes in land use histories. The NDVI response of each site over time is a function of several factors, most notably changes in the seasonal and annual weather pattern at each site and also changes in ecosystem response functions as a result of the effects of land use histories. NDVI differences on a temporal basis between (or among) sites having similar characteristics is hypothesized to be a function of the site's land use history.

In the California and Colorado Plateau data sets (Figures 3 and 4), the sites compared occur within the grassland formation type. The more disturbed grassland site in California has a much greater annual component in its composition while the less disturbed site has a greater perennial component. The NDVI response for the annual site shows a consistently steep greenup period, a rapid and concomitantly steep senescence, and an earlier date of peak greenness. The perennial grassland, on the other hand, responds with slower phenological changes, a broader green plateau, and a retarded date of peak greenness. It also has a more consistent peak than does the annual grassland. These observations are corroborated by other lines of research involving grassland ecology (e.g. Anderson and Inouye 1991; Chapman 1992; Rickard et al. 1988).

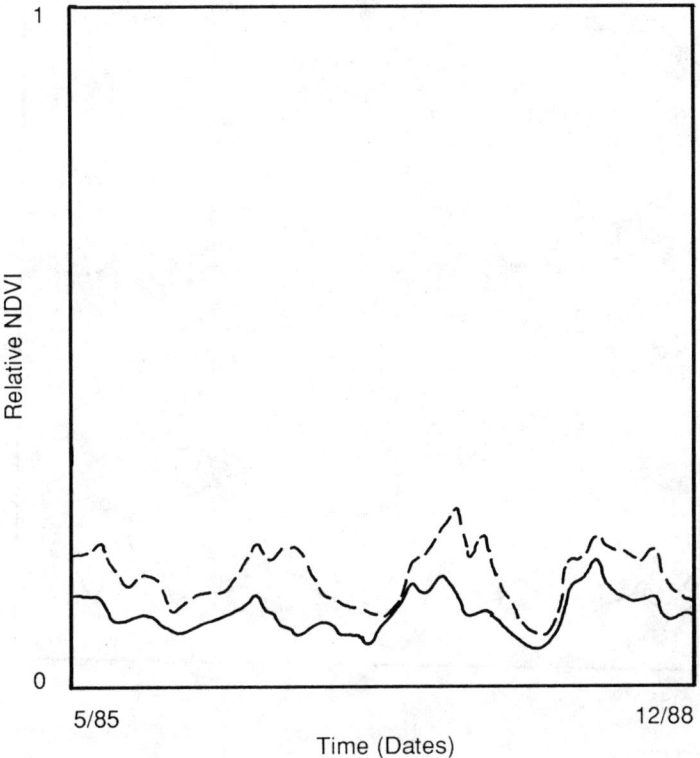

Figure 4. NDVI time-series (1985-1988) response curves for two grassland sites in southeast Utah (the more degraded grassland is indicated by the solid line while the less degraded grassland is indicated by the dashed line).

The grasslands depicted by the NDVI curves for a portion of the Colorado Plateau within the state of Utah (Figure 4) are predominantly perennial with a minor admixture of succulents, shrubs, annual grasses and forbs. In one site, an area has been protected from destructive land uses as a result of protection (i.e. fencing) resulting from the land's status as a park. In the other site, a variety of range and recreation uses has been allowed to continue with little interruption. As a result, the two areas have differing NDVI response curves. These follow, predictably, the curves illustrated for the California grassland sites, with the more protected grassland having a more consistent year-to-year variation and the less protected a less predictable NDVI response.

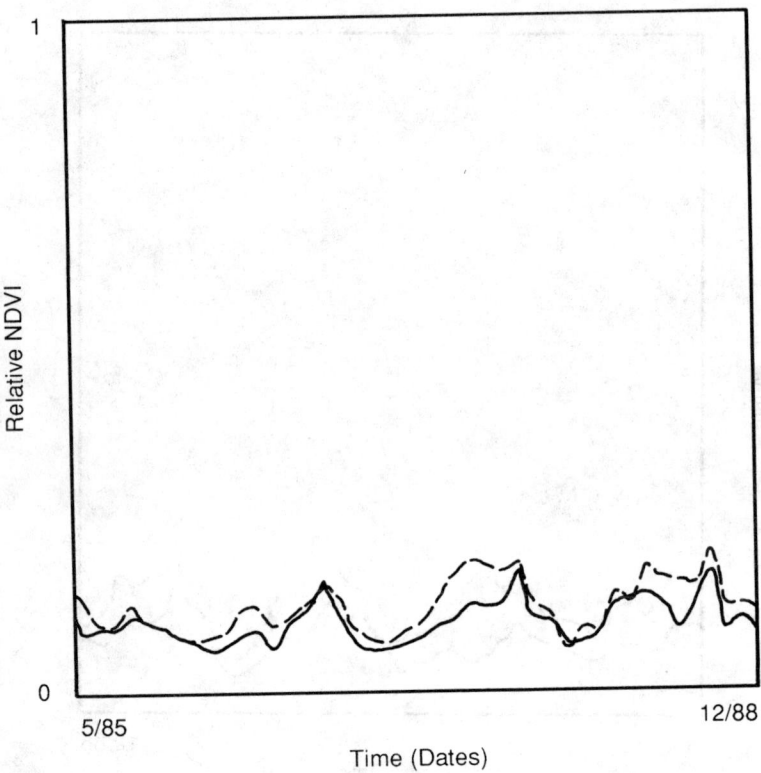

Figure 5. NDVI time-series (1985-1988) response curves for two grassland sites in the vicinity of the Jornada LTER in southern New Mexico (the more degraded grassland is depicted by a solid line while the less degraded grassland is depicted by a dashed line).

The data set examined for southern New Mexico includes a region encompassing the 200,000 ha Jornada Long Term Ecological Research site and includes barren areas (the White Sands Missile Range), shrubland, grassland and degraded grassland sites, woodland, and agricultural areas. The two NDVI response curves (Figure 5) correspond to a degraded grassland and a less degraded grassland. The two curves illustrate the variable (annually) and bimodal precipitation which occurs in southern New Mexico. The degraded grassland is comprised largely of mesquite and has a relatively small component of grasses. The less degraded grassland has an NDVI response which somewhat smooths out the bimodal precipitation distribution and also has higher overall amplitudes.

CONCLUSIONS

It appears, as a result of the examination of coarse spatial resolution satellite (Advanced Very High Resolution Radiometer - AVHRR) vegetation index (Normalized Difference Vegetation Index - NDVI) response patterns for ecosystems having diverging land use histories, that healthier systems have different response patterns than do less healthy ones. The ability to predict health, however, on the basis of these response patterns is still subject to additional research. Nevertheless, in all examples examined by the author, the healthier system has a more consistent year-to-year variation, has a slower greenup and senescence period, and has a broader green peak amplitude. In most instances, the healthier system has a retarded date of maximum or peak greenness. There are many explanations for these differences. Greater stresses on individual plants, changes in species composition toward annuals, increased soil erosion, lower moisture availability, and lower nutrient turnover and availability all contribute to these changing response patterns. Ecological health stress tests may be established on the basis of these response patterns with increased knowledge of the ecosystem dynamics and careful knowledge of the land use histories.

The use of techniques, including remote sensing, to assess, identify, and monitor the status of ecosystem health will greatly improve our ability to locate and manage areas in early stages of stress. The use of the techniques described previously must be considered in the context of the system limitations. Knowledge of the ecosystem dynamics and land use histories combined with accounting for limitations in the sensor data must be key elements in a strategy to assess ecosystem health with NDVI. Registration of hypertemporal data sets, radiometric and atmospheric calibration, and other factors obfuscate the potential use of these techniques. The examples extracted from the literature and from the author's observations of NDVI data sets illustrate a strong potential utility for examining ecosystem health from the context of time-series vegetation index (NDVI) data sets.

ACKNOWLEDGEMENTS

The author would like to thank the Environmental Protection Agency and the National Aeronautics and Space Administration for providing partial support for this effort. The material described in this paper comes partially from an EPA research project on ecosystem health. Thanks also go to Chris Elvidge and Tim Ball with the Desert Research Institute, David Rapport with the University of Ottawa, Wally Jansen with WTJ Software Services, and Sue Franson and Bruce Jones of the Environmental Protection Agency.

REFERENCES

Anderson G.L., Hanson J.D., Haas R.H. (1993) Evaluating Landsat Thematic Mapper Derived Vegetation Indices for Estimating Above-Ground Biomass on Semiarid Rangelands. Remote Sensing of Environment 45(2):165-176

Anderson J.E., Inouye R. (1991) Vegetation Dynamics and the Distribution of Cheatgrass at the Idaho National Engineering Laboratory. Department of Energy Program Review, Las Vegas, June 11, 1991

Burgan R.E., Hartford R.A. (1993) Monitoring vegetation greenness with satellite data. Ogden, Utah, U.S. Department of Agriculture, Forest Service, Intermountain Research Station, General Technical Report INT-297 14 pp.

Chapman G.P. (ed.) (1992) Desertified Grasslands: Their Biology and Management. Academic Press, London

Costanza R., Norton B.G., Haskell B.D. (eds.) Ecosystem Health: New Goals for Environmental Management. Island Press, Washington DC

Eidenshink J.C., Haas R.H. (1992) Analyzing vegetation dynamics of land systems with satellite data. Geocarto International 7(1):53-69

Elvidge C.D., Jansen W.T. (1991) Hypertemporal remote sensing: display and analysis of Global Vegetation Index data. Earth Observations and Global Change Decision Making: A National Partnership, ERIM Fall Conference, Washington DC, October 22-23, 1991

Gholz H.L. (1982) Environmental limits on aboveground net primary production, leaf area, and biomass in vegetation zones of the Pacific Northwest. Ecology 63(2):469-481

Graetz R.D. (1990) Remote sensing of terrestrial ecosystem structure: An ecologist's pragmatic view. pp:5-30. In Hobbs R.J., Mooney H.A. (eds.) Remote Sensing of Biosphere Functioning. Springer-Verlag, New York

Hargrove E. (1992) Environmental therapeutic nihilism. In Costanza R., Norton B.G., Haskell B.D. (eds.) Ecosystem Health: New Goals for Environmental Management. Island Press, Washington DC pp:124-134

Holben B.N. (1986) Characteristics of maximum-value composite images from temporal AVHRR data. International Jour. of Remote Sensing 7:1395-1416

Hobbs R.J., Mooney H.A. (eds.) (1990) Remote Sensing of Biosphere Functioning. Springer-Verlag, New York 312 pp.

Huete A.R., Tucker C.J. (1991) Investigation of soil influences in AVHRR red and near-infrared vegetation index imagery. International Journal of Remote Sensing 12(6):1223-1242

Justice C.O., Townshend R.G., Holben B.N., Tucker C.J. (1985) Analysis of the phenology of global vegetation using meteorological satellite data. International Journal of Remote Sensing 6:1271-1318

Lieth H. (1974) Phenology and Seasonality Modeling. Springer- Verlag, New York

Loveland T.R., Merchant J.W., Ohlen D.O., Brown J.F. (1991) Development of a Land-Cover Characteristics Database for the Conterminous U.S. Photogrammetric Engineering and Remote Sensing 57(11):1453-1463

Malingreau J.P., Belward A.S. (1992) Scale considerations in vegetation monitoring using AVHRR data. International Journal of Remote Sensing 13(12):2289-2307

Malo A.R., Nicholson S.E. (1990) A study of rainfall and vegetation dynamics in the African Sahel using normalized difference vegetation index. Jounal of Arid Environments 19:1-24

Mouat D.A., Fox C.A., Rose M.R. (1992) Ecological indicator strategy for monitoring arid ecosystems. In McKenzie D.H., Hyatt D.E., McDonald V.J. (eds.) Ecological Indicators. Elsevier Applied Science, London pp:717-737

Myneni R.B., Ganapol B.D, Asrar G. (1992) Remote sensing of vegetation canopy photosynthetic and stomatal conductance efficiencies. Remote Sensing of Environment 42:217-238

NOAA (1986) NOAA Polar Orbiter User's Guide, National Oceanic and Atmospheric Administration, National Environmental Satellite, Data and Information Service, National Climatic Data Center, Satellite Data Services Div., Washington, DC

Peters A.J., Reed B.C., Eve M.D., Havstad K.M. (1993) Satellite assessment of drought impact on native plant communities of southeast New Mexico, U.S.A. Journal of Arid Environments 24:305-319

Reed B.C., Brown J.F., VanderZee D., Loveland T.R., Merchant J.W., Ohlen D.O. (1994) Variability of land cover phenology in the United States. Journal of Vegetation Science (in press).

Reining P. (ed.) (1978) Handbook on Desertification Indicators. Washington, D.C. American Association for the Advancement of Science 141 pp.

Rickard W.H., Rogers L.E., Vaughan B.E., Liebetrau S.F. (1988) Shrub-Steppe: Balance and Change in a Semi-Arid Terrestrial Ecosystem. Elsevier, Amsterdam 272 pp.

Running S.W. (1990) Estimating terrestrial primary productivity by combining remote sensing and ecosystem simulation. In: Hobbs R.J., Mooney H.A. (eds.) Remote Sensing of Biosphere Functioning. Springer-Verlag, New York pp:65-86

Schlesinger W.H., Reynolds J.F., Cunningham G.L., Huenneke L.F., Jarrell W.M., Virginia R.A., Whitford W.G. (1990) Biological feedbacks in global desertification. Science 247:1043-1048

Sellers P.J. (1985) Canopy reflectance, photosynthesis and transpiration. International Journal of Remote Sensing 8:1335-1372

Sellers P.J. (1987) Canopy reflectance, photosynthesis and transpiration II - the role of biophysics in the linearity of their interdependence. Remote Sensing of Environment 21:143-183

Sellers P.J., Berry J.A., Collatz G.J., Field C.B., Hall F.G. (1992) Canopy reflectance, photosynthesis, and transpiration. III - a reanalysis using improved leaf models and a new canopy integration scheme. Remote Sens. Environ. 42:187-216

Spanner M.A., Pierce L.L., Running S.W., Peterson D.L. (1990) The seasonality of AVHRR data of temperate coniferous forests: relationship with leaf area index. Remote Sens. Environ. 33:97-112

Tappan G.G., Tyler D.J., Wehde M.E., Moore D.G. (1992) Monitoring rangeland dynamics in Senegal with Advanced Very High Resolution Radiometer data. Geocarto International 7(1):87-98

Waring R.H., Emmingham W.H., Gholz H.L., Grier C.C. (1978) Variation in maximum leaf area of coniferous forests in Oregon and its ecological significance. Forest Science 24:131-140

Wharton R.A., Wigand P.E., Rose M.R., Reinhardt R.L., Mouat D.A., Klieforth H.E., Ingraham N.L., Davis J.O., Fox C.A., Ball J.T. (1990) The North American Great Basin: A sensitive indicator of climatic change. In Osmond C.B., Pitelka L.F., Hidy G.M. (eds.) Plant Biology of the Great Basin. Ecological Studies 80:323-359 Springer-Verlag, Berlin

RAPPORTEUR'S REPORT

Connie Gaudet
Environment Canada
351 St. Joseph Blvd.
Hull, Quebec, K1A 0H3
Canada

SUMMARY OF PRESENTATIONS

The range of ecosystems (from desert to marine), the spatial and temporal variability within these systems, and the ambiguities surrounding the definition of a "healthy" system pose significant methodological challenges in evaluating and monitoring ecosystem health.

Magnuson addressed the importance of understanding and evaluating the natural temporal and spatial variability of different parameters across a diversity of ecological systems. In his work describing general patterns exhibited across a range of Long Term Ecological Research (LTER) sites, he emphasizes the need to "*stand way back and look at general properties of disparate systems*". Variability is highly dependent on the scale of the data set and Magnuson cautions that in comparative studies, it is important not to confound spatial and temporal variability with measures taken at different scales. He further emphasized the importance of comparing the same level of aggregation or organization. Methodologically, such comparative studies are challenging, with the length of the time series required to get a reference variation itself varying with the property under examination. Magnuson concludes however that four to five years is adequate for most observations. He also cautions that because of spatial heterogeneity, single point measures cannot be relied on.

Discussants pointed out that in marine systems, variance also increases as stress increases, though in an extremely stressed system (e.g. Sudbury) there may be no variance because the system is biologically "dead". The relationship between variance and stress is likely not linear i.e. variance is low at extremes of low and high stress and maximum at intermediate levels of stress. Also, multivariate analysis shows that signals are detectable across genera and family and that it is not necessary to go to the species level of measurement in comparative studies. However, Magnuson responded that as long as comparisons were made within the same level of aggregation, then interpretation would not be confounded, cautioning that though aggregated data may give more stable results, how much information is lost?

Sherman described a systems approach to measuring responses to changing ecological states ("health") of large marine ecosystems (LMEs) based on indices of diversity, resilience, stability, productivity, and yield. At this time, there is enough known about marine systems to propose core measures of "health" such as productivity and that the greatest challenge is in appropriate scaling in evaluating parameters. Indicators may be relevant from the molecular to the population level, and monitoring must address different spatial and temporal scales.

Monitoring of large transboundary systems such as LMEs poses not only significant scientific challenges, but also challenges in integrating and coordinating efforts and data across a number of different programs and nations. A holistic, ecosystem-wide approach to broad-area coastal and ocean marine resource assessment is both a necessity and a means for fostering international cooperation and support in linking scientific and societal needs to support long-term, broad-area studies.

Mouat discussed the use of remote sensing for assessing ecosystem condition. Using desertification as an example, Mouat emphasizes the need for measures of sensitivity to desertification risk such as grazing pressure, erosion hazard, percent of exotics, and the change in "greenness" as expressed by NDVI. The response of the NDVI, based on remote sensing imagery, is particulary promising as a methodological tool for evaluating large-scale systems. Data is available over large spatial areas and time series, and NDVI can serve as a measure of health that integrates aspects of the other parameters (such as grazing pressure). Using Advanced Very High Resolution Radiometer (AVHRR) imagery, it has been found that AVHRR response can distinguish sites based on their relative degree of degradation. Mouat cautions however, that though remote sensing is potentially an important methodological tool for assessing and monitoring the status of ecosystems, the use of such techniques must be considered in the context of other information such as a knowledge of ecosystem dynamics and land use histories combined with accounting for limitations in the sensor data.

From a logistical perspective, it was acknowledged that though AVHRR is a cheap and portable tool, there are extensive manpower needs in interpreting the data. It was also emphasized by discussants that interpretation of results is difficult in the absence of a process driven conceptual model, and in making comparisons across disparate systems (i.e. what does greenness mean in Amazonian rainforest as compared to desert?). Notwithstanding the importance of remote sensing tools, there is a need to support interpretation of results with more hypothesis testing and development of alternative conceptual models in order that indices such as NDVI can be convincingly linked to actual "health" or risk.

DISCUSSION

The discussion focussed on three areas 1) the need for expert systems to support monitoring and evaluation; 2) the need for indicators and "ecometers"; and 3) uncertainty.

Expert Systems

The major challenge in monitoring and evaluating health is linking theory to things that might be measurable, especially in evaluating ecosystem level properties such as stability, resilience and connectiveness. At present there is no single theory with regard to changing ecosystem states sufficient to drive monitoring and evaluation strategies, but there is a large body of expertise in the scientific and management community. There is a need to develop an "expert system" built on this core of expertise and experience to address common environmental problems. This expert system was conceptualized in the form of a matrix (Figure 1) supported by expert guidance and data from comparative case studies.

**CUMULATIVE EXPERIENCE
EXPERT GUIDANCE**
▼

System	Stressors	Indicators	Monitoring Strategy	Management Strategy
coral reef estuaries mud flats etc.	i.e. particular to a system	how can they assess problems?	how do you measure changes in state? Type II errors are critical.	how does this link with management strategies?

▼
MANAGEMENT

Figure 1. The expert matrix as an approach to evaluating and monitoring ecosystem health.

Indicators and Ecometers

Though rapid technological advances in areas such as remote sensing hold promise for increasing our ability to monitor and report the status of systems over large scales, it was emphasized that it is not the "methods" that are needed as much as validated indicators of ecosystem health (or ill health). The term "ecometer" was introduced to describe tools and approaches to monitor, integrate and communicate health status of a an ecosystem, much in the same way the dials in the cockpit of a plane inform the pilot of the condition of the plane's system and any impending problems (such as loss of altitude, low pressure or fuel). However simple and enticing the concept of an "ecometer" may be, their use must be based on a comprehensive understanding of the underlying system. If designed to measure an aspect of the system that is not an indicator of "health" (i.e. to track trends with no underlying understanding of what these trends mean in terms of stress and response or overall "health" goals for the system), or if used in the absence of an underlying conceptual model, then no matter how accurate the measure, and how much data it generates, it may actually impart very little information for evaluating the health or status of the system.

In implementing and verifying the usefulness of an indicator there are basic signs of success or failure:

- can you implement the measure?
- can you verify the indicator?
- is the indicator supported at the scientific, social and political level?

Science may verify ecological adequacy but political and public support will lend acceptance, status and power to the indicator. It was therefore considered important not only to validate methods but to develop tools to inform policy makers and the public of the results of the evaluation in order to effect action.

The "amoeba", used in The Netherlands, provides an integrated summary of the status of several indicators relevant to the overall health of a system. It was brought forward as one example of an ecometer or "ecoclock" with both a high diagnostic and communication power. Importantly, it is an approach firmly grounded in an understanding of underlying ecosystem dynamics, clearly defined ecological and societal goals for the "healthy" system, and based on indicators of both scientific and socio-economic relevance.

The need to have the scale of science correspond to the political scale was also emphasized. In short, the larger the scale of degradation, the larger the political scale and public scale of involvement and support required to deal effectively with the issue. In this regard, early warning, diagnostic indicators of health are critical to a cost-effective, preventative approach to sustaining ecosystem health. Without early warnings of impairment and loss of ecosystem function, resilience and services, not only will large scale crisis occur, such as fisheries collapse, they will occur at a time when public and political interest and awareness lag far behind. At the same time the socio-economic and remedial costs may be enormous.

The need for early warning indicators was discussed in the context of marine systems. Early warning indicators of, for example, eutrophication, are critically important for early diagnosis and prevention of severe, and perhaps irreversible health effects. Fish, benthos and zooplankton have potential as indicators of major disturbance. However, can these indicators serve as early warnings that the system is "at risk" before it is noticeably impaired? For example, could we have identified indicators that would have prevented over-fishing and the collapse of the Atlantic fishery, especially when some of the traditional measures of status, such as zooplankton abundance, did not change over the period that the fishery crashed?

One method for improving such diagnostic capability is comparative case studies. For example, increasing eutrophication has been observed in many seas. Is it possible to construct a series of stress/response curves based on the observations within systems at different degrees of degradation (e.g. Black Sea, Baltic Sea, North Sea, North East Atlantic) that could be used to predict where the less degraded systems are heading? In concluding, discussants cautioned that in developing diagnostic protocols and expert systems, we must avoid being single minded in focussing on one major stress (e.g. eutrophication) and response (e.g. fishery) and that a holistic ecosystem approach must be emphasized.

Uncertainty

Though indicators are needed for monitoring and evaluating health, it was recognized that there is a great deal of uncertainty in interpreting data and making decisions. Indicators of ecosystem health are extrapolated, therefore immediately introducing uncertainty. The delay between a stress and response leads to uncertainty (e.g. nitrate in groundwater: delay of 10-50 years in linking stress and response). The bigger the scale, the less the certainty. The more dissimilar systems are (e.g. comparing geographic regions), the less the certainty in the relevance of the indicators.

Using remote sensing as an example, major classes of uncertainty were identified as:
- leaving out major considerations
- too much information
- unnecessary and ambiguous indicators
- sensor system issues
- interpretation of data (e.g. greenness may not be a good proxy for...)
- historical issues
- lack of an alternative conceptual framework (i.e. how else might the results be interpreted)

The criteria for success or failure were identified as:
- feedback loops
- management appraisal

- statistical tests (Type II errors may be more important than Type I)
- expert opinion

There is little immediate resolution to uncertainty, except to acknowledge that we must deal with it. However, research emphasizing hypothesis testing to establish predictive stress/response relationships and validated indicators of "health" or "risk" such as loss of resilience, comparative case studies and the building of expert systems, are all important to reducing this uncertainty.

CONCLUSIONS

1. Critical scientific methodological issues to be addressed in monitoring and evaluating the health of large-scale systems are:

    Variability - understanding natural variability in a system (spatial and temporal, and across different levels of aggregation and organization).
    Stress/Response - integration of stress and response in interpreting results and in designing monitoring strategies.
    Scaling - different measures/indicators may be relevant at different spatial and temporal scales.
    Hypothesis Testing - more research is needed to develop basic conceptual and predictive models required to support interpretation of health.

2. Though technological capability is important, selection, validation, and communication of **early-warning indicators** of ecosystem health is a major goal for evaluating and monitoring the health of large-scale ecosystems. Such indicators would be most useful if they enabled early identification of systems "at risk" (for example loss of resilience) before impairment and associated socio-economic costs are apparent and recovery is either expensive or impossible.

3. The goal of monitoring and selection of indicators should be to provide the information necessary to practice cost-effective, **preventative health care** of ecosystems.

4. Evaluation of health should be based on policy relevant, integrated indicators of health that have **scientific, public** and **political** support.

5. **Expert systems** which build on and make available the collective expertise within the scientific and management community should be developed and maintained to provide the support necessary to implement effective monitoring and evaluation of large-scale ecosystems.

WORKSHOP SUMMARY

Peter Calow and David Rapport

CONCLUSIONS

- There is a necessary separation between social values which determine what is a healthy state, and the science of assessment which determines whether you have it or not. The science is needed to describe, understand, predict, and manipulate change in ecological systems and to address questions such as: can we ascribe change to anthropogenic causes? This effort clearly must draw upon the ecological sciences, as well as the health sciences, economics, ethics, law etc.

- To measure change, a suite of indicators is required. These indicators need to be both biologically and physically based (the physical parameters may dominate in confirming that a recovery is possible and has begun).

- Variability in systems is an integral part of them and needs to be described at various scales, and sorted by temporal and spatial coherence.

- To determine if a change is significant, it needs to be measured against a value system: this involves both science and the values determined by politicians, public interest groups, policy makers, and pressure groups.

- The challenge will be to catalyze a meaningful dialogue between scientists and other interested parties.

RECOMMENDATIONS

1. Future Research Programs

- There is a need to identify criteria in which we describe and measure change - positive and negative, and in terms of structure or function. The scale of the discussion is also important. In any event, we need to follow a rigorous approach in which hypotheses are formulated and tested in critical experimental observational circumstances.

- There is a need to develop a more systematic approach to diagnosis - that is a rational decision-making process which identifies probable causes of observed change, and also probable consequences of observed change involving both retrospective and prospective approaches.

- There is also a need to develop for example more useful risk assessment procedure which are transparent and based directly on significant ecosystem features and public values.

- There needs to be relevant to new legislative programs, for example, the program on remedying environmental damage by the EC (European Community). Environmental damage means "any significant physical, chemical, or biological deterioration of the environment". The objective is environmental protection at the level of quality that society determines. Where environments are damaged below that standard, restoration is the only environmentally sound remedy.

2. Integrated Case Studies

- It will be important to compile case studies which illustrate the involvement of science, policy and, and public pressure in environmental protection and remediation.

3. International Society and Journal

- An international society and companion journal could be established that encourages a transdisciplinary approach.

4. Follow-up Workshop

- There could be a follow-up workshop - this current workshop was designed to set the topology for the development of the field of assessing and monitoring the health of large scale ecosystems. Thus its objectives were broad based and the contributions reflect these global goals. The discussions were wide-ranging, and emphasized the relevance and interconnectedness of all levels and scales of investigation. Clearly, each topic warrants a more focused and in-depth analysis.

- Of the topics the three inviting most immediate attention are: 1) diagnosis; 2) rehabilitation; and 3) the interface between information and policy development. Therefore we recommend a series of workshops to address these topics. To maintain the momentum and integration of the approach. it will be important for these kinds of events to networked and capitalize on what we have started here. The society and journal (recommended above) are an important step towards this.

The ASI Series Books Published as a Result of
Activities of the Special Programme on Global Environmental Change

This book contains the proceedings of a NATO Advanced Research Workshop held within the activities of the NATO Special Programme on Global Environmental Change, which started in 1991 under the auspices of the NATO Science Committee.

The volumes published as a result of the activities of the Special Programme are:

Vol. 1: **Global Environmental Change.**
Edited by R. W. Corell and P. A. Anderson. 1991.
Vol. 2: **The Last Deglaciation: Absolute and Radiocarbon Chronologies.**
Edited by E. Bard and W. S. Broecker. 1992.
Vol. 3: **Start of a Glacial.** Edited by G. J. Kukla and E. Went. 1992.
Vol. 4: **Interactions of C, N, P and S Biogeochemical Cycles and Global Change.**
Edited by R. Wollast, F. T. Mackenzie and L. Chou. 1993.
Vol. 5: **Energy and Water Cycles in the Climate System.**
Edited by E. Raschke and D. Jacob. 1993.
Vol. 6: **Prediction of Interannual Climate Variations.**
Edited by J. Shukla. 1993.
Vol. 7: **The Tropospheric Chemistry of Ozone in the Polar Regions.**
Edited by H. Niki and K. H. Becker. 1993.
Vol. 8: **The Role of the Stratosphere in Global Change.**
Edited by M.-L. Chanin. 1993.
Vol. 9: **High Spectral Resolution Infrared Remote Sensing for Earth's Weather and Climate Studies.**
Edited by A. Chedin, M.T. Chahine and N.A. Scott. 1993.
Vol. 10: **Towards a Model of Ocean Biogeochemical Processes.**
Edited by G. T. Evans and M.J.R. Fasham. 1993.
Vol. 11: **Modelling Oceanic Climate Interactions.**
Edited by J. Willebrand and D.L.T. Anderson. 1993.
Vol. 12: **Ice in the Climate System.** Edited by W. Richard Peltier. 1993.
Vol. 13: **Atmospheric Methane: Sources, Sinks, and Role in Global Change.**
Edited by M. A. K. Khalil. 1993.
Vol. 14: **The Role of Regional Organizations in the Context of Climate Change.**
Edited by M. H. Glantz. 1993.
Vol. 15: **The Global Carbon Cycle.**
Edited by M. Heimann. 1993.
Vol. 16: **Interacting Stresses on Plants in a Changing Climate.**
Edited by M. B. Jackson and C. R. Black. 1993.
Vol. 17: **Carbon Cycling in the Glacial Ocean: Constraints on the Ocean's Role in Global Change.**
Edited by R. Zahn, T. F. Pedersen, M. A. Kaminski and L. Labeyrie. 1994.
Vol. 18: **Stratospheric Ozone Depletion/UV-B Radiation in the Biosphere.**
Edited by R. H. Biggs and M. E. B. Joyner. 1994.
Vol. 19: **Data Assimilation: Tools for Modelling the Ocean in a Global Change Perspective.**
Edited by P. O. Brasseur and J. Nihoul. 1994.

Vol. 20: **Biodiversity, Temperate Ecosystems, and Global Change.**
Edited by T. J. B. Boyle and C. E. B. Boyle. 1994.
Vol. 21: **Low-Temperature Chemistry of the Atmosphere.**
Edited by G. K. Moortgat, A. J. Barnes, G. Le Bras and J. R. Sodeau. 1994.
Vol. 22: **Long-Term Climatic Variations – Data and Modelling.**
Edited by J.-C. Duplessy and M.-T. Spyridakis. 1994.
Vol. 23: **Soil Responses to Climate Change.**
Edited by M. D. A. Rounsevell and P. J. Loveland. 1994.
Vol. 24: **Remote Sensing and Global Climate Change.**
Edited by R. A. Vaughan and A. P. Cracknell. 1994.
Vol. 25: **The Solar Engine and Its Influence on Terrestrial Atmosphere and Climate.**
Edited by E. Nesme-Ribes. 1994.
Vol. 26: **Global Precipitations and Climate Change.**
Edited by M. Desbois and F. Désalmand. 1994.
Vol. 27: **Cenozoic Plants and Climates of the Arctic.**
Edited by M. C. Boulter and H. C. Fisher. 1994.
Vol. 28: **Evaluating and Monitoring the Health of Large-Scale Ecosystems.**
Edited by D. J. Rapport, C. L. Gaudet and P. Calow. 1995.